The Last Deglaciation: Absolute and Radiocarbon Chronologies

NATO ASI Series

Advanced Science Institutes Series

A series presenting the results of activities sponsored by the NATO Science Committee, which aims at the dissemination of advanced scientific and technological knowledge, with a view to strengthening links between scientific communities.

The Series is published by an international board of publishers in conjunction with the NATO Scientific Affairs Division

A Life Sciences B Physics	Plenum Publishing Corporation London and New York
C Mathematical and Physical Sciences D Behavioural and Social Sciences E Applied Sciences	Kluwer Academic Publishers Dordrecht, Boston and London
F Computer and Systems Sciences G Ecological Sciences H Cell Biology I Global Environmental Change	Springer-Verlag Berlin Heidelberg New York London Paris Tokyo Hong Kong Barcelona Budapest

NATO-PCO DATABASE

The electronic index to the NATO ASI Series provides full bibliographical references (with keywords and/or abstracts) to more than 30000 contributions from international scientists published in all sections of the NATO ASI Series. Access to the NATO-PCO DATABASE compiled by the NATO Publication Coordination Office is possible in two ways:

- via online FILE 128 (NATO-PCO DATABASE) hosted by ESRIN, Via Galileo Galilei, I-00044 Frascati, Italy.
- via CD-ROM „NATO-PCO DATABASE" with user-friendly retrieval software in English, French and German (© WTV GmbH and DATAWARE Technologies Inc. 1989).

The CD-ROM can be ordered through any member of the Board of Publishers or through NATO-PCO, Overijse, Belgium.

Series I: Global Environmental Change, Vol. 2

The Last Deglaciation: Absolute and Radiocarbon Chronologies

Edited by

Edouard Bard

Laboratoire de Géosciences de l'Environnement
URA CNRS 132
Université d'Aix-Marseille III, Saint Jérome
F-13397 Marseille Cedex 13, France

Wallace S. Broecker

Lamont-Doherty Geological Observatory
Columbia University
Palisades, NY 10964, USA

Springer-Verlag
Berlin Heidelberg New York London Paris Tokyo
Hong Kong Barcelona Budapest
Published in cooperation with NATO Scientific Affairs Division

Proceedings of the NATO Advanced Research Workshop on The Last Deglaciation: Absolute and Radiocarbon Chronologies held at Erice, Sicily (Italy), December 9–13, 1990

Library of Congress Cataloging-in-Publication Data
The last deglaciation : absolute and radiocarbon chronologies / edited by Edouard Bard and Wallace S. Broecker.
(NATO ASI series. Series I, Global environmental change ; vol. 2)

ISBN-13: 978-3-642-76061-7 e-ISBN-13: 978-3-642-76059-4
DOI: 10.1007/978-3-642-76059-4

1. Glacial epoch–Congresses. 2. Radiocarbon dating–Congresses. 3. Climatic changes–Congresses. I. Bard Edouard. II. Broecker, Wallace S., 1931- . III. Series. QE697.L292 1992 551.7'92–dc20

Softcover reprint of the hardcover 1st edition 1992

Typesetting: Camera ready by author
31/3145 - 5 4 3 2 1 0 - Printed on acid-free paper

TABLE OF CONTENTS

PROLOGUE

LIST OF AUTHORS AND PARTICIPANTS

I- RADIOCARBON AND ABSOLUTE CHRONOLOGIES

II- COSMONUCLIDE PRODUCTION CHANGES DURING THE PAST

III- CLIMATIC CHANGES DURING THE LAST DEGLACIATION

PROLOGUE

The last deglaciation was a period of intense and rapid climatic changes which affected the global climate of the Earth from about 20,000 to 5000 years before present. Paleoclimatologists present during the Erice meeting, have reconstructed global variations of the chemical composition of the atmosphere (30% increase in CO_2 and CH_4 and decrease in dust content etc), temperature of the atmosphere and surface of the ocean (mean global change of about +4°C), major reorganization of the ocean circulation and a sea level rise of about 120 meters followed by a slow rebound of the continents below the ice caps.

All these events, described during the workshop, seem to have been triggered by a maximum in north hemisphere radiation centered at about 11,000 years BP as calculated by astronomers. In addition the last deglaciation is characterized by a major transient oscillation. This so-called Younger Dryas event cannot be attributed to a direct extraterrestrial forcing but is interpreted as a symptom of the complex relations existing between atmosphere, biosphere, cryosphere and oceans. Our climate is influenced by a large number of mechanisms which can act as direct causes but also as amplifiers of the primary forcings (so-called positive feedbacks). In particular, our capability of forecasting the near future climate relies on our understanding of these subtle phenomena.

The main focus of the workshop was to discuss the timing of the last deglaciation in order to provide a common absolute chronological framework for the numerous researchers in the field. Results obtained in different scientific disciplines and relevant for solving the chronological problem have been discussed during the meeting: glaciology, varved clay sedimentology, geophysics, paleomagnetism, dendrochronology and radiochronology (^{14}C and U/Th). We then discussed the implications of the timing on the identification of the climatic forcings. The information obtained from different methods have been compared in order to obtain a better understanding of the climatic changes: palynology, deep sea paleontology and geochemistry of sediment, coral, wood and ice cores.

The workshop has been sponsored by the Scientific Affairs Division of the North Atlantic Treaty Organization with additional contributions from the US National Science Foundation and from the Ettore Majorana Center for Scientific Culture of Erice in Sicily where the meeting was held (9-13 December 1990). For their financial and logistical supports we wish to thank all these organizations

The Editors

LIST OF AUTHORS AND PARTICIPANTS

B. Ammann, Universität Bern, Systematisch-Geobotanisches Institut, Altenbergrain 21, CH-3013 Bern, Switzerland

M. Arnold, Centre des Faibles Radioactivités, CNRS-CEA, av. de la Terrasse, 91198, Gif-sur-Yvette, France

E. Bard, Centre des Faibles Radioactivités, CNRS-CEA, av. de la Terrasse, 91198, Gif-sur-Yvette, France and Géosciences de l'Environnement, URA CNRS 132, Université d'Aix-Marseille III, Saint Jérome, 13397, Marseille Cedex, France

J.M. Barnola, Laboratoire de Glaciologie et Géophysique de l'Environnement, BP 96, 38402 St Martin dHères Cedex, France

N.I. Barkov, Arctic and Antarctic Research Institute, Beringa Street 38, 199226 Leningrad, USSR

B. Becker, Institut für Botanik, Universität Hohenheim 210, Garbenstr. 20, 7000 Stuttgart 70, Germany

J. Beer, Environmental Physics, ETH-Zürich, Institute for Aquatic Sciences and Water Pollution Control, c/o EAWAG, CH-8600 Dübendorf,Switzerland

S. Björck, Department of Quaternary Geology, University of Lund, Tornavägen 13, S-223 63 Lund, Sweden

G. Bonani, Institut für Mittelenergiephysik, ETH-Hönggerberg, CH-8093 Zürich, Switzerland

N. Bonhommet, Département de Géotectonique, Université de Paris VI, 4 place Jussieu, Tour 25, F-75230 Paris Cédex 05, France

W. S. Broecker, Lamont-Doherty Geological Observatory, Columbia University, Palisades, NY10964, USA

L. Brunnberg, Dept. of Quaternary Research, Stockholm University, Odengatan 23, S-113 22 Stockholm, Sweden

I. Cato, Geological Survey of Sweden, Box 670, S-751 28, Uppsala, Sweden

J. Chappelaz, Laboratoire de Glaciologie et Géophysique de l'Environnement, BP 96 38402 St Martin dHères Cedex, France and NASA/Goddard Institute for Space Studies, 2880 Broadway, New York, NY10025, USA

P. Ciais, Laboratoire de Géochimie Isotopique, DSM/CEA, CEN Saclay, 91191 Gif-sur-Yvette, France

J. D'addezio, Istituto Nazionale di Geofisica, Ria di Villa Ricotti, Roma, Italy

W. Dansgaard, Geophysical Isotope Laboratory, University of Copenhagen, Haraldsgade 6, DK-2200 Copenhagen N, Denmark

M. Dulinski, Institute of Physics and Nuclear Techniques, Academy of Mining and Metallurgy, 30-059 Krakow, Poland

J.-C. Duplessy, Centre des Faibles Radioactivités, CNRS-CEA, av. de la Terrasse, 91198, Gif-sur-Yvette, France

J. Duprat, Dept. de Géologie et Océanographie, Université de Bordeaux 1, av. des Facultés, 33405 Talence, France

H. Erlenkeuser, C14-Labor, Christian-Albrechts-Universität, Olshausenstr. 40, D-2300 Kiel, Germany

R. G. Fairbanks, Lamont-Doherty Geological Observatory, Columbia University, Palisades, NY10964, USA

R.C. Finkel, Center for Accelerator Mass Spectrometry, Lawrence Livermore National Laboratories, Livermore, California 94550, USA

D. A. Fisher, Glaciology Section, Terrain Sciences Div., Geological Survey of Canada, 601 Booth St.Ottawa, Ontario K1A 0E8, Canada

A. Flatoy, Department of Geology sec. B, University of Bergen, Allégt. 41, N- 5007 Bergen, Norway

J.-Ch. Fontes, Laboratoire d'Hydrologie et de Géochimie Isotopique, Université de Paris-Sud, Bat 504, F-91405, Orsay Cedex, France

F. Gasse, Laboratoire d'Hydrologie et de Géochimie Isotopique, Université de Paris-Sud, Bat 504, F-91405, Orsay Cedex, France

T. Goslar, Radiocarbon Laboratory, Institute of Physics, Silesian Technical University, Ul Boleslawa Krzywoustego 2, PL-44-100 Gliwice, Poland

I. Hajdas, EAWAG Umweltphysik, CH-8600 Dübendorf,Switzerland

B. Hamelin, Géosciences de l'Environnement, URA CNRS 132, Université d'Aix-Marseille III, Saint Jérome, 13397, Marseille Cedex, France

C. U. Hammer, Geophysical Isotope Laboratory, University of Copenhagen, Haraldsgade 6, DK-2200 Copenhagen N, Denmark

B. Haverkamp, Abt. Geophysik, Universität Münster, Corrensstr. 24, D-4400 Münster, Germany

J. Heinemeier, Institute of Physics, University of Aarhus, DK-8000 Aarhus C, Denmark

E. Jansen, Department of Geology sec. B, University of Bergen, Allégt. 41, N- 5007 Bergen, Norway

S. Johnsen, Science Institute, University of Iceland, Dunhaga 3, IS-107 Reykjavik, Iceland, and Geophysical Isotope Laboratory, University of Copenhagen, Haraldsgade 6, DK-2200 Copenhagen N, Denmark

J. Jouzel, Laboratoire de Géochimie Isotopique, DSM/CEA, CEN Saclay, 91191 Gif-sur-Yvette, France and Laboratoire de Glaciologie et Géophysique de l'Environnement, BP 96, 38402 St Martin dHères Cedex, France

A. Juillet-Leclerc, Centre des Faibles Radioactivités, CNRS-CEA, av. de la Terrasse, 91198, Gif-sur-Yvette, France

L.D. Keigwin, Woods Hole Oceanographic Institution, Woods Hole, MA 02543, USA

V.M. Kotkyakov, Institute of Geography, Staromonetny per 29, 109017 Moscow, USSR

B. Kromer, Institut für Umweltphysik der Universität Heidelberg, Im Neuenheimer Feld 366, D-6900 Heidelberg, Germany

T. Kuc, Institute of Physics and Nuclear Techniques, Academy of Mining and Metallurgy, 30-059 Krakow, Poland

L.D. Labeyrie, Centre des Faibles Radioactivités, CNRS-CEA, av. de la Terrasse, 91198, Gif-sur-Yvette, France

C. Laj, Centre des Faibles Radioactivités, CNRS-CEA, av. de la Terrasse, 91198, Gif-sur-Yvette, France

D. Lal, Scripps Institution of Oceanography, Geological Research Division, La Jolla, California 92093-0220, USA

C.C. Langway, Ice Core Laboratory, Dept. of Geology, State University of New-York at Buffalo, Amherst, NY 14226, USA

C. Lorius, Laboratoire de Glaciologie et Géophysique de l'Environnement, BP 96, 38402 St Martin dHères Cedex, France

A. Lotter, EAWAG Umweltphysik, CH-8600 Dübendorf,Switzerland

A. Mazaud, Centre des Faibles Radioactivités, CNRS-CEA, av. de la Terrasse, 91198, Gif-sur-Yvette, France

J.F.W. Negendank, Abt. Geologie, Universität Trier, Postfach 3825, D-5500 Trier, Germany

E. Nelson, Archaeology Department, Simon Fraser University, Burnaby, B.C. Canada V5A1S6

H. Oeschger, Physics Institute, University of Bern, CH 3012 Bern, Switzerland

D. Pantosti, Istituto Nazionale di Geofisica, Ria di Villa Ricotti, Roma, Italy

M. F. Pazdur, Radiocarbon Laboratory, Institute of Physics, Silesian Technical University, Ul Boleslawa Krzywoustego 2, PL-44-100 Gliwice, Poland

W. R. Peltier, Department of Physics, University of Toronto, Ontario, M5S 1A7, Canada

D. Peteet, NASA/Goddard Institute for Space Studies, 2880 Broadway, New York, NY10025, USA and Lamont-Doherty Geological Observatory, Columbia University, Palisades, NY10964, USA

J.R. Petit, Laboratoire de Glaciologie et Géophysique de l'Environnement, BP 96, 38402 St Martin dHères Cedex, France and Laboratoire de Géochimie Isotopique, DSM/CEA, CEN Saclay, 91191 Gif-sur-Yvette, France

V.N. Petrov, Arctic and Antarctic Research Institute, Beringa Street 38, 199226 Leningrad, USSR

G. Raisbeck, Centre National de Spectrométrie Nucléaire et de Spectrométrie de Masse, IN2P3-CNRS, bat. 104-108, 91405 Orsay Campus, France

D. Raynaud, Laboratoire de Glaciologie et Géophysique de l'Environnement, BP 96, 38402 St Martin dHères Cedex, France

C. Ritz, Laboratoire de Glaciologie et Géophysique de l'Environnement, BP 96, 38402 St Martin dHères Cedex, France

K. Rozanski, International Atomic Energy Agency, Isotope Hydrology Section, Wagrammerstrasse 5, POB 100, A-1400, Vienna, Austria

N. Rud, Institute of Physics, University of Aarhus, DK-8000 Aarhus C, Denmark

J.S. Salis, Département de Géotectonique, Université de Paris VI, 4 place Jussieu, Tour 25, F-75230 Paris Cédex 05, France

M. Sarnthein, Geologisch-Paläontologisches Institut, Christian-Albrechts-Universität, Olshausenstr. 40, D-2300 Kiel, Germany

J. Southon, Archaeology Department, Simon Fraser University, Burnaby, B.C. Canada V5A1S6 and Center for Accelerator Mass Spectrometry, Lawrence Livermore National Laboratories P.O. Box 808, Livermore, California 94550, USA

B. Stauffer, Physics Institute, University of Bern, CH 3012 Bern, Switzerland

B. Strömberg, Dept. of Physical Geography, Stockholm University, S-106 91, Stockholm, Sweden

M. Stuiver, Department of Geological Sciences, University of Washington, Seattle, Washington 98195, USA

M. Sturm, EAWAG Umweltphysik, CH-8600 Dübendorf,Switzerland

M. Suter, Paul Scherrer Institute, CH-5232 Villigen, Switzerland

A. E. Sveinbjörnsdottir, Science Institute, University of Iceland, Dunhaga 3, IS-107, Reykjavik, Iceland

E. Tric, Dept. de Géomagnétisme, Université de Paris-Sud, Bat 504, F-91405, Orsay Cedex, France

T. Veum, Department of Geology sec. B, University of Bergen, Allégt. 41, N- 5007 Bergen, Norway

J. Vogel, Archaeology Department, Simon Fraser University, Burnaby, B.C. Canada V5A1S6 and Center for Accelerator Mass Spectrometry, Lawrence Livermore National Laboratories P.O. Box 808, Livermore, California 94550, USA

E. Vogelsang, Geologisch-Paläontologisches Institut, Christian-Albrechts-Universität, Olshausenstr. 40, D-2300 Kiel, Germany

A. Walanus, Radiocarbon Laboratory, Institute of Physics, Silesian Technical University, Ul Boleslawa Krzywoustego 2, PL-44-100 Gliwice, Poland

M.S. Weinelt, Geologisch-Paläontologisches Institut, Christian-Albrechts-Universität, Olshausenstr. 40, D-2300 Kiel, Germany

W. Woelfli, Institut für Mittelenergiephysik, ETH-Hönggerberg, CH-8093 Zürich, Switzerland

F. Yiou, Centre National de Spectrométrie Nucléaire et de Spectrométrie de Masse, IN2P3-CNRS, bat. 104-108, 91405 Orsay Campus, France

B. Zolitschka, Abt. Geologie, Universität Trier, Postfach 3825, D-5500 Trier, Germany

PART I:

RADIOCARBON AND ABSOLUTE CHRONOLOGIES

Tree-ring ^{14}C Calibration at 10.000 BP

Bernd Kromer
Institut für Umweltphysik derUniversität Heidelberg
Im Neuenheimer Feld 366
D-6900 Heidelberg
Germany

Bernd Becker
Institut für Botanik, Universität Hohenheim 210
Garbenstr. 20
D-7000 Stuttgart 70
Germany

Abstract:
Calibration of ^{14}C dates between 9000 and 10.000 BP is proposed using two tree-ring chronologies, the absolutely dated German oak series and a floating pine series covering the Preboreal and late Younger Dryas. The floating pine series is matched to the oak master by the ^{14}C pattern providing a minimum absolute age for its placement. The YD/Preboreal transition is defined by ^{13}C and ^{2}H as measured in the pine series. A minimum age for the transition of 10.970 cal BP is obtained.

Dendrochronologically dated tree-ring chronologies provide the most precise source for calibration of the radiocarbon time scale. Through high precision ^{14}C analyses of tree-ring series several ^{14}C laboratories contributed to the calibration task. The status of the published calibration data sets is shown in figure 1. As is evident from this compilation the German oak series is the longest absolutely dated chronology extending from 9888 dendro BP (7938 BC) to present[1]. Its start is close to the immigration of the mixed oak forest into Central Europe replacing the early Holocene pine forest.

---

[1]In this paper we will use three time scales : (1) **Dendro years BP,** ie. solar years before AD 1950 (2) conventional radiocarbon years BP (before AD 1950), denoted **^{14}C BP** for clarity and (3) **cal BP** to denote years before AD 1950, calibrated by some means, eg. varve counting

NATO ASI Series, Vol. I 2
The Last Deglaciation: Absolute and Radiocarbon Chronologies
Edited by E. Bard and W. S. Broecker

In addition to the oak series we have established several floating oak and pine chronologies in the interval 12.400 to 8930 ^{14}C BP (fig. 2). Of particular interest is a 1604 yr Pine chronology which starts at 10.030 ^{14}C BP and ends at 8930 ^{14}C BP covering the entire Preboreal period. If it were linked dendrochronologically to the absolute oak series it immediately would provide tree-ring calibrated ages of the 10.000 ^{14}C BP time mark.

However, the suggested overlap of about one century (see below) between the oak and the Preboral pine series is insufficient to permit a cross-match based on the tree-ring pattern. Yet we can make use of the ^{14}C pattern at the linkage to provide constraints to the absolute position of the floating pine series. The argument is as follows:

The atmospheric ^{14}C level in the Holocene as derived from the tree-ring chronologies shows variations on time scales of centuries to millenia (Stuiver et al. 1991). The key to the linking of the pine series to the oak master are the century scale oscillations. They have been shown (Stuiver & Braziunas, 1989) to occur at regular intervals throughout the Holocene and their shape is rather uniform. The amplitude of the increase in the atmospheric ^{14}C level is 2 to 4 % and the duration is between 150 and 250 years. These oscillations are superimposed upon the long-term ^{14}C trend of the Holocene which decreased by about 10 % with a characteristic time scale of 10.000 years.

From the pattern of the ^{14}C age vs. tree-ring age at the respective ends of the pine and the oak chronology (fig. 3 b) it is obvious that they could mark a century type oscillation. The sharp decline of ^{14}C ages at 9100 to 8930 ^{14}C BP is caused by the rise in atmospheric ^{14}C whereas the almost constant ^{14}C age of 8800 ^{14}C BP at the beginning of the oak chronology reflects the return to the quasi-steady state level of the time prior to the oscillation. Using the information derived from all the Holocene oscillations we put narrow constraints on the absolute placement of the floating pine series:

(i) The lowest possible absolute position is obtained when we assume the gap of ^{14}C years between the youngest measured pine section (ring 1510 at 8930 ^{14}C BP) and the earliest oak section (9825 dendro BP, 8800 ^{14}C BP) to be zero in true years.

(ii) For a more realistic estimate we choose the absolute age placement of the pine series in such a way that an oscillation of the century type results (fig. 3a, C1/C2). This procedure leads to an estimated calibration curve (^{14}C vs. dendro age), fig. 3b, and to the level of atmospheric ^{14}C from 10.000 to 8500 ^{14}C BP as shown in fig 3a.

From this figure several observations are noted :

(1) If the assumption about the shape of the oscillation at the link of the chronologies is accepted then there is room for only slight shifts of the floating pine series. Any lower absolute placement beyond the statistical error of the ^{14}C analyses (30 years) would conflict with the ^{14}C levels obtained from the absolutely dated oak series; in view of the long-term trend of the ^{14}C baseline, a shift to higher absolute ages by as much as a few centuries cannot be excluded. However, for a symmetrical shape of the oscillation the range would be limited to about one century (1% in delta ^{14}C). In any case, the minimum age of the beginning of the pine chronology is fixed to 11.370 dendro BP.

(2) There are three 'plateaus' of apparent constant ^{14}C age, ie. at 10.000, 9600 and 8800 ^{14}C BP, covering true time intervals of several centuries in each case. At these intervals precise conversion of^{14}C dates is not possible.

(3) With respect to the chronology of the Younger Dryas the ^{14}C age plateau at 10.000 BP does not allow a precise *absolute* dating of the end of this event by ^{14}C. In the tree rings measured up to now the plateau extends over 250 dendro years already.

A solution to the dating precision dilemma may come from analyses of stable isotopes (^{13}C,^{2}H) in the tree rings of the pine chronology published elsewhere (Becker et al, 1991). Both isotopes show a coincident and distinct transition starting at 10.970 dendro BP and lasting for 200 (^{13}C) and 450 (^{2}H) dendro years, respectively. This tree-ring isotopic signal is interpreted as evidence for the climatic change at the transition from the Younger Dryas to the Preboreal. If this interpretation of the stable isotope record of the pine chronology holds true we can a minimum date for the climatic transition marking the end of the Younger Dryas in Central Europe rather precisely at 10.970 dendro BP.

Discussion and Conclusion:

Several authors have attempted the calibration of the radiocarbon time scale around 10.000 ^{14}C BP. A full account is given by Stuiver et al (1986) and Stuiver et al. (1991); a summary is presented in table 1.

Table 1. Absolute age at the YD/Preboral transition

reference	dating method	absolute age YD/PB transition [cal BP]
Hammer et al, 1986	ice cores, stable isotopes	10.720 ± 150
Strömberg et al, 1985	varves	10.700 +50 -150
Stuiver et al, 1986	tree rings, varves, ^{14}C matching	10.970 ± 110
Stuiver et al, 1991	tree rings, varves, ^{14}C matching	11.300
Bard et al, 1990, and this volume	U/Th dated corals	>= 11.090
Kromer & Becker, this work	tree rings, ^{14}C matching, stable isotopes	>= 10.970

From this compilation it can be seen that ^{14}C analyses on the Hohenheim tree-ring chronologies are important in several respects :

(1) The tree-ring sections provide sufficent material to allow ^{14}C analyses of high resolution (< =10 years) and high precision (1 sigma error in ^{14}C < = 30 years). Therefore even with floating chronologies *relative* ^{14}C variations can be determined very accurately. We have identified two long 'plateaus' of apparent constant ^{14}C ages at 9600 and 10.000 BP.

(2) The pine series provides constraints on the absolute age of the YD/Preboreal transition. It is higher (by more than two centuries) than the estimates based on ice cores and on the Swedish varve chronology. To align the pine series with these estimates, the pine series is required to overlap with the absolute oak series by at least two centuries which is immediately contradicted by the incompatible ^{14}C ages as already measured at the respective ends of the tree-ring series.

(3) Using the pattern of Holocene ^{14}C variations and taking into account the evidence from the stable isotopes ^{13}C and ^{2}H as analysed in the pine series, the minimum age of theYD-Preboreal transition is 10.970 dendro BP.

(4) Recently, Bard et al. (1990) expanded the calibration range well into glacial times by AMS ^{14}C analyses on U/Th-dated corals. For the Holocene part of the coral record overlapping with the oak chronology, an excellent agreement was found (see Bard, this volume). For the time interval covered by the pine series only one coral section exists, which is dated at 11.090 cal BP by U/Th. Its ^{14}C age is 9700 ± 110 (1 sigma) ^{14}C BP (Bard, this volume), confirmed by several conventional and AMS ^{14}C dates. The absolute age is close to the rapid transition from 10.000 to 9600 ^{14}C BP; its ^{14}C age confirms our argument about the minimum position of the pine series.

Figure captions:

Fig. 1 : Present state of the ^{14}C calibration data base as of end 1989. Most of the data are published in Radiocarbon (calibration issue, volume 28, No. 2B); the latest results will be published in a second issue.

Fig. 2 : ^{14}C ages of tree-ring chronologies as established in the Hohenheim laboratory in the interval 8000 to 12400 ^{14}C BP. The German oak chronology is dendrochronologically absolutely dated, the remaining series or single trees (oak and pine between 8.900 BP and 9200 BP, pine elsewhere) are floating. They are arranged according to their ^{14}C age. The time axis is expanded between 9500 and 10.000 BP to indicate the two long 'plateaus' at 9600 ^{14}C BP and 10.000 ^{14}C BP.

Fig. 3 : delta ^{14}C vs. dendro age (a) and ^{14}C age vs. dendro age (b) for the absolute German oak chronology (open circles) and the floating YD/Preboreal pine chronology. The pine series is matched to the oak series to result in a century-type ^{14}C oscillation (see text). Four strong oscillations are indicated (A,B,C1/C2,D) leading to 'plateaus' of apparent constant ages. The oscillation marked C1/C2 is the result of the age matching of the floating pine series to the absolute oak chronology.

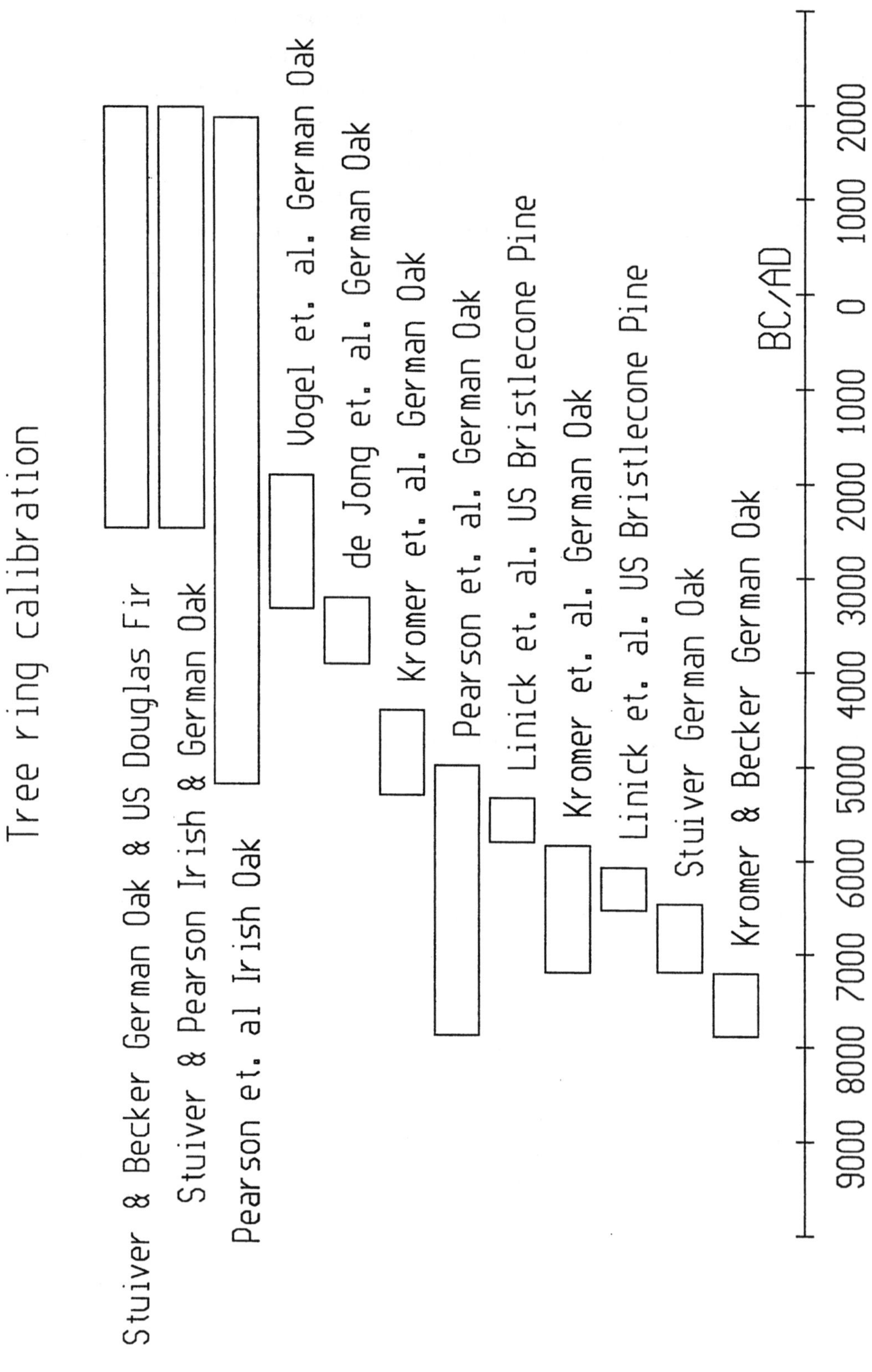
Tree ring calibration
Stuiver & Becker German Oak & US Douglas Fir
Stuiver & Pearson Irish & German Oak
Pearson et. al Irish Oak
Vogel et. al. German Oak
de Jong et. al. German Oak
Kromer et. al. German Oak
Pearson et. al. German Oak
Linick et. al. US Bristlecone Pine
Kromer et. al. German Oak
Linick et. al. US Bristlecone Pine
Stuiver German Oak
Kromer & Becker German Oak
BC/AD
9000 8000 7000 6000 5000 4000 3000 2000 1000 0 1000 2000

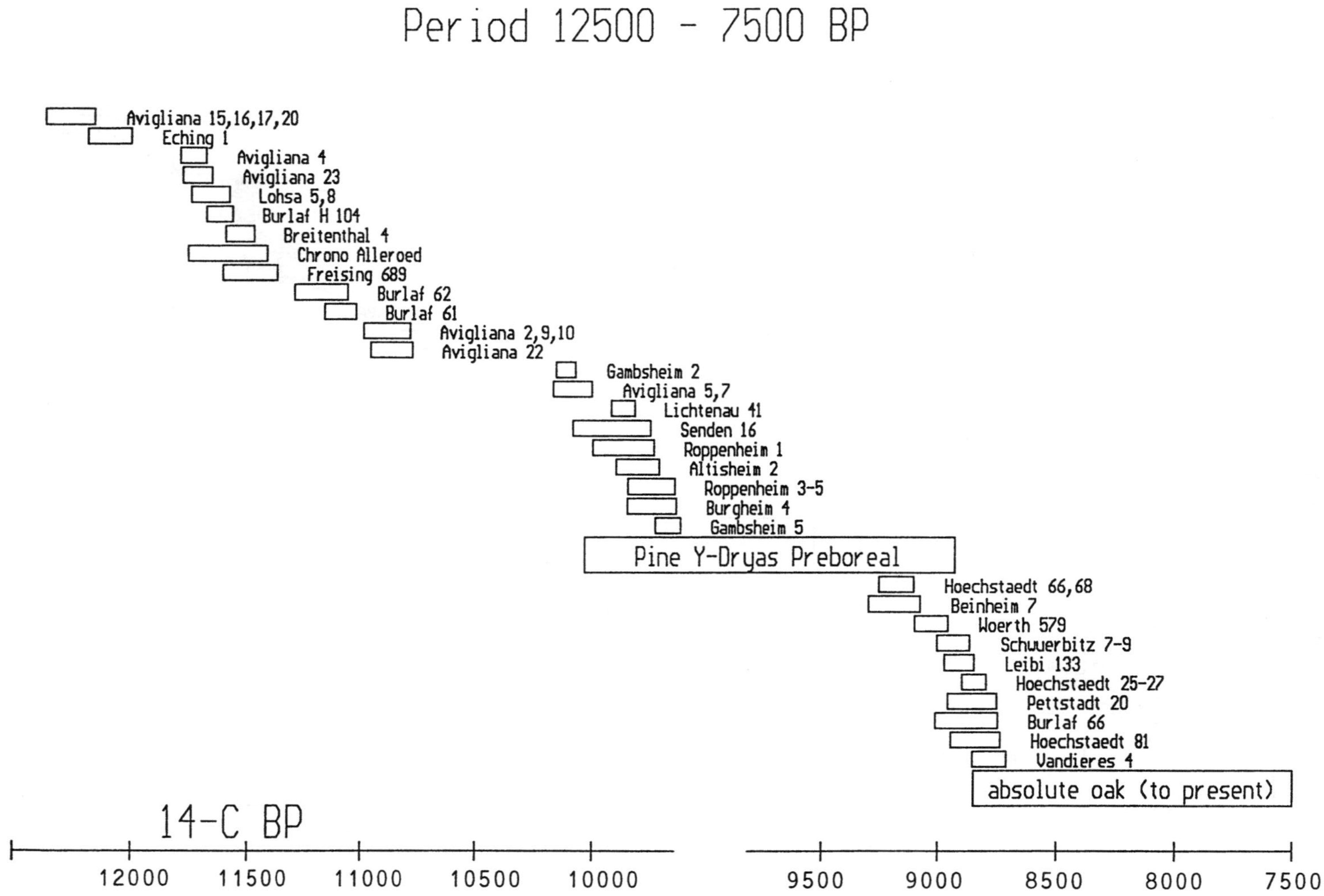
Period 12500 - 7500 BP
Avigliana 15,16,17,20
Eching 1
Avigliana 4
Avigliana 23
Lohsa 5,8
Burlaf H 104
Breitenthal 4
Chrono Alleroed
Freising 689
Burlaf 62
Burlaf 61
Avigliana 2,9,10
Avigliana 22
Gambsheim 2
Avigliana 5,7
Lichtenau 41
Senden 16
Roppenheim 1
Altisheim 2
Roppenheim 3-5
Burgheim 4
Gambsheim 5
Pine Y-Dryas Preboreal
Hoechstaedt 66,68
Beinheim 7
Woerth 579
Schwuerbitz 7-9
Leibi 133
Hoechstaedt 25-27
Pettstadt 20
Burlaf 66
Hoechstaedt 81
Vandieres 4
absolute oak (to present)
14-C BP
12000
11500
11000
10500
10000
9500
9000
8500
8000
7500

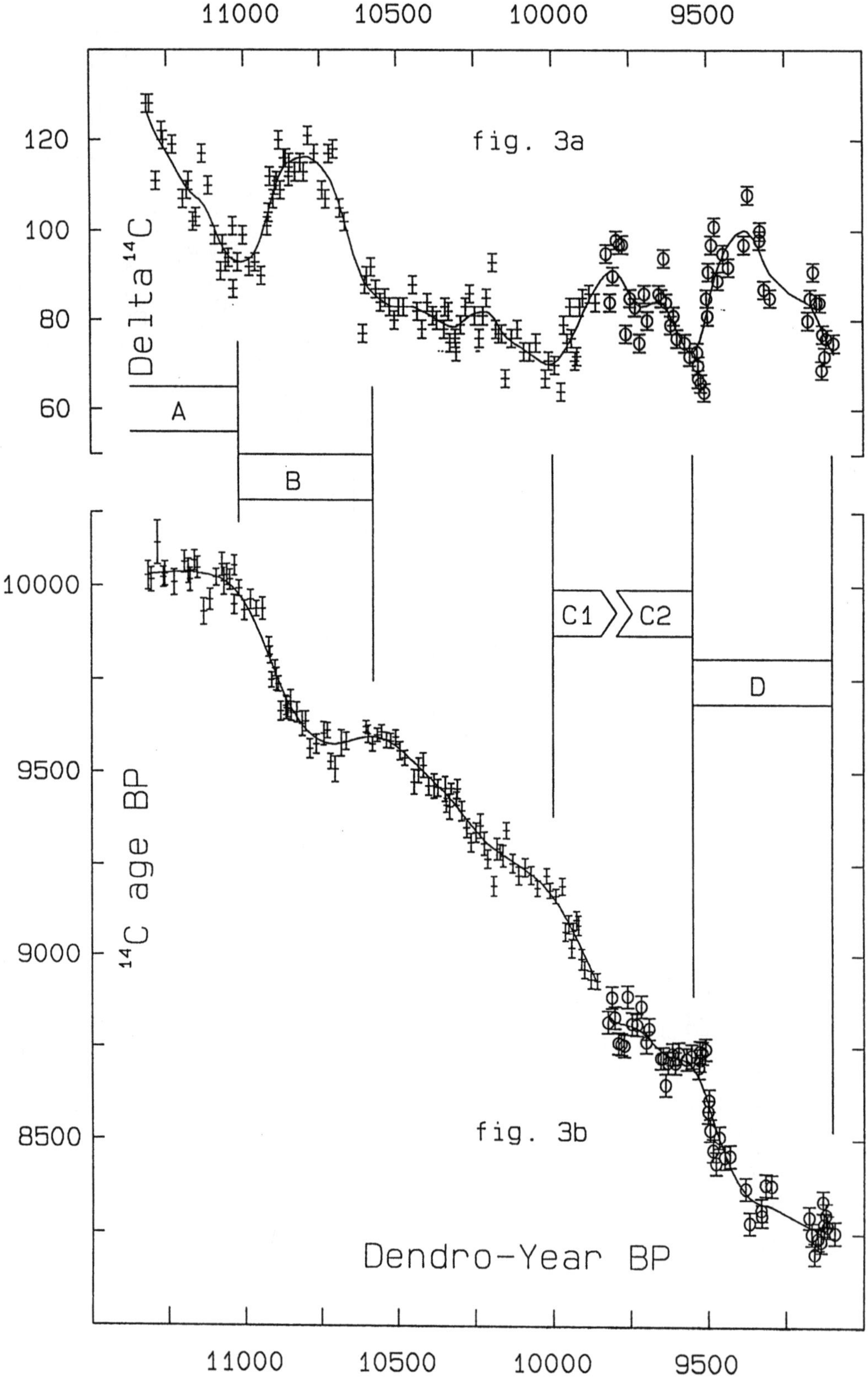

11000
10500
10000
9500
fig. 3a
Delta^{14}C
120
100
80
60
A
B
C1
C2
D
10000
9500
9000
8500
^{14}C age BP
fig. 3b
Dendro-Year BP
11000
10500
10000
9500

References :

Bard, E., Hamelin, B., Fairbanks, R.G., Zindler, A. (1990) Calibration of the ^{14}C timescale over the past 30.000 years using mass spectrometric U-Th ages from Barbados corals. Nature 345:405-410

Becker, B., Kromer, B., Trimborn, P. (1991) A stable isotope tree-ring time scale of the Late Glacial/Holocene boundary. Nature, in press.

Hammer, C.U., Clausen, H.B., Tauber, H. (1986) Ice-core dating of the Pleistocene/Holocene boundary applied to a calibration of the ^{14}C time scale. Radiocarbon 28:2A, 284-291

Strömberg, B. (1985) Revision of the Late-Glacial Swedish varve chronology. Boreas 14:101-105

Stuiver, M., Kromer, B., Becker, B., Ferguson, C.W. (1986) Radiocarbon age calibration back to 13.300 years BP and the ^{14}C age matching of the German oak and US Bristlecone Pine chronologies. Radiocarbon 28:2B, 969-979

Stuiver, M. and Braziunas, Th. F. (1989) Atmospheric ^{14}C and century-scale solar oscillations. Nature 338: 405-408

Stuiver, M., Braziunas, Th. F., Becker, B., Kromer, B. (1991) Climatic, solar, oceanic and geomagnetic influences on Late-Glacial and Holocene atmospheric $^{14}C/^{12}C$ change. Quaternary Research 35:1-24

ON FLOW MODEL DATING OF STABLE ISOTOPE RECORDS FROM GREENLAND ICE CORES

S.J. Johnsen
Science Institute
University of Iceland
Dunhaga 3
IS-107 Reykjavik
Iceland

and

W. Dansgaard
Geophysical Institute
University of Copenhagen
Haraldsgade 6
DK-2200 Copenhagen N
Denmark

Abstract

In 1988, a Nordic expedition recovered a continuous surface-to-bedrock ice core from the summit of the small separate ice cap on the Renland peninsula in Scoresbysund Fjord, East Greenland. The 324 m long core has been dated by a new ice flow model, based on the assumption that the ice close to the bottom is of Eemian origin. The flow model takes into account (i) the special ice flow at the Summit, which is characterized by a lack of shear stresses, and (ii) the enhanced flow in a relatively thin layer of soft silty ice at the bedrock. Applying the same dating technique on the Camp Century and the Dye 3 deep cores leads to good correlations between the $\delta^{18}O$ profiles along the three cores. ^{10}Be and volcanic ash spikes in the two deep cores are dated in essential agreement with independent datings.

Introduction

Profiles of stable isotope and chemical composition through the polar ice sheets are perhaps the most important palaeoclimatic continental complement to the climatic records from marine sediments. Ice cores contain a great variety of information on past environments, on atmospheric chemistry and physics, including climatic conditions, for example mean air surface temperature, which is recorded by the stable isotopic composition of the ice. However, the difficulty of independent dating of very old ice is a draw back.

Ice cores have previously been dated (i) by counting annual layers downward from the surface; this is an accurate method, but so far it is applicable only in postglacial time, at best, (Hammer et al. 1986); (ii) by radioactive isotopes, in particular ^{14}C (Andrée et al. 1984), but even with AMS the range is limited to 30,000 years B.P.; (iii) by correlation with other dated records, using e.g. reference horizons (Dansgaard et al. 1982); and (iv) by ice flow model

NATO ASI Series, Vol. I 2
The Last Deglaciation: Absolute and Radiocarbon Chronologies
Edited by E. Bard and W. S. Broecker

calculations (Haefeli 1961; Dansgaard and Johnsen 1969; Paterson and Waddington 1984; Reeh et al. 1985; Lorius et al. 1985; Reeh 1989; Dahl-Jensen 1989; Schøtt et al.).

Extending model calculations to depths approaching the ice bedrock interface has not been very successful in the past. This is mainly due to (a) the actual vertical strain rates at great depths in the ice being poorly known; (b) lack of knowledge about past accumulation rates; (c) temporal variation of the strain rate profiles along the cores; and (d) disturbed stratigraphy close to bedrock at locations far from ice divides.

A combination of the above mentioned methods (iii) and (iv) will be applied for dating the Greenland cores from Camp Century, Dye 3 and Renland. The dating method includes an improved version of a widely used flow model (Dansgaard & Johnsen 1969). It accounts for the enhanced flow in the bottom layers of silty ice, which is known to be softer than glacier ice of low impurity content (Shoji & Langway 1984; Gundestrup and Hansen 1984). This enhances the vertical strain rate in the silty, which results in much lower ages of the ice resting immediately on top of it.

All available glaciological data, and three obvious time markers in the $\delta^{18}O$ records are used as inputs to the ice flow model, which calculates the in situ annual layer thickness versus depth relationship and, by numerical integration, a time scale along the cores. As to the Camp Century and Renland cores, the basic assumption is that the high δ layers found on top of silty ice close to bedrock were deposited during the last interglacial.

The new flow model

The original flow model (Dansgaard & Johnsen 1969) assumed uniform horizontal velocity u from the surface to a distance h above the bedrock, and at greater depths u-values proportional to the distance z to the bedrock, cp. the heavy dashed u-profile in Fig. 1. The new version differs from the old one for $z < h$ by taking into account the "lubricating" effect of a relatively soft bottom layer of silty ice, which corresponds to a small, but very important sliding along the bedrock. Furthermore, an "s" shape of the vertical strain rate profile is introduced in order to account for special flow conditions close to ice divides, where shear stresses are negligible (Raymond 1983). The following expression for u is therefore used for $z < h$ (cp. the full heavy curve in Fig. 1):

$$u(z) = f_b \cdot u_o + (1-f_b) \cdot u_o \cdot z/h - u_s \cdot \sin(2\pi z/h), \qquad (1)$$

f_b being the ratio between the "sliding" and the surface velocity u_o, and $u_s = f_s \cdot u_o$ being the amplitude of a harmonically varying deviation from the linear trend of u for $z < h$.

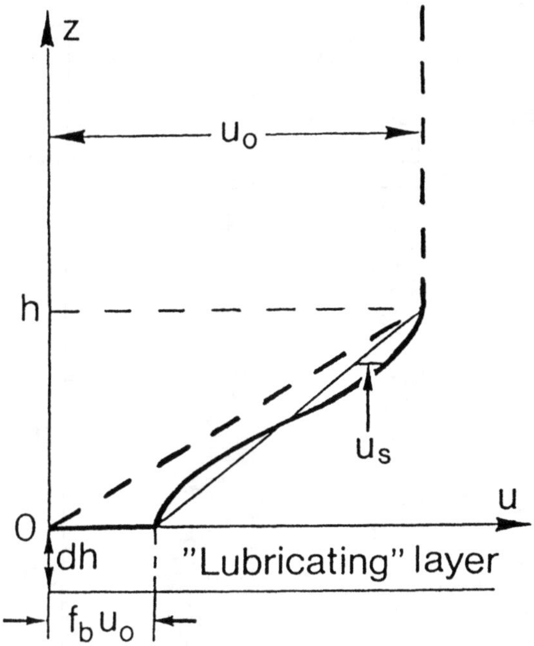

Fig. 1. Horizontal velocity profiles along an ice core frozen to the bedrock. The heavy dashed curve is the original profile used by Dansgaard and Johnsen (1969). The full heavy curve takes into account the enhanced deformation in a soft silty bottom layer, which acts as a small but important sliding over the bedrock.

If we assume a "sandwich type" of flow (i.e. ice thickness H and surface accumulation rate λ_H independent of time and geographical position), use of the continuity equation leads to an expression for the vertic al velocity as a function of z, independent of u_o and numerically identical with the in situ annual layer thickness $\lambda(z)$ (Dansgaard and Johnsen 1969). The age difference Δt between two layers z_1 and z_2 meters above bedrock is

$$\Delta t = \int_{z_1}^{z_2} dz/\lambda(z) \,. \tag{2}$$

Since H, h, f_s, and f_b are all independent of λ_H, r times less accumulation rate during the formation of the layer between z_1 and z_2 will make Δt r times higher.

Far from ice divides (Camp Century and Dye 3) we have $u_s=0$, i.e. the original u-profile (Dansgaard & Johnsen 1969) is maintained.

In case of bottom temperatures at the pressure melting point, we have uniform horizontal velocity profile along the entire core, which corresponds to $h = 0$.

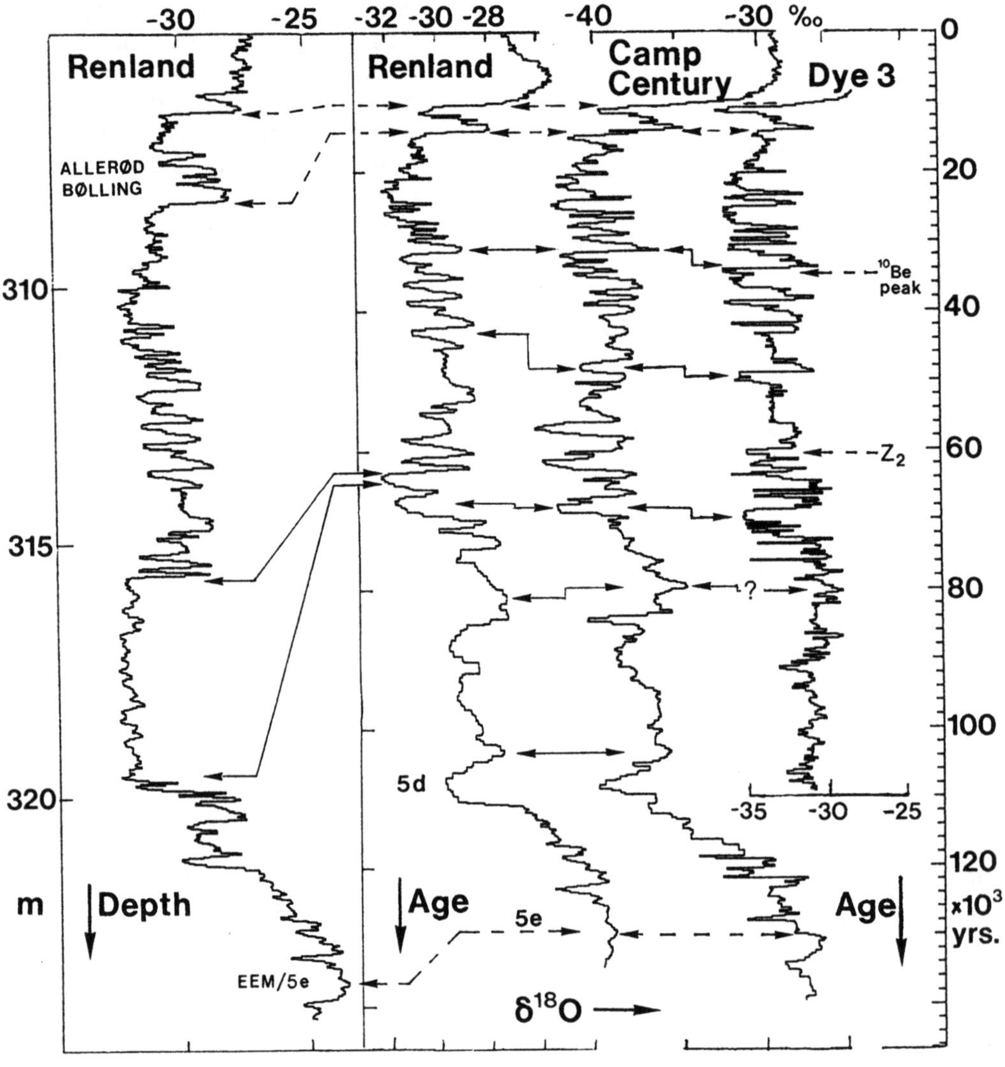

Fig. 2. *Left hand section:* $\delta^{18}O$ profile along the deepest 20 meters of a 325 m long ice core drilled through the small, isolated ice cap on the Renland peninsula in the Scoresby Sund Fjord, East Greenland. The dashed arrows point at the chosen time scale fix points listed in the bottom section of Table 1. Most of the 4 m thick, homogeneous increment between the full drawn arrows is disregarded in the time scale calculation (see the text).
Right hand section: $\delta^{18}O$ profiles (plotted on a calculated linear time scale) along the Greenland ice cores from Renland and Camp Century with scales on top, and Dye 3 with scale below the curve. Tentative reconciliation is indicated by arrows. The Dye 3 record is hardly continuous beyond 80,000 years B.P.

The Greenland ice cores

A previous publication (Dansgaard et al. 1984) showed $\delta^{18}O$ profiles plotted on a linear depth scale along the deepest few hundred meters of the ice cores from Dye 3, Southeast Greenland, and Camp Century, Northwest Greenland. The glacial to Holocene transition is marked by an abrupt increase in δ 252 and 230 m above bedrock, respectively. More than 180 m of the ice below this horizon originates from the last glaciation, as indicated by the very low δ values.

In the Dye 3 core, the low δ region extends all the way down to a 25 m thick layer of silty ice close to bedrock. The silty ice itself has δ's higher than present day snow at Dye 3, and it may therefore have been formed during the last interglacial, Eem or isotope stage 5e. Another possibility is that the silty ice at Dye 3 originates from a succeeding period (stage 5d ?) of glacier build up at low elevations, following a full or partial disintegration of the ice sheet in South Greenland during Eem (cp. Koerner 1989). It should be borne in mind, however, that the deepest ice in the Dye 3 core has travelled several tens of kilometers over a mountainous bedrock from its site of deposition. The silt, including pebbles of considerable size, may therefore be material crushed off from obstacles on the bedrock. Regardless of the origin of the silt, the layer sequence in the deepest part of the Dye 3 ice is not considered continuous (Hammer et al. 1985).

On the other hand, going downward through the glacial ice in the Camp Century core, the low δ's increase to values equal to or higher than present day δ's already 22 m above a 17 m thick layer of silty ice, which has considerably *lower* δ than present day snow at Camp Century. The high δ ice was therefore probably deposited during the Eem interglacial, and the silty ice in a cooler period preceding Eem.

The 325 m long core from the small separate ice cap on the Renland peninsula in the Scoresby Sund Fjord, East Greenland, was drilled in 1988 close to the summit (2340 m a.s.l.) by a shallow drill (Johnsen et al. 1980). The core reached very close to the bedrock, to judge from radar sounding from the surface, but approximately 1 m of silty ice may be missing. The bottom temperature is -15°C. The mean annual surface temperature is -18°C, and only little melting occurs in the summer time. The mean annual accumulation is 0.50 m of ice equivalent. The δ profile along the deepest 20 m is shown to the left in Fig. 2.

In spite of the short distance to the bedrock, where the stratigraphy might be expected to be totally destroyed by strong local stresses around scattered obstacles, many features in the Renland δ record are recognizable from those along the deep Greenland ice cores, e.g. the late glacial interstadial Bølling-Allerød; the numerous fast oscillations during most of the last glaciation; and the details in the deepest 5 m of ice with higher δ's than present day snow on the Renland ice cap.

From a climatological point of view, the essentially homogeneous layer from 315.6 to 319.6 m depth must be a fake - unchanged climate through a very long period of time is not compatible with any other climatic record. One possibility is that this layer consists of ice from the main ice sheet, deposited on Renland during an advanced stage. However, (i) in view of the vertical strain since deposition, the 4 meter layer would originally have been several hundred meters thick, and therefore not homogeneous; (ii) the high dust content indicating a very cold period of deposition (Hammer et al., 1985), makes the δ value (-32 $^{o}/_{oo}$) incompatible with an origin inland, where glacial δ values are considerably lower; (iii) both dust content and δ value of the homogeneous layer are close to those of other ice layers deposited on Renland in the cold periods later in the glaciation; (iv) an overflow by inland ice would have removed most of the originally several hundred meters of previously deposited Renland ice, which is now thinned into the 5 meters thick layer below the homogeneous ice; (v) a possible overflow would probably have been repeated several times later in the glaciation, and at least during the glacial maximum 20,000 years ago; and (vi) the Renland peninsula is almost surrounded by deep fjords that function as efficient conveyors of ice, thereby protecting the high elevation Renland plateau from being overflowed by ice created elsewhere.

Another, and perhaps more likely possibility is that the homogeneous layer was created by the Boudinage effect (Staffelbach et al. 1988), which makes particularly soft layers (i.e. those of high dust content; Gundestrup and Hansen 1984; Dahl-Jensen and Gundestrup 1984) grow or shrink irregularly, depending on coincidental, strong local stresses around scattered obstacles on the bedrock. If so, the homogeneous layer must have thickened through several tens of millennia, during which the adjoining layers underwent normal thinning and, in addition, local thinning due to the growth of the Boudinage layer.

The only way to settle the problem is to drill and analyze additional ice cores to bedrock close to the first one. Regardless of the mode of creation, however, there is no sign of disturbed stratigraphy above, nor below the homogeneous layer. We therefore remove most of it from the δ record, and re-establish a continuous δ record over the entire core length by stretching the remaining parts of the glacial record to fill in the gap, thereby compensating for the additional thinning: 9.2 m of the record above the gap (up to the shift to the relatively rigid Holocene ice), and 1.8 m below the gap (down to the shift to the relatively rigid Eemian ice) are stretched uniformly and hooked together at 318.3 m depth. This procedure does not influence the age calculation outside the 306.8 to 321,6 m depth interval.

Table 1

	Camp Century	Dye 3	Renland
Measured:			
H m ice equiv.	1369.4	2013	306.6
λ_H m ice/year	0.38	0.49	0.50
dh meters	17	25	(1)
$f_b = u_b/u_o$	0.135	0.16	
Derived:			
h meters	500 ± 60	300 ± 40	250 ± 40
$f_b = u_b/u_o$			0.010 ± 0.0015
$f_s = u_s/u_o$			0.056 ± 0.09
Accumulation rate (% of present) in the period (years B.P.):			
0 - 10,720	100	100	100
10,720 - 14,000	20	30	40
14,000 - 70,000	20	28	22
70,000 - 120,000	30	(42)	30
120,000 - 130,000	120		120
Time scale fix points (years B.P.)			
1. End of Younger Dryas	10,720	10,720	10,720
2. Onset of Bølling	14,000	14,000	14,000
3. ^{10}Be peak		35,000	
4. Culmination of Eem	130,000		130,000

Dating the cores

The parameter values used as input to the model calculation of the Greenland time scales are listed in the two upper sections of Table 1. Some of them were measured in the field (top section), others were derived (mid section) by tuning the model so as to reproduce the well known dates of three of the four easily identifiable features in the δ records shown in the bottom section of the table. This puts rather narrow limits on the choice of input parameters, cp. the ± values referring to the partial changes of the individual parameters that would cause a 5 % deviation of the calculated ages from those of the events listed as time scale fix points in the bottom section of the table:

1. the end of the Younger Dryas cold spell; age 10,720 ± 150 years (Hammer et al. 1987);
2. onset of the Bølling/Allerød interstadial; age 14,000 years;
3. a remarkably high ^{10}Be concentration peak, first demonstrated in the Vostok core at an age of 35,000 years in the Vostok chronology (Raisbeck et al.

1987); it has later been identified in the Camp Century core (Beer et al. this volume); according to the close correlation between the Camp Century and the Dye 3 δ records, the peak is to be found at z = 137 m in the Dye 3 core;

4. the culmination of the Eemian interglacial, or isotope stage 5e; age 130,000 years (Lorius et al. 1985).

The three Greenland δ records are plotted on the common linear time scale in the right hand section of Fig. 2. The Dye 3 and the Camp Century records have been shortened by omitting the silty ice. The Dye 3 core is believed to reach no more than 80,000 years back in time, at least not in a continuous sequence.

Discussion

A tentative interpretation/reconciliation is indicated in Fig. 2 by a few simultaneous δ oscillations set out in black, and by double arrows pointing at layers of presumably simultaneous deposition.

The validity of the dating procedures is supported by a number of check points:

(i) There is a close correlation, even in details, between the Greenland δ records as far as they reach, in spite of the completely different flow conditions in the areas, where the cores were recovered. With the suggested interpretation, the three time scales deviate less than 8 % throughout the range of the records, except for a 13 % difference between the Camp Century and the Renland time scales around 50,000 years B.B. The applied compensation for the Boudinage layer in the Renland core is supported by the generally excellent δ correlation with the Camp Century record in between the two time scale fix points 14,000 and 130,000 years B.P. This correlation is obvious in spite of diffusive smoothing of the Renland record (due to the extreme thinning of the deepest layers), cp. the fast oscillations, the general trends, and the details during the end of the Eemian period. As to the fast oscillations, they do not occur as regularly, nor as violently in the Antarctic records (Johnsen et al. 1972; Lorius et al. 1979; 1985). Hence, the Greenland oscillations seem to reflect a regional, probably North Atlantic climatic instability with two quasi-stable modes of atmospheric and oceanic circulation.

(ii) all of the long term variations within the last glacial cycle go in parallel in the Camp Century and the Renland records. Both of the records suggest that the 5e to 5d transition, and thereby the early stage of the build-up of new great ice sheets following the warm 5e, took 10,000 to 15,000 years and proceeded in several steps. This is also suggested by the Vostok record (Lorius et al. 1985).

(iii) the ^{10}Be peak dated at 35,000 years in the Vostok core appears at 32,500 years B.P. in the Camp Century chronology (Beer et al.; this volume).

(iv) the 64,700 ± 3,500 years B.P. volcanic zone Z_2 found in North Atlantic deep sea sediments (Ruddiman and Glover 1975), seems to appear as a visible dust layer 88 m above the bottom in the Dye 3 core. This layer was first believed to consist of bottom material (Hammer et al. 1985), but it has later been shown to be of volcanic origin (Ram and Gayley 1991). The calculated Dye 3 chronology dates it at 60,500 years B.P., in surprisingly good agreement with the measured age of Z_2, considering the strong bottom undulations under the Dye-3 flowline (Dahl-Jensen 1985), which might be expected to strongly modify the thickness of annual layers in the deepest part of the South Greenland ice sheet. If the interpretation of these horizons is correct, it excludes the possibility of the high δ layers close to bedrock being much younger than assumed.

A second pronounced dust peak from the middle of the Younger Dryas period in the Dye 3 core (cf. Fig. 14 in Hammer 1989) might also provide an important fix point for the Dye 3 time scale. Preliminary investigations of the dust particles in the peak show them to be of volcanic origin, most probably correlating with the volcanic zone 1 (Ruddiman and Glover 1975) and the Vedda ash, which has been dated at 10,600 ^{14}C years B.P. (Mangerud et al. 1984). Microprobe and PIXE analysis of the dust particles is under progress for further identification of the eruption.

In addition to time scales, the models provide informations on past accumulation rates. In Greenland, for example, they suggest that accumulation rates in glacial times were only between 20 and 30 % of the present values. These estimates must be taken as lower limits, however, considering the fact that present day values of h are lower than normal due to the relatively thin layer of soft glacial ice. Higher values of h during the glaciation (when most of the ice sheet consisted soft ice due to the high dust content, Hammer et al. 1985), enhanced the thinning of the layers from that period. On the other hand, the estimate on glacial accumulation rates is not far from another estimate based on the same ice cores (Reeh 1990).

Conclusion

Simple ice flow model calculations based on glaciological parameters and a few time scale fix points are adequate for dating of deep ice cores to a few meters above bedrock. This method is useful as long as the extremely laborious counting of annual layers is the only completely independent method for dating of ice older than 30,000 years.

Like the Camp Century ice core, the new 325 m long Renland ice core reaches back into the Eemian interglacial. Both of the δ records show that the glacial climate in the North Atlantic region was unstable, dominated by frequent shifts between mild and cold stages. They also show that the transition from stage 5e to stage 5d took at least 10,000 years, and that this cooling event was interrupted several times by shorter periods of relatively stable climatic conditions.

Acknowledgements

The Camp Century ice core was drilled in 1966 by the U.S. Army Cold Region Research Laboratory. The logistics for the recovery of the Dye 3 ice core was funded by the U.S. National Science Foundation. The Danish-Icelandic-Swedish expedition to Renland in 1988 was funded by the Nordic Council of Ministers. Chartered airplanes from The Royal Swedish Air Force, Flugfelag Nordurlands, and Grønlandsfly did the transport of personnel and cargo to Renland. The rest of this work was funded by The Danish Natural Science Research Council, The Commission for Scientific Research in Greenland, and The Icelandic Natural Science Council.

References

Andrée M, Moor E, Beer J, Oeschger H, Stauffer B, Bonani G, Hofman HJ, Morenzoni E, Nessi M, Suter M, Wolfli W (1984) ^{14}C dating of polar ice. Nucl. Instr. and Methods Phys. Res. 233: 380-384.

Beer J, Johnsen SJ, Finkel RC, Langway CC, Oeschger H, Stauffer B, Suter M, Woelfli W (1991) ^{10}Be peaks as time markers in Arctic ice cores. This volume.

Dahl-Jensen D (1985) Determination of the flow properties at Dye 3, South Greenland, by bore hole tilting measurements and perturbation modeling. J Glaciol. 31: 92-98.

Dahl-Jensen D (1989) Two-dimensional thermo-mechanical modelling of flow and depth-age profiles near the ice divide in Central Greenland. Ann. Glaciol. 12: 31-36.

Dansgaard W, Clausen HB, Gundestrup N, Johnsen SJ, Rygner C (1985) Dating and climatic interpretation of two deep Greenland ice cores. In: Langway CC Jr, Oeschger H, Dansgaard W (eds.) Greenland Ice Cores: Geophysics, Geochemistry and Environment. Am. Geophys. Un. Monograph 33: 71-76.

Dansgaard W, Johnsen SJ (1969) A flow model and a time scale for the ice core from Camp Century, Greenland. J Glaciol. 8: 215-223.

Dansgaard W, Clausen HB, Gundestrup N, Hammer CU, Johnsen SJ, Kristinsdottir PM, Reeh N (1982) A new Greenland deep ice core. Science 218: 1273-1277.

Gundestrup NS, Hansen BL (1984) Bore-hole survey at Dye 3, South Greenland. J Glaciol. 30: 282-288.

Haefeli R (1961) Contribution to the movement and the form of ice sheets in the Arctic and Antarctic. J Glaciol 3: 1133-1151.

Hammer CU (1989) Dating by physical and chemical seasonal variations and reference horizons. In: Oeschger H, Langway CC Jr (eds.) The environmental record in glaciers and ice sheets, Dahlem Konferenzen 1988. Phys., Chem., and Earth Sci. Res. Rep.: 99-121.

Hammer CU, Clausen HB, Dansgaard W, Neftel A, Kristinsdottir P, Johnson E (1985) Continuous impurity analysis along the Dye 3 deep core. In: Langway CC Jr, Oeschger H, Dansgaard W (eds.) Greenland Ice Core: Geophysics, Geochemistry and Environment, Am. Geophys. Un. Monograph 33: 90-94.

Hammer CU, Clausen HB, Taüber H (1986) Ice-core dating of the Pleistocene-/Holocene boundary applied to a calibration of the ^{14}C time scale. Radiocarbon 28: 284-291.

Johnsen SJ, Dansgaard W, Clausen HB, Langway CC Jr (1972) Oxygen isotope profiles through the Antarctic and Greenland ice sheets. Nature 235: 429-434 and 236: 249.

Johnsen SJ, Dansgaard W, Gundestrup N, Hansen SB, Nielsen JO, Reeh N (1980) A fast lightweight core drill. J Glaciol. 25: 169-174.

Koerner RM (1989) Ice core evidence for extensive melting of the Greenland ice sheet in the last interglacial. Science 244: 964-968.

Lorius C, Merlivat L, Jouzel J, Pourchet M (1979) A 30,000 yr isotope climatic record from Antarctic ice. Nature 280: 644-648.

Lorius C, Jouzel J, Ritz C, Merlivat L, Barkov NI, Korotkevich YS, Kotlyakov VM (1985) A 150,000 year climatic record from Antarctic ice. Nature 316: 591-596.

Mangerud J, Lie SE, Furnes H, Kristiansen IL, Lømo L (1984) A Younger Dryas ash bed in Western Norway, with possible correlations to the Norwegian Sea and the North Atlantic. Quat. Res. 21: 85-104.

Paterson WBS, Waddington ED (1984) Past precipitation rates derived from ice core measurements: Methods and data analysis. Rev. Geophys. Space Phys. 22:123-130.

Raisbeck GM, Yiou F, Bourles D, Lorius C, Jouzel J, Barkov N (1987) Evidence for two intervals of enhanced ^{10}Be deposition in Antarctic ice during the last glaciation period. Nature 326: 273-276.

Ram M, Gayley RI (1991) Long range transport of volcanic ash to the Greenland ice sheet. Nature 349: 401-404.

Raymond CF (1983) Deformation in the vicinity of ice divides. J Glaciol. 29: 357-373.

Reeh N (1989) The age-depth profile in the upper part of a steady-state ice sheet. J Glaciol 35: 406-417.

Reeh N (1990) Past changes in precipitation rate and ice thickness as derived from age-depth profiles in ice-sheets; application to Greenland and Canadian Arctic ice core records. In: Bleil U and Tiede J (eds) Geological History of the polar oceans (Nato series edited by Kluwer Academic Press): 255-271.

Reeh N, Johnsen SJ, Dahl-Jensen D (1985) Dating the Dye 3 deep ice core by flow model calculations. In: Langway CC Jr, Oeschger H, Dansgaard W (eds) Greenland Ice Cores: Geophys., Geochim., and Environment. Am. Geophys. Un. Monograph 33: 57-65.

Ruddiman WF, Glover LK (1975) Subpolar North Atlantic Circulation at 9,300 yr B.P.: Faunal Evidence. Quat. Res. 5: 361-389.

Ice Cores: Geophys., Geochim., and Environment. Am. Geophys. Un. Monograph 33: 57-65.

Ruddiman WF, Glover LK (1975) Subpolar North Atlantic Circulation at 9,300 yr B.P.: Faunal Evidence. Quat. Res. 5: 361-389.

Shoji H, Langway CC Jr (1984) Flow behavior of basal ice as related to modeling considerations. Ann. Glaciol. 30: 141-148.

Staffelbach T, Stauffer B, Oeschger H (1988): A detailed analysis of the rapid changes in ice core parameters during the last ice age. Ann. Glaciol. 10: 167-170.

Schøtt C, Waddington ED, Raymond CF (to be published) Predicted time scales for GISP2 and GRIP boreholes at Summit, Greenland. J Glaciol.

Errata: In figure 1 the thick line at z=0 should not be drawn from (0,0) but from (0,-dh).

THE CLAY-VARVE BASED SWEDISH TIME SCALE AND ITS RELATION TO THE LATE WEICHSELIAN RADIOCARBON CHRONOLOGY

Svante Björck, Ingemar Cato [1], Lars Brunnberg [2] and Bo Strömberg[3]
Department of Quaternary Geology
Lund University
Tornav. 13
S-223 63 LUND
Sweden

INTRODUCTION

The Swedish Time Scale is mainly associated with two Swedish geologists: the pioneer of varve chronology, Gerard De Geer and one of his pupils, Ragnar Lidén.

The varve chronology was born a little more than one hundred years ago when, for the first time in history, De Geer made correlations between three clayey varve series taken at different localities in Stockholm (De Geer 1884). He thereby proved and pursued a four-year-old idea of his own, i.e. that the varves of the glacial clay corresponded to an annual sedimentation in front of the inland ice and that this could probably be used as a method of dating the recession of the inland ice.

According to the classical view, suspended sediment was brought into the Baltic basin from the melting inland ice. The silt, being heaviest, settled out first, but the clay particles remained in suspension, settling out gradually during autumn and winter after spring and summer melting had ceased. The low temperature and hence relatively great density of the fresh water were important factors in delaying settlement of the clay particles. This gradual separation caused the observed gradation from the light coloured coarse layer to the dark coloured fine grained layer making up one varve. With resumption of melting the next spring, new sediment entered the lake, and the coarse fraction, settling out rapidly, produced the sharp contact between the top of the older varve and the base of the younger

---

1) Geological Survey of Sweden, Box 670, S-751 28 UPPSALA, Sweden

2) Department of Quaternary Research, Stockholm University, Odengatan 23, S-113 22 STOCKHOLM, Sweden

3) Department of Physical Geography, Stockholm University, S-106 91 STOCKHOLM, Sweden

NATO ASI Series, Vol. I 2
The Last Deglaciation: Absolute and Radiocarbon Chronologies
Edited by E. Bard and W. S. Broecker

one. The younger varve was not only deposited above the older varves but also over recently exposed bottom areas due to the annually retreat of the inland ice. The process was repeated each sedimentary year and resulted in varves covering each other like tiles on a roof from the southern Baltic to the northern central Sweden, i.e. during the entire subaqutic melting period of the inland ice. The sedimentary year commences with the spring.

De Geer developed a method of measuring varves, where each varve in a sequence plotted against its thickness. Connected with each other they form a curve. From the slope of the curve the relative thickness appears, the significant parameter, since absolute thickness varies from one locality to another. Curves from different localities, placed side by side,can then be moved up or down until a correlation appears. The curves from a number of localities are then combined into a chronology longer than that implied at any one locality. However, the connections are not only based on the relative varve thicknesses and changes, but also on the individual feature of several "unique" varves. By this way both the Swedish and Finnish varve chronologies were built.

De Geer's discovery and connection of varve series from Stockholm was a sensation, but it still took another 20 years until, with great intensity and the aid of young geologists, he was able to continue the work of obtaining a chronology for the last part of the Ice Age. This was an admirable work since at this time only open exposures, digged by hand, were used for varve studies. During the XIth International Geological Congress in Stockholm 1910 De Geer was then able to present the outlines of a glacial time scale (De Geer 1912), but the investigations proceeded until in 1940 in the classical work "Geochronologia Suecica" he published a "complete" time scale for the recession of the inland ice in Sweden from Skåne in the south to Indalsälven in the north (De Geer 1940). Supplements, connections and extensions, which will be commented below, have since then been produced by various scientists (see, Fig. 1 and Strömberg 1985b, 1989), but the main features of De Geer's chronology still remain.

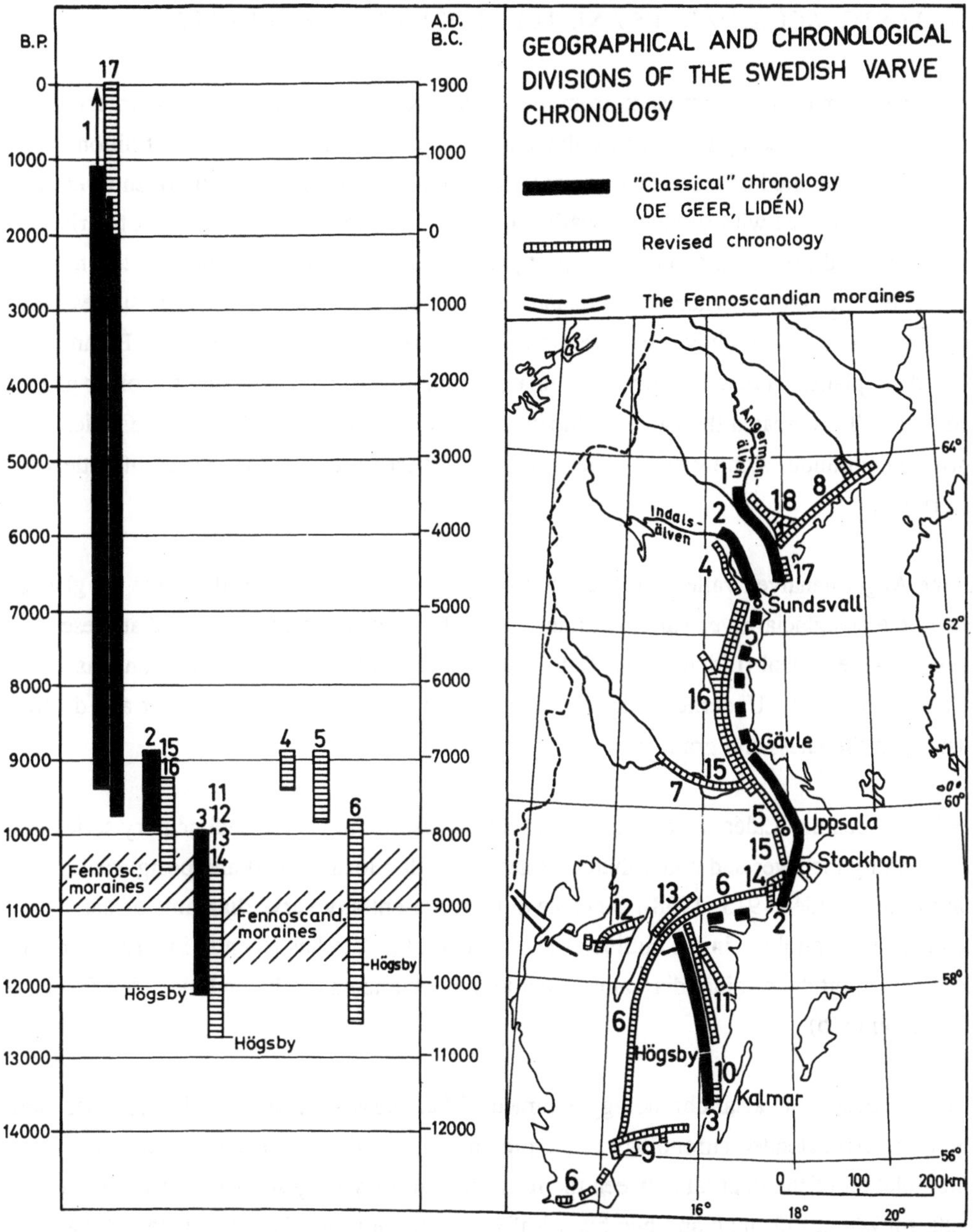

Fig 1. The Swedish time scale in 1989 (from Strömberg 1989). 1 = Lidén 1913 and 1938. 2 = De Geer 1940. 3 De Geer, unpublished. 4. Borell & Offerberg 1955. 5 = Järnefors 1963. 6 = Nilsson 1968. 7 = Fromm 1964. 8 = Hörnsten 1964 and Bergström 1968. 9 = Ringberg 1979 and Ringberg & Rudmark 1985. 10 Rudmark 1975. 11 = Kristiansson 1982 and 1988. 12 = Caldenius 1944, Strömberg 1977, 1985 and 1989. 13 = Perhans unpublished. 14 = Brunnberg 1989. 15 = Strömberg 1989. 16 = Forsmark, unpublished. 17 = Cato 1985 and 1987. 18 = Fözö 1980.

THE POSTGLACIAL VARVES AND THE LINK WITH THE PRESENT

After the ice-margin had retreated from the interior portions of the fjords in northern Sweden, and receded up to the areas above the highest coast line (HK, see definition in Cato 1982), the mud of the glacial rivers was carried out into the fjords by open rivers in which it was mixed with sediments washed out from the ice-free land areas. At the river mouths the sediment was deposited as postglacial deltas. Through the land uplift, which proceeded after the recession of the inland ice, the outlets of the rivers moved successively from the HK at 230 m above sea level down to the present level of the Gulf of Bothnia. New delta planes were constantly accumulated upon lower levels covering the older glacial sediments. This created the valley sediments of northern Sweden with a surface which forms a continuous sloping delta plane from the highest shore line to the present outlets of the rivers.

In the Ångermanälven valley, northern Sweden (Fig. 1), Lidén found that, like the glacial clays, the postglacial river valley sediments were laminated and he assumed that these laminae were annual (Lidén 1911). Cato`s studies showed that this presumption was valid (Cato 1985, 1987). Underneath the delta surface the varves overlap each other and decrease distally in thickness and grain size.

Early this century Lidén devoted many years to measure the varved sedimentary series in bluffs along the river and could thereby establish both a local glacial and postglacial chronology (Lidén 1913, 1938). Since he did not present any varve diagrams of the postglacial chronology in his outline article from 1938, Cato has recently incorporated and commented on the original diagrams found in the documents left behind by Lidén (Lidén (†) & Cato in prep).

Lidén`s local postglacial chronology embraces 7,522 varves measured at 19 localities along the river. An extended chronology to the present was roughly achieved by Lidén by interpolation of the interval between the formation of his youngest measured varve on the delta surface at Prästmon and two historically datable shore levels in the valley of the Ångermanälven river: (Lidén 1938). However, several authors had emphasized Lidén`s weak connection of the postglacial time scale with the present, since it was based mainly on extrapolated shore displacement data (Wenner 1968, Fromm 1970, Tauber 1970).

Several attempts have been made to get a more exact extension of the Swedish Time Scale to the present (G De Geer 1912, Caldenius 1924, E H De Geer 1933, Kullenberg & Fromm

1944, Fromm unpubl., and Granar 1956) but no researcher progressed further than Lidén (1938) until recently when Cato presented an exact connection with the present time (Cato 1985, 1987).

In order to establish a definitive link between the Swedish Time Scale and historical time, Lidéns chronology had to be supplemented with new varve sequences that extend to the present. In the years 1978-1983 Cato collected 25 sediment cores, amounting to a total lenght of 200 m, from the rivermouth and 20 km upstreams in the valley of the Ångermanälven Over 10,000 varves were measured and counted before all the connections were clear within the barely 15 km long stretch of river (Fig 2). The calendar year affinity of the recent varves formed in the non-tidal estuary of the river was proved by several different methods (Cato 1987). They all showed that the varves are annual and that the current sedimentation and unique varve formation are still in progress in the area.

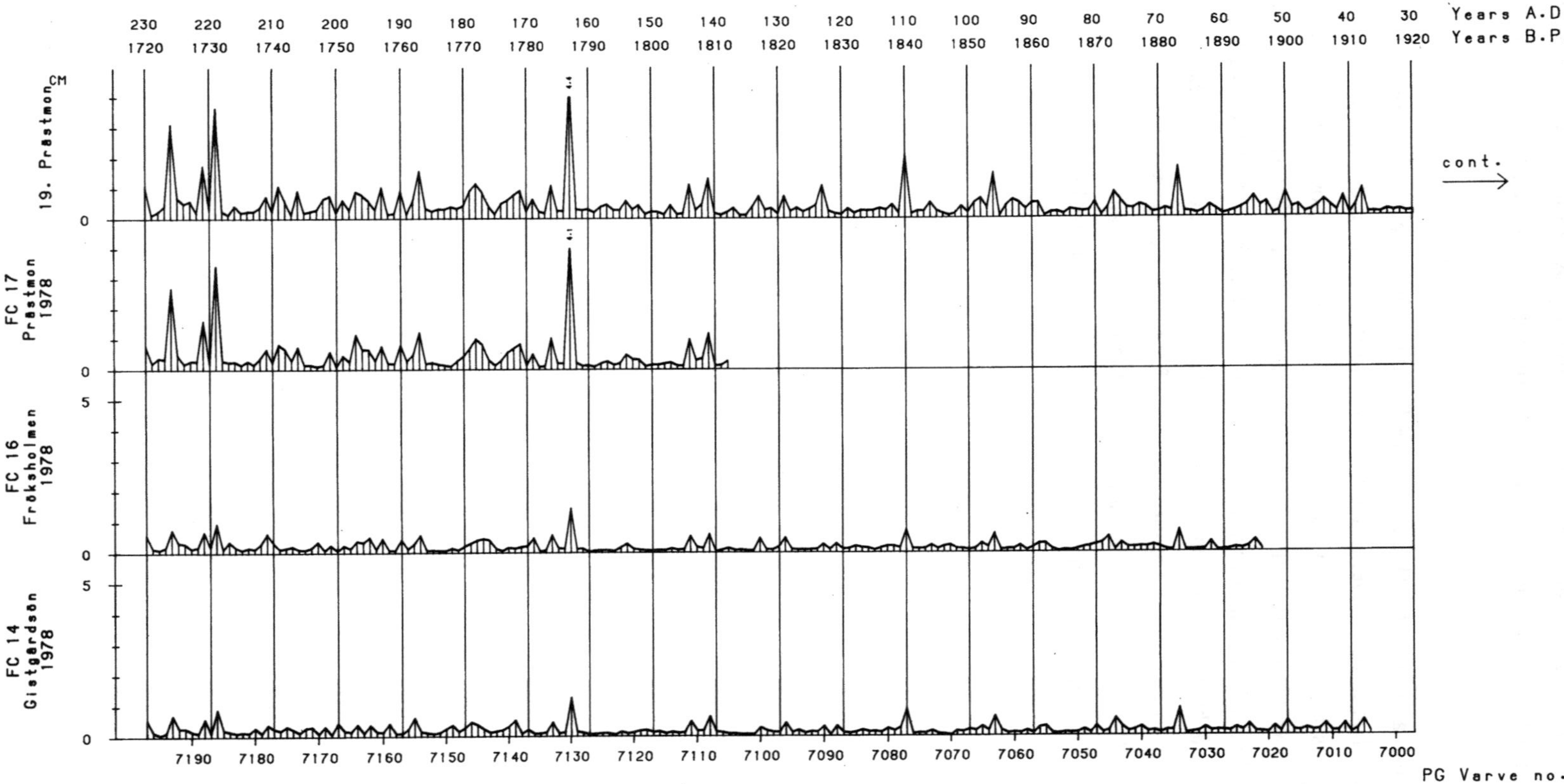

Fig 2. Varve diagrams showing the connection between Cato`s (FC14, FC16, FC17) and Lidén`s (19. Prästmon) varve series in the Ångermanälven valley 1720 to 1920 B.P. (from Cato 1987).

The combination of Cato`s varve series (Cato 1987) with Lidén`s (Lidén (†) & Cato in prep) yielded a complete varve chronology embracing 8,945 varves for the entire postglacial period (up to the fieldwork of Cato in early 1979) in Ångermanland, i. e. 8916 varves up to 1950 = BP. Here it must be stressed that the change from glacial to postglacial sedimentation is not synchronous in different valleys, which implies that the age relation must be determined by varve connections and not by varve numbers, since different local time scales were used in different valleys, e.g. by Lidén (1913) and Caldenius (1924).

THE SWEDISH ZERO-YEAR

The zero-year, which since long denotes the boundary between finiglacial and postglacial time in the Swedish Time Scale was first defined by De Geer (1940) as the varve corresponding to an extreme drainage of melt water into the Indalsälven river (Fig 1). The varve was originally thought to represent a bipartition of the remaining Scandinavian ice sheet (an event discussed and questioned however, by inter alios Lundqvist 1959, 1973, E H De Geer 1959, Caldenius 1960, Fözö 1980).

However, several different drainage varves were deposited in the Indalsälven valley, and Borell & Offerberg (1955) found that De Geer (1940) had made a confusion. In this correlation between the 'finiglacial' and the 'postglacial' time scales in the Indalsälven and Ångermanälven valleys he had used a drainage varve which was c. 80 years too young.

The zero-year varve was previously dated by Nilsson (1960, 1968) on the basis of Lidén`s provisional connection with the present. Now it is dated by Cato (1985) to 9,238 varve years B.P., as a consequence of his connection of the Swedish Time Scale to present time. Nilssons chronological calculations were recently checked and confirmed by Fromm (1985) and Strömberg (1989, p 5-10).

The total number of postglacial varves in the Swedish Time Scale is according to Cato (1985) 9,266 (including the zero-year varve) up to recent time (early 1979). The margin of error has for the total period been estimated at +10,-180 years (Cato 1987).

THE FINIGLACIAL VARVES

The finiglacial varves were deposited during the deglaciation of east-central Sweden from the Stockholm area in the south to the Sundsvall area in the north (Fig. 1). This part of the Swedish Time Scale ends with the zero-varve mentioned above.

A varve chronology for the entire finiglacial period was first presented by G De Geer (1940). The chronology was later revised by Järnefors (1963) and recently also by Strömberg (1989). The revision of Järnefors implied an extension of De Geer`s Time Scale for deglaciation between Uppsala and Sundsvall by 19 varve years. Strömberg`s revision, based on more than 200 varve measuring sites, led to a correction and a further extension with 64 varve years north of Uppsala and 35 varve years between Stockholm and Uppsala. According to Strömberg`s varve correlations De Geer`s recession line -1073, just south of Stockholm, is now dated -1191, i.e. 10,429 varve years B.P. An estimated margin of error for the finiglacial part of the Swedish Time Scale is +-25 years and for the entire period from the recession line -1191 to 1950 A.D. the error is estimated to +35,-205 years (Strömberg 1989, Cato 1987).

THE CONNECTION BETWEEN THE SWEDISH AND FINNISH VARVE CHRONOLOGIES

A connection between a lateglacial part of the Swedish and Finnish varve chronologies have recently been carried out by Strömberg (1990). His connection is based on varve measurements from east-central Sweden via Åland in the Baltic to south-western Finland. According to Strömberg, the Finnish varve 500 corresponds to the Swedish varve -905, i.e. the Finnish varve can be dated to 10,143 B.P. and the Finnish +-0 varve (Sauramo 1923) to 10,643 BP as a consequence of Cato`s connection of the Swedish Time Scale with the present (Cato 1985, 1987). A maximum margin of error was by Strömberg estimated at +25, -195 varve years without regard to possible errors in the Finnish Time Scale. Strömberg`s connection is strengthened by the fact that the so called spot zone with limestone fragments is found not only in east-central Sweden but also in a corresponding stratigraphical position at Åland and in Finland.

THE GOTIGLACIAL VARVES

The gotiglacial varves were deposited during the deglaciation ofsouthern Sweden, from the province of Skåne to approximately the Stockholm area (Fig. 3). According to the ^{14}C deglaciation chronology of e.g. Björck e al. (1988), more or less all of these varves are of Late Weichselian age, i.e. older than 10,000 ^{14}C years B.P. As been mentioned above the pioneer works on these varves were carried out in the early years of this century by G. De Geer and his collaborators. With the exception for the local varve chronology of NE Skåne and W Blekinge (Antevs 1915) the varve diagrams were never published, although a map published by De Geer showed e.g. that the outermost coast of Blekinge was deglaciated 4,000 years before De Geer's zero varve. Although the original diagrams measured along the Swedish south-eastern coast were never published by De Geer they were, among others, used by Nilsson (1968) to create a calendar year based (in years B.C.) deglaciation chronology of Southern Sweden, which Ringberg (1971) used to date the deglaciation of Blekinge. His detailed varve chronology showed that the southern half of Blekinge was deglaciated between c. and 12,200 and 12,050 varve years B.P. The criticism of Nilsson's (1968) time scale by e.g. Mörner (1969) and Lundqvist (1975) led, however, Ringberg (1979) to regard his chronology as a local time scale, valid only for NE Skåne and Blekinge.

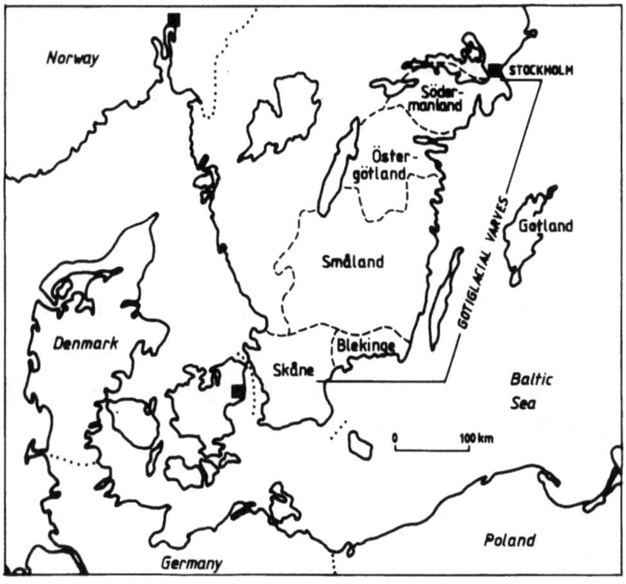

Fig 3. Map of South Sweden showing the provinces along the east coast where (Late Weichselian) varves are found, are preliminary connected with the present, and have been correlated to the ^{14}Ctime scale.

The two other rather recently published varve chronological works along the Swedish east coast south of Stockholm, Rudmark (1975) and Kristiansson (1986), did also result in local varve chronologies without any connection to a time scale. Rudmark's (1975) chronology was short and actually seems to be synchronous with Ringberg's (1971) youngest (Ringberg & Rudmark 1985) and Kristiansson's (1986) oldest varves (Björck & Möller 1987). Based on those assumptions, combined with correlations of varve stratigraphic data, Björck & Möller (1987) actually made a preliminary connection between Ringberg's (1979) and Kristiansson's (1986) local varve chronologies. This important connection will be further evaluated in a coming project by Björck and Wohlfarth.

The most critical part of the gotiglacial varve sequence is the varve stratigraphical relationship between Kristiansson's (1986) younger varves and Brunnberg's (in prep.) gotiglacial-finiglacial varve sequence. Another critical task is to connect Caldenius' (1944), Nilsson's (1968), Strömberg's (1985a), and Perhans' (unpublished) varve series further west with the varve series of Kristiansson/Brunnberg. These connections are necessary to be able to connect the varve series of the Younger Dryas ice marginal zone with the present time. Although a provisional connection already exists (Strömberg 1985a, 1989) the connection has to be more firm by the connection to a Kristiansson/Brunnberg time scale. Below follows a short summary of the gotiglacial varves along the South Sweden east coast, geographically subdivided into three groups: 1) The Stockholm-Södermanland varves (Brunnberg in prep), 2) the Östergötland-Eastern Småland varves (Kristiansson 1986) and 3) the SE Småland-Blekinge-NE Skåne varves (Ringberg 1971, 1979; Ringberg & Rudmark 1985; Rudmark 1975).

The Stockholm-Södermanland varves

This varve series consists of two main units, situated below the Red Glaciomarine Marine Clay unit of the finiglacial clay: the upper Gray Clay and the Greyish Brown Clay (Brunnberg 1988). The youngest of these youngest gotiglacial varves is the varve preceding the -1191 varve, called the 10,430 varve year.

The Grey Clay is an unusually well-developed varved clay with distinct varve boundaries and a well-developed graded bedding in the fine silt-clay fraction. It consists of c. 310 varves and was deposited between 10,430 and 10,740 varve years B.P. The lower boundary to the Greyish Brown Clay is very distinct and most likely represents an important change in sedimentation environment. This change may be related to the final drainage of the Baltic

Ice Lake. The most recent ^{14}C age of that event is 10,300-10,400 ^{14}C years B.P. (Björck & Digerfeldt 1989, Svensson 1989).

The Greyish Brown Clay is a relatively thin-varved silty clay with varve thicknesses between 2 and 10 mm. Altogether 600 varves (years) of this type have been found in the area and this unit spans between 10,740 and 11,340 varve years B.P. According to radiocarbon based deglaciation chronologies the varves should have been deposited in the Younger Dryas and perhaps also during the very end of the Alleröd Chronozone as defined by Mangerud et al. (1974).

The Östergötland-Småland varves

Kristiansson's (1986) local chronology covers a distance of 180 km along the coasts of the provinces of Östergötland and the northern half of Småland. The chronology, which is mainly based on his own, Nilsson's (1968) and De Geer's unpublished diagrams, contains c. 2300 varve years. The most well defined part of the chronology covers 150 km, which was deglaciated between the local years 2825 and 1339, i.e. a mean ice recession rate during these 1486 years of c. 100 m/year. An important result of Kristiansson's (1986) study was that he managed to relate his varve chronology to the Younger Dryas stadial, i.e. when the ice margin oscillated in the Middle Swedish end moraine zone. The beginning of this relative still stand was dated to the local varve year 1760 and lasted for 400-500 varve years. Most of the varve sequences in this zone are disturbed (by the oscillating ice margin) and the disturbances occurred between the years 1660 and 1460. The ice recession rate south of the YD varied between 80 and 280 m/year, while the mean rate within the zone is 30-40 m/year. The rate just north of the zone is estimated to 150-200 m/year (Kristiansson 1986). The equicesses in the whole area display a NNE-SSW direction.

In this context one should mention that we suspect that Kristiansson's (1986) Alleröd-Younger Dryas varve complex may contain a few hundred varves (years) more than presented in his work. This assumed "weak" part of the chronology is for the moment being investigated and it may result in a slightly extended time scale than at present.

Unfortunately the varves cored in this region do not display the distinct color and lithology changes that Brunnberg has found to be so typical and useful in the Stockholm-Södermanland area. This may owe to the fact that Brunnbergs corings have mainly been performed in lake basins where the cored sediments have been situated below the ground water level since deglaciation. Kristianssons's (1986) varves have been measured, however, on clays sampled with the foil piston corer on the clay plains or from open sections. A fluctuating ground water level may cause secondary color changes that are useless for correlations.

The SE Småland-Blekinge-NE Skåne varves

Rudmark (1975) established a 119 year long varve chronology in a 40 km long and 8-12 km wide area along the coast of southeasternmost Småland. The chronology was based on 25 sites with 19-76 varves in each varve series. He found that the equicesses are more or less parallel to the north-south stretching coast, and the rate of ice recession (from east to west) was estimated to 125-150 m/year with locally even higher rates between the mainland and the island of Öland. Since Rudmark (1975) found an almost perfect connection to one of Ringberg's (1971) northeasternmost sites he used Ringberg's calendar year chronology (based on Nilsson's time scale) to date the deglaciation of the coast area of SE Småland to c. 12,150-12,050 varve years B.P.

Ringberg's (1971) studies in the eastern part of the province of Blekinge were based on 40 sites of which 22 were originally measured (but unpublished) by De Geer and his co-operators. Apart from the easternmost part of the province, where the equicesses turn north, the equicesses are more or less parallel to the coast, i.e. stretch west-east. The Blekinge time scale, when connected to Nilsson's (1968) chronology, covered approximately the years 12,150-11,750 varve years B.P. The deglaciation from today's coastline (disregarding the archipelago and peninsulas) to the highest shoreline, a distance of 12-15 km, occurred between c. 12,150 and 12,030 varve years B.P. with a mean deglaciation rate of 90 m/year.

Except for the youngest and most distal varves the varves in Blekinge are usually quite distinct with sharp boundaries between both varves and summer and winter layers. A special type of varves are the so-called 60-85 varves with a diffuse appearance (Ringberg 1979), dated to c.11,965-11,880 varve years B.P. (Ringberg 1971). They are rather thick, 1-3 cm, with generally thicker summer layers and the diffuse appearance is explained by the mixed grain-size of the summer layer, containing both clay and silt laminea. Björck & Möller (1989) related these varves to a rather warm period at c. 12,300-12,400 ^{14}C years B.P. creating a climax of stragnant ice melting in Blekinge and southern Småland and the formation of large deltas in Blekinge c. 10 m below the highest shore line. Ringbergs (1979) new chronology for NE Skåne and Blekinge was based on more than 185 sites. Most of these are from western Blekinge and NE Skåne and were originally presented by Antevs (1915). Based on the criticism of Nilsson's (1968) time scale e.g. by Mörner (1969), Lundqvist (1975), and Berglund (1976), Ringberg (1979) started out from Antev's (1915) thick varve no -100, whose equicess runs approximately along the Blekinge south coast, to create a floating local chronology for NE Skåne and Blekinge. The deglaciation of the area below the highest shore line occurred between the local years -325 and +20, and as a whole the chronology covers 640 years, i.e. -325 to +315 in the local chronology. Although local varved clays have been found and measured south and southwest of the Kristianstad Plain in NE Skåne no reliable connections have been made between these and Ringberg's (1979) floating chronology. This probably owes to the fact that these rather scattered occurrences of varved clay in south Skåne were deposited in small and restricted glacial lakes, while the above mentioned gotiglacial varves were deposited in the same basin, i.e. the Baltic Basin. The criticism of Nilsson's (1968) time scale was based on the same type of reasoning, since he mainly based his gotiglacial time scale on connections between large but still restricted glacial lake basins in southern Sweden.

ATTEMPTS TO RELATE THE GOTIGLACIAL VARVES WITH THE ^{14}C TIME SCALE

Even before the Swedish Time Scale was reliably connected to the present (Cato 1985) attempts were made to estimate the differences between the varve chronology and the radiocarbon time scale (cf. Mörner 1969, 1970; Tauber 1970). Comparisons between clay-varve and ^{14}C dated deglaciation chronologies have, for natural reasons, been the most obvious way to find differences between the time scales. When Cato's (1985) connection to the present was presented, Strömberg (1985b) estimated that the amount of varve years exceed ^{14}C years with 200-400 years at the end of the younger Dryas. When Lundqvist (1986) corrected the clay-varve chronology according to Cato's (1985) findings (+365 years) and compared it with a possible ^{14}C based deglaciation chronology he found that there were no significant differences between the chronologies. There are, however, great uncertainties when time scale comparisons are based on differences between these two types of deglaciation chronologies since, (a) correlations between ice marginal zones along the west coast and ice recession lines along the east coast are often hypothetical, (b) there exist so many different ^{14}C based deglaciation chronologies for southern Sweden, and (c) that these chronologies are often based on different types of radiocarbon dated material (cf. Björck et al. 1988) that are neither very reliable nor date the deglaciation event in the same precise way that the clay-varves do. Marine molluscs and mammals and the low organic lake sediments are examples of such radiocarbon dated material.

As a consequence of the above described uncertainties, one of Björck & Sandgren's (1986) main aims behind establishing a Late-Weichselian magnetostratigraphy was to create possibilities to make direct connections between the two time scales. The basic idea was that (a) the Late Weichselian secular variations should be synchronous within southern Sweden and (b) that these variations could be documented both in radiocarbon dated lake sediments and in clay-varve dated clays. It was soon found (Björck & Sandgren 1986, Sandgren 1986) that lake sediments from the late Alleröd to early Preboreal was characterised by a gradual and significant declination change with an unusually rapid change between c. 10, 300 and 10,000 ^{14}C years B.P. This was very fortunate since the younger part of the gotiglacial varves was, at that time, the only part that was connected to the Swedish Time Scale in a satisfactory way (Brunnberg 1986 and pers comm.). Detailed magnetic measurements (double parallel sections) on one of Brunnberg's and one of Kristiansson's (1986) sites resulted in absolute declination curves that were dated with the Swedish Time Scale and with a local varve chronology, respectively. They did not only seem correlatable between themselves but also with the radiocarbon dated absolute declination curves from Blekinge (Björck & Sandgren 1986) and Skåne (Sandgren 1986). By using a specially designed statistic analysis the clay-varve dated curves were correlated with the most complete and well-dated declination record in lake sediments, namely Lake Halsjön in Blekinge (Björck 1981).

The correlations with Brunnberg's dated varved clay showed that the Swedish Time Scale contains 500-600 more years than the ^{14}C time scale at 10,200 ^{14}C years B.P. (Björck el al. 1987). By correlating Brunnberg's varves with Kristiansson's (1986) by means of declination changes followed by a correlation to the ^{14}C dated declination changes Björck et al. (1987) also found it possible to make correlations between the time scales at 11,000 ^{14}C years B.P. This owes to the fact that Kristiansson (1986) has been able to relate a certain local varve year (1760) to the onset of Younger Dryas. The result of these correlations was that 11,000 ^{14}C years correspond to 11,400 - 11,500 varve years. The basic reasoning behind these correlations is illustrated in Björck et al. (1987, Fig. 3). Furthermore, Björck & Möller (1987) correlated the abruptly increasing varve thickness at Kristiansson's (op. cit.) varve 2560 with the climatic amelioration at the beginning of the Alleröd. This makes the Alleröd 800 varve years long (2560-1760), while the radiocarbon dated biostratigraphy indicated that it is 1000-1100 ^{14}C years long (Björck 1984, Björck & Möller 1987). The difference between the two time scales thus became reduced to c. 200 years at 12,000 ^{14}C years B.P.

Another direct correlation tool between the time scales is to find significant paleoenvironmental events recorded in both varved clay and in directly or indirectly radiocarbon dated sediments. The ingression of marine water into the Baltic, i.e. the beginning of the finiglacial varve series at varve -1191 dated to 10,429 varve years B.P., is typically an event that can be traced from the west coast to the Baltic (Björck et al. 1990). It has been recorded and dated in lake basins in the Göta älv River Valley (Svedhage 1985), in the area west of Mt. Billingen (Björck & Digerfeldt 1986), and in the Baltic proper (Svensson 1989). The ^{14}C age of this short (100-150 years) event has in all those, above mentioned, studies been estimated to 9900-10,000 ^{14}C years B.P. Based on this correlation the difference between the time scales at 10,000 ^{14}C years B.P. is c. 500 years, which is in agreement with the paleomagnetic correlations (Björck et al. 1987).

Finally it should be mentioned that serious attempts have been made to ^{14}C date remains of Portlandia arctica as well as fish remains, found in the lower part of the finiglacial varved clay deposited during the marine ingression and dated to c. 10,430 varve years B.P. This project (Brunnberg 1986, Brunnberg & Possnert in prep.) has been based on AMS dates since the submitted samples have been very small. Altogether 12 dates have been performed and the rather puzzling results were as follows (without marine correction): eight dates on carbonate shell remains yielded ages between 11,600 and 10,700 ^{14}C years B.P., three dates on perostraka and one age on fish remains gave ages between 9300 and 8300 ^{14}C years B.P. As often is the case with ^{14}C dates on marine organisms the results are more confusing than elucidating, and once again point to the danger of using marine dates for creating a reliable chronology.

CONCLUSIONS

We can conclude that the Swedish Time Scale now covers 11,340 varve years, representing the post-, and finiglacial varve series and the last c. 900 varves of the gotiglacial series (the Stockholm-Södermanland varves). Varves older than that are only preliminary connected with the Time Scale, but will hopefully be definitely connected in a very near future. The preliminary connection indicates that the gotiglacial varve series contains c. 2400 varve years along the Swedish southeast coast down to NE Skåne. This means that the length of the Swedish Time Scale, from the earliest deglaciation of the Kristianstad Plain in NE Skåne up to 1950 A.D., is c. 12,800 varve years, or perhaps even more.

Different attempts to correlate between the varve chronology (the Swedish Time Scale) and the ^{14}C time scale suggest that the varve chronology contains c. 500 more years at the end of the Younger Dryas. Correlations further back in time, based on the preliminary connected Time Scale, indicate that the difference between the time scales gradually decreases to c. 200 years at c. 12,000 ^{14}C years B.P. When Kristiansson's (1986) varve chronology is firmly connected with the Swedish Time Scale it may, however, result in a significantly larger difference. A definitive connection between the older parts of the gotiglacial varve series and the Swedish Time Scale combined with a planned project on AMS datings of terrestrial macrofossils in the gotiglacial varved clay will hopefully be the basis for a reliable connection between the two time scales between 10,000 and c. 13,000 varve years B.P. Then the Swedish varve chronology will be the longest calendar year chronology connected with the present, embracing roughly 13,000 years with each year represented.

ACKNOWLEDGEMENTS

Partly, these studies received financial support from the Swedish Natural Science Research Council and the Geological Survey of Sweden

REFERENCES

Antevs E (1915) Landisens recession i nordöstra Skåne. Geol. fören. förh. Stockh. 37, 345-366
Berglund B E (1976) The deglaciation of southern Sweden: Presentation of a research project and a tentative radiocarbon chronology. University of Lund, Dept. of Quaternary Geol. Report 10, 1-67.
Bergström R (1968) Stratigrafi och isrecession i södra Västerbotten. Sver. geol. unders. Ser C 634, 76 pp. Stockholm.
Björck S (1981) A stratigraphic study of Late Weichselian deglaciation, shore displacement and vegetation history in southeastern Sweden. Fossils and Strata 14, 1-93.
Björck S (1984) Bio- and chronostratigraphic significance of the Older Dryas Chronozone on the basis of new radiocarbon dates. Geol. fören. förh. Stockh. 106, 81-91.
Björck S, Berglund B E, Digerfeldt G (1988) New aspects on the deglaciation chronology of South Sweden. Geographica Polonica 55, 37-49.
Björck S, Dennegård B, Sandgren P (1990) The marine stratigraphy of the Hanö Bay, SE Sweden, based on different sediment stratigraphic methods. Geol. fören. förh. Stockh. 112, 265-280.
Björck S, Digerfeldt G (1986) Late Weichselian-Early Holocene shore displacement west of Mt. Billingen, within the Middle Swedish end-moraine zone. Boreas 15, 1-18.
Björck S, Digerfeldt G (1989) Lake Mullsjön - a key site for understanding the final stage of the Baltic Ice Lake east of Mt. Billingen. Boreas 18, 209-219.
Björck S, Möller P (1987) Late Weichselian environment history in southeastern Sweden during the deglaciation of the Scandinavian ice-sheet. Quaternary Research 28, 1-37.
Björck S, Sandgren P (1986) A 2000 year geomagnetic record from two Late Weichselian sequences in southeast Sweden. Geol. fören. förh. Stockh. 108, 21-29.
Björck S, Sandgren P, Holmquist B (1987) A magnetostratigraphic comparison between ^{14}C years and varve years during Late Weichselian, indicating significant differences between the time scales. Journal of Quaternary Science 2, 133-140.
Borell R, Offerberg J (1955) Geokronologiska undersökningar inom Indalsälvens dalgång mellan Bergforsen och Ragunda. Sver. geol. unders. Ser Ca 31, 24 pp. Stockholm.
Brunnberg L (1986) ^{14}C - daterade lervarv från Stockholmsområdet. 17 Nordiska Geologmötet Abstracts, Helsinki, 19.
Brunnberg L (1988) Diatoms in glacial clay in the Södertörn area. Stockholm Region. Proceedings of Nordic Diatomist Meeting, Stockholm 1987. University of Stockh. Dept. of Quaternary Research Report 12, 7-9.
Caldenius C (1924) Ragundasjöns stratigrafi och geokronologi. Sver. geol. unders. Ser Ca 12, 91 pp. Stockholm.
Caldenius C (1944) Baltiska issjöns sänkning till Västerhavet. En geokronologisk studie. Geol. fören. förh. Stockh. 66, 366-382. Stockholm.
Caldenius C (1960) The Zero-Year Varve of the Swedish Geological Time Scale. A Rectification. Geol. fören. förh. Stockh. 82, 126-133.
Cato I (1982) Highest coastline. In Schwarts, M.L. (ed): The Encyclopedia of Beaches and coastal Environments, 461-462. Hutchinson Ross Publ Comp. Pennsylvania.
Cato I (1985) The definitive connection of the Swedish Geochronological Time Scale with the present and the new date of the zero-year in Döviken, northern Sweden. Boreas 14, 117-122. Oslo.

Cato I (1987) On the definitive connection of the Swedish Time Scale with the present. Sver. geol. unders. Ser Ca 68, 55 pp.Uppsala.

De Geer E H (1933) Teleconnection of geochronology and historic time. Report of XVI International geological congress 1933, Washington, 203-211.

De Geer E H (1959) Das wahre Null-Warw der Zeitskala von G De Geer. Eiszeitalter und Gegenwart 10, 113-117.

De Geer G (1884) Autoreport. Geol. fören. förh. Stockh., 7, 3. Stockholm.

De Geer G (1912) A Geochronology of the last 12000 years. Compte rendu 11 Congr. Geol. Intern., Stockholm 1910, 241-253.

De Geer G (1924) Om den definitiva förbindelsen mellan den svenska tidskalans senglaciala och postglaciala del. Geol. fören. förh. Stockh. 46, 493-494. Stockholm.

De Geer G (1940) Geochronologia Suecica, Principles. Kungl Sv. Vet. Akad. Handl. Bd 18:6 367 pp. Stockholm.

Fromm E (1964) the Ice Recession through Southern Dalecarlia. Geol. fören. förh. Stockh. 86, 173-174. Stockholm.

Fromm E (1970) An estimation of errors in the Swedish varve chronology. Nobel Symp. 12, Radiocarbon variations and absolute chronology, 163-172. Stockholm.

Fromm E (1985) Chronological calculation of the varve zero.Boreas 14, 123-125. Oslo.

Fözö I (1980) Ådalen. Den geologiska utvecklingen vid slutet av istiden. Länsstyrelsen i Västernorrlands län. 65 pp. Sundsvall.

Granar L (1956) Dating of recent fluvial sediments from the estuary of the Ångerman River. Geol. fören. förh. Stockh. 78, 654-658. Stockholm.

Hörnsten Å (1964) Ångermanlands kustland under isavsmältningsskedet. Prel. medd. Geol. fören. förh. Stockh. 86, 181-205. Stockholm

Järnefors B (1963) Lervarvskronologien och isrecessionen i östra Mellansverige. Sver. geol. Unders. C 594. 67 pp.

Kristiansson J (1982) Varved sediments and the Swedish time scale. Geol. fören. förh. Stockh. 104, 273-275. Stockholm.

Kristiansson J (1986) The ice-recession in the southeastern part of Sweden, a varve-chronological time scale for the latest part of the Late Weichselian. University of Stockh. Dept. of Quaternary Research Report 7, 1-132.

Kullenberg B, Fromm E (1944) Nya försök att upphämta långa sedimentprofiler från havsbottnen. Geol. fören. förh. Stockh. 66, 501-510. Stockholm.

Lidén R (1911) Om isafsmältningen och den postglaciala landhöjningen i Ångermanland. Ref. av föredrag. Geol. fören. förh. Stockh. 33, 271-280. Stockholm.

Lidén R (1913) Geokronologiska studier öfver det finglaciala skedet i Ångermanland. Sver. geol. unders. Ser Ca 9, 39 pp. Stockholm.

Lidén R (1938) Den senkvartära strandförskjutningens förlopp och kronologi i Ångermanland. Geol.fören. förh. Stockh. 60, 397-404. Stockholm.

Lidén R (+), Cato I (in prep.) Ragnar Lidéns postglacial varve chronology from the Ångemanälven valley, northern Sweden. Sver. geol unders. Ser. Ca.

Lundqvist J (1959) Issjöar och isavsmältning i östra Jämtland. Sver. geol. unders. Ser. C 573, pp. Stockholm.

Lundqvist J (1973) Isavsmältningens förlopp i Jämtlands län. Sver. geol. unders. Ser. C 681, 187 pp, Stockholm.

Lundqvist J (1975) Ice recession in central Sweden, and the Swedish Time Scale. Boreas 4, 47-54.
Lundqvist J (1986) Late Weichselian Glaciation and Deglaciation in Scandinavia. In V Sibrava, DQ Bowen, GM Richmond (eds) Quaternary Glaciations in the Northern Hemisphere. Quaternary Science Reviews 5, 269-292.
Mangerud J, Andersen S T, Berglund B E, Donner JJ (1974) Quaternary stratigraphy of Norden, a proposal for terminology and classification. Boreas 3, 109-127.
Mörner N-A (1969) The Late Quaternary History of the Kattegatt Sea and the Swedish West Coast. Deglaciation, shorelevel displacement, chronology, isostasy and eustasy. Sver. geol. unders. Ser. C 640, 1-487.
Mörner N-A (1970) The Younger Dryas Stadial. Geol. fören. förh. Stockh. 92, 5-20.
Nilsson E (1960) Södra Sverige i senglacial tid. Geol. fören. förh. Stockh. 82, 134-149. Stockholm.
Nilsson E (1968) Södra Sveriges senkvartära historia. Geokronologi, issjöar och landhöjning. Kungl. Sv. Vet. Akad. Handl. 4, 12, 1, 1-117.
Ringberg B (1971) Glacialgeologi och isavsmältning i östra Blekinge. Sver. geol. unders. C 661, 1-174.
Ringberg B (1979) Varve chronology of the glacial sediments in Blekinge and northeastern Skåne, southeastern Sweden. Boreas 8, 209-215.
Ringberg B, Rudmark L (1985) Varve chronology based upon glacial sediments in the area between Karlskrona and Kalmar, southeastern Sweden. Boreas 14, 107-110.
Rudmark L (1975) The deglaciation of Kalmarsund, southeastern Sweden. Sver. geol. unders. C 713, 1-88.
Sandgren P (1986) Late Weichselian palaeomagnetic secular variations from the Torreberga basin, south Sweden. Physics of Earth and Planetary Interior 43, 160-172.
Sauramo M (1923) Studies on the Quaternary varve sediments in southern Finland. Bulletin de la Commission Géologique de Finlande 60. 164 pp.
Strömberg B (1977) Einige Bemerkungen zum Rückzug des Inlandeises am Billingen (Västergötland, Schweden) und den Ausbruch des Baltischen Eisstausees. Zeitschrifte fur Geomorphologie, Neue Folge, Suppl. Bd. 27, 89-111.
Strömberg B (1985a) New varve measurements in Västergötland, Sweden, Boreas 14, 111-116.
Strömberg B (1985b) Revision of the late glacial Swedish varve chronology. Boreas 14, 101-105. Oslo.
Strömberg B (1989) Late Weichselian deglaciation and clay varve chronology in east-central Sweden. Sver. geol. unders. Ser Ca 73, 70 pp. Uppsala.
Strömberg B (1990) A connection between the clay varve chronologies in Sweden and Finland. Ann. Acad. Sci. Fennice A. III. 154, 32 pp.
Svedhage K (1985) Shore displacement during Late Weichselian and early Holocene in the Risveden area, SW Sweden. Chalmers tekniska Högskola/Göteborgs Universitet, Geol. Inst. 51, 1-111
Svensson N-O (1989) Late Weichselian and early Holocene shore displacement in the Central Baltic, based on stratigraphical and morphological records from eastern Småland and Gotland, Sweden. Lundqua 25, 1-195.

Tauber H (1970) The Scandinavian varve chronology and C-14 dating. In I U Olsson (ed) Radiocarbon Variations and Absolute Chronology. Nobel Symposium 12, 173-196 Wiley & Sons New York.

Wenner C-G (1968) Comparison of varve chronology, pollen analysis and radiocarbon dating. Including an investigation of A0 as a synchronous level in Sweden. Acta Univ. Stockh., Stockh. Contrib. in Geol. 18:3, 76-97. Stockholm.

A STEP TOWARDS AN ABSOLUTE TIME-SCALE FOR THE LATE-GLACIAL: ANNUALLY LAMINATED SEDIMENTS FROM SOPPENSEE (SWITZERLAND)

A.F. LOTTER, B. AMMANN[1], J. BEER, I. HAJDAS & M. STURM

EAWAG
Umweltphysik
CH-8600 Dübendorf
Switzerland

"Die Zeit ist ein sonderbares Ding", sagte ich einmal, "es gibt sie, und gibt sie auch wieder nicht -"
MAX FRISCH, 1954: Bin oder die Reise nach Peking.

Abstract

AMS radiocarbon stratigraphies on terrestrial plant remains from Switzerland reveal three distinct phases of constant radiocarbon age (plateaux) during the Late-Glacial and the early Holocene, centred at 12,700, at 10,000, and at 9500 ^{14}C yr B.P. These results are correlated with a floating varve chronology from Soppensee (Switzerland) based on palynostratigraphy. First estimates for the age-plateaux at 10,000 and 9500 yr B.P. yield a duration of ca. 400 varves. Emphasis is laid on a strict distinction between biozones and chronozones. Preliminary absolute age-estimates for the late-glacial and early Holocene chrono- and biozones are given. A comparison of the varve chronology and the ^{14}C time-scale indicates high atmospheric $^{14}C/^{12}C$ ratios between 11,000 to 12,000 varve yr B.P. and decreasing values between 12,000 to 13,000 varve yr B.P.

---

[1] Universität Bern, Geobotanik, Altenbergrain 21, CH-3013 Bern, Switzerland

NATO ASI Series, Vol. I 2
The Last Deglaciation: Absolute and Radiocarbon Chronologies
Edited by E. Bard and W. S. Broecker

1. Introduction

Time is a major factor in Quaternary science. Without a trustworthy chronometer any interpretation of changes in proxy data of stratigraphical origin is on weak ground. In fact, any attempt at a sound reconstruction of timing and rates of past climatic change as well as the response of the biosphere can only be achieved on the basis of a reliable chronology. Moreover, all correlations and comparisons through time on a continental or global scale depend heavily on the reliability of the time-scale used. Therefore the establishment of an absolute time-scale is a fundamental goal. In this contribution we refer to the term "absolute time-scale" as a time-scale consisting of ages determined on the basis of sidereal years.

Traditional stratigraphical methods of absolute dating include the Swedish glacial varve chronology, already developed early in this century by De Geer (1912) and since then continuously improved (e.g. Strömberg 1985; Cato 1987). Unfortunately, however, a spatial correlation with other stratigraphies outside Fennoscandia is difficult.

The most widely used dating technique in Quaternary sciences is the radiocarbon method. However, due to changes in the concentration of the atmospheric $^{14}C/^{12}C$ ratio in the past (De Vries 1958; Suess 1970) the ^{14}C-ages have to be calibrated in order to obtain absolute ages. This is commonly done by analyzing tree rings that have been absolutely dated by dendrochronological methods. Nevertheless, the calibration curve reaches only back as far as ca. 10,000 radiocarbon years at present due to the lack of subfossil trees. This radiocarbon calibration therefore does not cover the important period of the Late-Glacial with its marked environmental changes. However, attempts have been made to assess an absolute late-glacial time-scale on the basis of different non-glacial varve chronologies (e.g. Welten 1944; Stuiver *et al.* 1986; Pazdur *et al.* 1987; Zolitschka 1989; Lotter 1991a). In the absence of a continuous dendrochronology these varve chronologies are a valuable instrument for an extension of the absolute time-scale into the Pleistocene.

2. The sites

Pollen analyses are available from two lakes: Rotsee (Lotter 1988) and Soppensee (Lotter 1991a). Both sites are located in the Canton of Lucerne on the Central Swiss Plateau close to the border of the Alps (Fig.1). Rotsee (420 m a.s.l., 8°20'E, 47°9'N) lies at the northern periphery of Luzern, and Soppensee (600 m a.s.l., 8°20'E, 47°5'30''N) is located 18 km to the NW. Both lakes are of glacial origin and were formed after the retreat of the Würmian glaciers. Their sediment includes the environmental history of the last 15,000 years, and due to the short geographical distance between the two sites their pollen records are very similar.

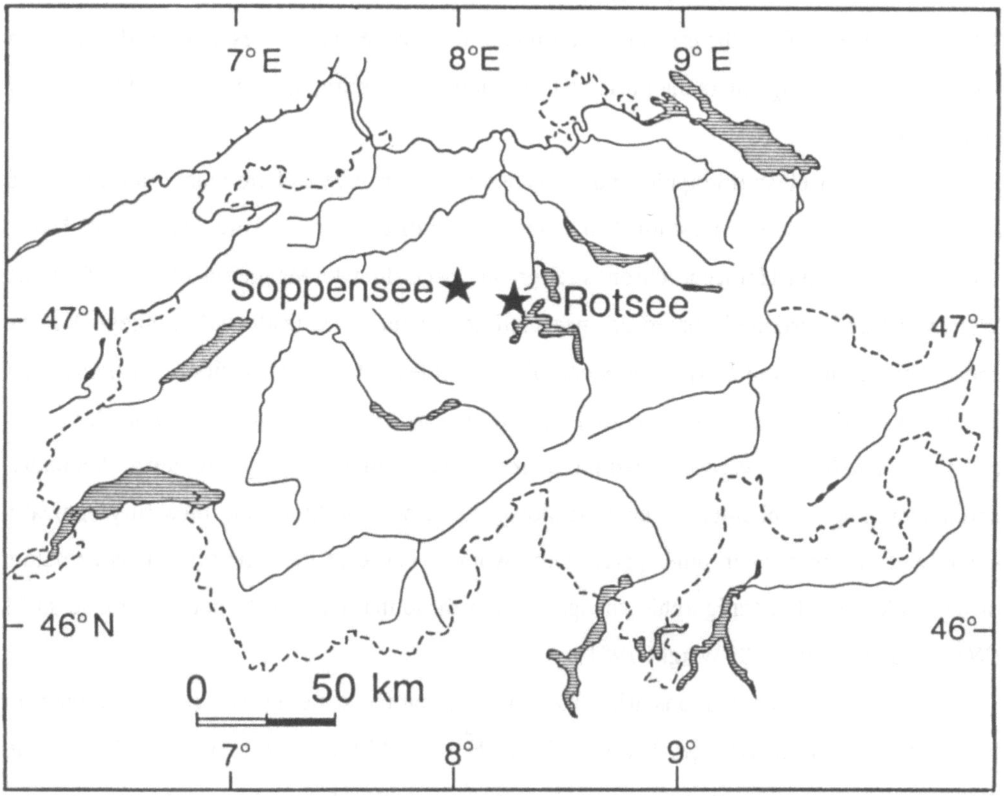

Fig.1 Map of Switzerland, indicating the location of the investigated sites

3. Zonation: biozones and chronozones

In stratigraphical studies the necessity is felt to separate clearly descriptions from subsequent steps such as interpretations, comparisons, and correlations. Descriptions are usually given as litho-, bio-, or/and chronostratigraphies for which a set of rules and a terminology is recommended by the *International Guide to Stratigraphy, Classification, Terminology and Usage* (Hedberg 1972a, 1972b) or the *North American Stratigraphic Code* (North American Commission on Stratigraphic Nomenclature 1983). Lithostratigraphy deals with the nature, distribution and relations of strata on the basis of their mere lithologic character. Chronostratigraphy, on the other hand, involves the correlation and classification of strata on the basis of their relative or absolute age, and biostratigraphy deals with the succession and classification of strata on the basis of their fossil content (see e.g. Geyh & Schleicher 1990). Nevertheless, the determination of zone boundaries always depends on several factors, among others the intrinsic resolution of the stratigraphic record and the density of sampling.

Like other biostratigraphies, pollen zones are descriptive and are most often presented as assemblage zones (e.g. Cushing 1967; for detailed discussion see Birks & Gordon 1985). The late-glacial and Holocene chronostratigraphy according to Mangerud *et al.* (1974) and Welten (1982) is originally based on measured zone boundaries (radiocarbon ages). Yet the concept of chronostratigraphy is quite complex for at least two reasons: firstly, *chronos* itself is not measured, but e.g. ^{14}C-concentrations and then radiocarbon ages are calculated. In such cases *radiocarbon stratigraphy* would therefore be a more appropriate term. Secondly, chronostratigraphic schemes set up at stratotypes can become litho- and biostratigraphically empty and meaningless in other parts of the world because of the time-transgressive nature of both litho- and biostratigraphic changes - even those that respond to major climatic shifts (Watson & Wright 1980; Wright 1981).

Additional complications arise from the fact that the same terms originally coined for biozones (e.g. *Bφlling, Allerφd, Younger Dryas*, Hartz & Milthers 1901; Iversen 1942; Firbas 1949) were redefined as chronozones (Mangerud *et al.* 1974). Full agreement between bio- and chronozones is given in the area for which these definitions were made (e.g. for Scandinavia, see Mangerud *et al.* 1974). However, discrepancies between the bio- and chronozonation become apparent with increasing distance from that area. Hence, further confusion could be prevented if a regular labelling of such terms for usage as bio- or chronostratigraphically was made (see Ammann & Lotter 1989). In recent literature as well as during the NATO Advanced Research Workshop in Erice, December 1990, it has become obvious that authors need to give their definition of terms such as the *Younger Dryas*, since there are so many concepts and criteria and hence durations (ranging from e.g. 350 to 1500 years for the Younger Dryas).

3.1. Biostratigraphy: late-glacial and early Holocene biozones on the Swiss Plateau

In Central Europe the Firbas (1949, 1954) system of pollen zonation is traditionally applied. It is a modification of the Jessen system (1935, 1938), thus originally a biostratigraphic zonation system derived from the Blytt-Sernander scheme of climatic periods. The Firbas-zones are considered here as strict biozones (i.e. regional pollen assemblage zones= regional PAZ) and must not be confused with the chronozones (Mangerud *et al.* 1974; Welten 1982), although they have the same names (see Tab.1, and Ammann & Lotter 1989).

Owing to numerous pollen-analytical investigations on the Swiss Plateau these regional biozones are well defined (see e.g. Lang 1985a; Zoller 1987). Criteria for the zone boundaries are given by Lotter (1988), Ammann (1989), and Ammann & Lotter (1989). For a better understanding we summarize briefly the main features of these biozones as well as the criteria used for the definition of their boundaries (see Tab.1 and Figs.2 and 3, for further details refer to literature cited):

Zone **Ia** (Oldest Dryas) can usually be subdivided into two subzones: a lower *Artemisia-Helianthemum* regional PAZ characterized by high values of non-arboreal pollen (NAP), mainly heliophilous herbs (e.g. *Artemisia, Helianthemum,* Caryophyllaceae, Chenopodiaceae), and grasses (Poaceae). These assemblages are commonly interpreted as representing tree-less steppe-tundra on open soils, consisting of a mosaic of arctic and alpine herbs. The upper subzone, the *Artemisia-Betula nana* regional PAZ, is distinguished by its increased birch values, originating dominantly from the dwarf birch (*Betula nana*). This increase in dwarf shrubs indicates the onset of pedogenesis as well as a denser vegetation cover. The upper boundary of the biozone Ia is defined by an increase of the juniper percentages (*Juniperus*

>10%). Stable-isotope analyses on bulk-sediment carbonates (e.g. Eicher & Siegenthaler 1976; Eicher 1987) have revealed a close correlation between the first change in the $\delta^{18}O$ values and this increase in juniper, associated with a decrease in heliophilous NAP. Zone **Ib/c** (Bϕlling-Older Dryas complex) is usually divided into two subzones. The *Juniperus-Hippophaë* regional PAZ represents a pioneer shrub phase of juniper, sea buckthorn and willow, with a decreasing amount of light-demanding herbs. At some sites pollen concentrations have revealed that the important event of reforestation by birch had already occurred during this PAZ (e.g. Gaillard 1985; Ammann 1989). At other sites the reforestation occurs with the increase in the birch percentages during the *Betula*-Poaceae regional PAZ (e.g. Rösch 1985; Lotter 1985, 1988). The inferred vegetation consisted of open birch forests with tall herbs. Depending on the sampling resolution and on the site elevation, a brief regressive phase in vegetation development (Ic, Older Dryas), characterized by an increase in heliophilous NAP and grasses as well as a depression in *Betula*, can sometimes be discerned during this second subzone. The upper boundary of the biozone Ib/c is given by the increase in pine (*Pinus* >20%).

Zone **II** (Allerϕd: *Pinus-Betula* regional PAZ) is characterized by increasing pine values associated with decreasing birch percentages. After the expansion of pine the inferred vegetation consisted of pine forests with some birch. The upper boundary of this zone is marked by the increase of NAP (*Artemisia* >2%).

Zone **III** (Younger Dryas: *Pinus*-Poaceae-NAP regional PAZ) represents a second regressive phase; it is characterized by an increase of heliophilous taxa (mainly *Artemisia* and *Juniperus*), indicating open pine forests on the Swiss Plateau. The upper boundary of zone III is given by the NAP decrease (*Artemisia* <2%). Stable-isotope investigations in several circumalpine sites show a close negative correlation between the $\delta^{18}O$ values and the relative amount of NAP (mainly heliophilous taxa such as *Artemisia, Helianthemum*). Furthermore, a comparison of $\delta^{18}O$ records from the Alps with the Greenland ice-core record suggests that the "Younger Dryas" event in Dye 3 corresponds to the decrease in the bulk-sediment oxygen-isotope values in the Alps (Siegenthaler *et al.* 1984).

During zone **IV** (Preboreal: *Pinus-Betula-Corylus* regional PAZ) the pine forest became dense again and birch played a more important role. Its upper zone limit is defined by the increase of hazel (*Corylus* >10%). The onset of zone **V** (Boreal: *Corylus*-QM regional PAZ) is marked by the rapidly increasing mesic trees (*Corylus, Ulmus, Quercus, Tilia*) and the decreasing pine values. This marks the ecologically important transition from late-glacial coniferous forest to Holocene deciduous forest.

Table 1. Compilation of zonation systems for the Swiss Plateau in relation to time and stratigraphic events.

$\delta^{18}O$	^{14}C Plateaux (ZBINDEN *et al.* 1989)	Time conventional ^{14}C yr B.P.	Chronozones (MANGERUD *et al.* 1974)	Chronozones (WELTEN 1982)	Biozones (JESSEN 1935,1938 FIRBAS 1949, 1954)	**Criteria for biozone boundaries on the Swiss Plateau** (LOTTER 1988, AMMANN1989, AMMANN & LOTTER 1989)	**Inferred vegetation**
			BO	**BO**			mixed decidous forest
		9000	**PB**	**PB**	**V** (Boreal)		hazel, elm and oak spread into pine forest
						Corylus rise (> 10%)	
					IV (Preboreal)		denser pine forest with increased birch stands
rise		10,000	**YD**	**DR3**		NAP decrease (*Artemisia* < 2%)	
					III (Younger Dryas)		open pine forest with some birch and juniper stands
de-crease						NAP increase (*Artemisia* >2%)	
		11,000	**AL**	**AL**	**II** (Allerød)		pine forest with some birch
			OD				pine expansion
		12,000	**BØ**	**BØ** (s.l.)		*Pinus* rise (> 20%)	
					I b/c (Older Dryas/ Bølling)		open birch forest with tall herbs reforestation by birch pioneer scrub of juniper, sea buckthorn and willow
rise						*Juniperus* rise (> 10%)	
		13,000		**DR1**	**I a** (Oldest Dryas)		dwarf shrubs and meadows with heliophilous, alpine and steppe herbs

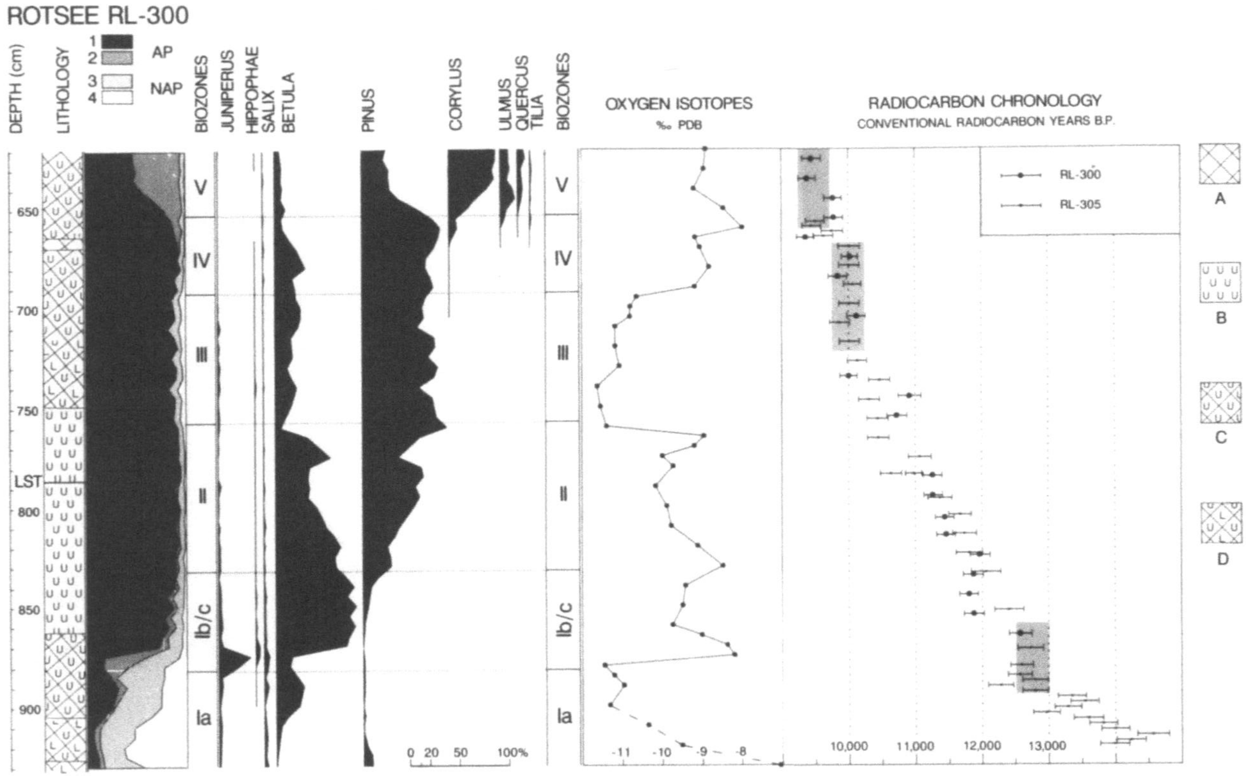

Fig.2 Rotsee core RL-300. Pollen, oxygen-isotope, and radiocarbon stratigraphy.
Cumulative summary pollen diagram: 1 trees; 2 shrubs; 3 NAP; 4 heliophilous NAP (mainly *Artemisia, Helianthemum*, Caryophyllaceae, Chenopodiaceae). 100% = ΣAP+ΣNAP. The biozones refer to Firbas (1949, 1954). The lithology is modified after Troels-Smith (1955): A fine detritus gyttja (Ld4); B lake marl (Lc4); C chalk gyttja (Lc3, Ld1); D clayey chalk gyttja (Lc2, Ld1, As1).
The three lowest $\delta^{18}O$ values of the bulk-sediment are influenced by allochthonous carbonate. The AMS ^{14}C-datings originate from cores RL-300 and RL-305, and include $\pm 1\sigma$ (Zbinden *et al.* 1989). The shaded areas highlight phases of constant radiocarbon age.

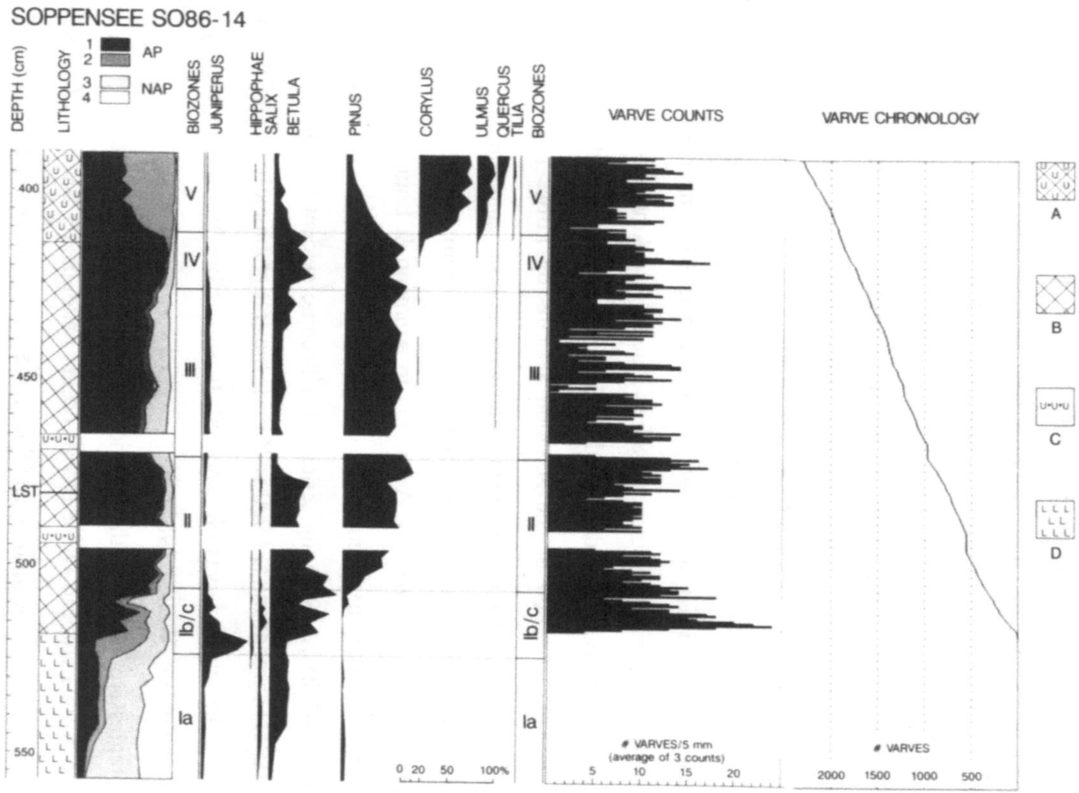

Fig.3 Soppensee core SO86-14. Pollen stratigraphy, average varve counts, and the resulting preliminary varve chronology.
Cumulative summary pollen diagram: 1 trees; 2 shrubs; 3 NAP; 4 heliophilous NAP (mainly *Artemisia, Helianthemum*, Caryophyllaceae, Chenopodiaceae). 100% = ΣAP+ΣNAP. The biozones refer to Firbas (1949, 1954).
The lithology is modified after Troels-Smith (1955): A calcareous gyttja (Ld2, Lc2); B fine detritus gyttja; C calcareous homogenite (Lc3, Ga1); D silt/clay (As4).

3.2. Radiocarbon stratigraphy: late-glacial and early Holocene chronozones

In connection with monographic studies of the environmental history of the last 15,000 years on the Swiss Plateau (Lang 1985b) the radiocarbon-dating technique was used to estimate the timing and velocity of environmental changes. At two sites, Lobsigensee near Bern (515 m a.s.l., Ammann 1989) and Rotsee near Luzern (Fig.1, 420 m a.s.l., Lotter 1988), two AMS dating series on over 90 samples yielded a ^{14}C stratigraphy for the late-glacial period (Andrée *et al.* 1986; Ammann & Lotter 1989; Zbinden *et al.* 1989). By dating exclusively plant remains of terrestrial origin (mainly fruits and catkin scales of *Betula*) we have avoided problems in connection with hard-water dating errors. Following Stuiver & Polach (1977), all ^{14}C dates are expressed as conventional radiocarbon years B.P. (i.e. before 1950).

The dates of the Rotsee AMS dating series range between 14,570 ±240 and 9450 ±140 yr B.P., thus encompassing the whole late-glacial period as well as the early Holocene. The high-resolution radiocarbon dating series from Rotsee and Lobsigensee, free of hard-water error, are not only the basis for the estimate of the duration of the late-glacial biozones and the age of the zone boundaries (Tab.1) but also for a detailed connection of bio- and chronozonation of the late-glacial period on the Swiss Plateau (see Ammann & Lotter 1989).

The chronozonation system widely used in Switzerland was introduced by Welten (1979, 1982). It is a combination of the zonation systems by Firbas (1949, 1954) and Mangerud *et al.* (1974). This fusion of a biozonation with a chronozonation has contributed once more to the confusion of terms. On the other hand, this system does not take account of the time-transgressive nature of biozones. According to Ammann & Lotter (1989) we use a somewhat modified relationship between the bio- and chronostratigraphy set up for the Swiss Plateau. On the basis of the AMS dating series it has become evident that the upper end of the Oldest Dryas (**DR1**) chronozone (13,000 yr B.P.) is biostratigraphically located during the dwarf birch phase of the Oldest Dryas biozone (Ia). It is therefore biostratigraphically older than previously thought (e.g. Welten 1982).

The first regressive phase in the palynostratigraphy after reforestation (Ic), often correlated with the north European Older Dryas, occurs undoubtedly before 12,000 yr B.P. According to Björck (1984) the Scandinavian Older Dryas is also older in age (12,250-12,100 yr B.P.) than originally thought. We therefore follow Welten (1982) by omitting the Older

Dryas (**DR2**: 12,000-11,800 yr B.P.) as a chronozone and merge the Bϕlling and the Older Dryas into a Bϕlling *sensu lato* chronozone (**Bϕ s.l.**: 13,000-12,000 yr B.P.).

The onset of the Allerϕd chronozone (**AL**: 12,000-11,000 yr B.P.) coincides with the rational limit of the pine curve. The Younger Dryas chronozone (**DR3**: 11,000-10,000 yr B.P.) begins litho- and chronostratigraphically with the Laacher See Tephra (LST), a volcanic ash layer deposited ca. 11,000 yr B.P. (Van den Bogaard & Schmincke 1985). Radiocarbon dates on terrestrial macrofossils from different sites have yielded the following age range: Rotsee (see Fig.2) 10,640 ±160, 10,990 ±120 B.P. (new dating of the same sample), 11,260 ±140 B.P., Lobsigensee 10,860 ±130 B.P., 10,900 ±130 B.P., 11,060 ±140 B.P. (all except the new dating are listed in Ammann & Lotter 1989), Schleinsee 11,080 ±155 (Clark *et al.* 1989).

Due to the 10,000 yr B.P. plateau of constant radiocarbon age the onset of the Preboreal chronozone (**PB**: 10,000-9000 yr B.P.) cannot be assigned to a specific event in the biostratigraphy or in the stable-isotope curve.

4. Phases of constant radiocarbon age

The high-resolution AMS dating series revealed three periods of constant radiocarbon age: at 12,700, 10,000, and 9500 yr B.P. These age plateaux coincide with highly dynamic phases of environmental change. The plateau at 12,700 B.P. concurs with the reforestation and the first major change in the oxygen-isotope record (Fig.2). The plateau at 10,000 B.P. includes the last major shift in the $\delta^{18}O$ record as well as the transition from the Younger Dryas (III) to the Preboreal (IV) biozone. First evidence for this plateau was provided by Oeschger *et al.* (1980) from the Wachseldorn peat bog. The 9500 B.P. plateau, which is only present as its lower part in the Rotsee record (Fig.2), coincides with the ecologically important transition from coniferous forest to deciduous forest. Dendrochronological studies (Becker & Kromer 1986, 1990) on subfossil pine trees confirm the existence of the two younger plateaux.

Plateaux of constant radiocarbon age cause a number of critical complications:

(1) They preclude more refined absolute dating by ^{14}C within the plateau phases.

(2) They prevent reliable estimation of accumulation rates and rates of change during these plateau phases on the basis of ^{14}C datings only (see e.g. Lotter *et al.* 1991).

(3) They act as "attractors" in attempts to correlate specific events between different sites. More radiocarbon ages of samples fall into these periods than would be expected. Hence, we consider events as synchronous that only seem so within the plateau.

(4) It is not possible to assign a specific stratigraphic event (e.g. changes in pollen, or stable-isotopes) to the transition from the chronozone Younger Dryas (DR3) to Preboreal (PB). Hence the important transition from Pleistocene to Holocene remains still ambiguous (see e.g. Fairbridge 1983).

These difficulties are especially perplexing as they affect periods of unusually rapid environmental and ecological change for which estimates of rates of change would be useful to understand the underlying processes. Any estimation of rates of change, however, is critically dependent on a reliable absolute chronology. Without such an absolute chronology apparent periods of rapid change cannot be distinguished from changes in the atmospheric $^{14}C/^{12}C$ ratio (Lotter *et al.* 1991). These problems can only be overcome by ^{14}C-independent dating techniques such as dendrochronology or/and varves. Although an extension of the radiocarbon calibration curve further into the Pleistocene is needed, it will not solve the problems connected with these radiocarbon age plateaux.

5. Varve chronology

During a geochemical sediment survey in 1986 laminated sediments were discovered in the deepest part of Soppensee (Fig.1). The laminations proved to be of annual character, i.e. non-glacial biogenic varves (Lotter 1989, 1991b). The Holocene varves are very distinct and include a light calcite layer precipitated during summer and a dark lamina consisting of an autumn chrysophyte cyst layer, a winter layer of organic fine-detritus, and a spring diatom layer. The late-glacial varves are poor in carbonate and are therefore macroscopically not conspicuous. They consist of a dark organic fine-detritus layer deposited during autumn and winter and a light spring and summer layer of diatoms and chrysophyte cysts with very few calcite crystals. The formation of these late-glacial varves started as the minerogenic input from the catchment decreased substantially, after the birch expansion of Bϕlling (see Fig.3), ca. 12,500 yr B.P.

Triplicate varve counts were performed on overlapping sediment thin-sections 11 cm long and 2 cm wide (Merkt 1971) using a polarizing microscope. The mean number of varves per 5 mm was estimated. The root mean squared error for the triplicate counts (±2σ) as well as the sums of the maximum and minimum deviations from the mean counts ($^{+}/_{-}$) are given in the text.

Preliminary counts yielded ca. 6000 varves between Bϕlling and mid-Atlantic for this floating varve chronology (Lotter 1991a). Further varve counts and measurements are at present in progress on other Soppensee sediment cores.

The AMS dating series of Rotsee has been integrated into the depth scale of the SO86-14 core by correlating the Rotsee RL-300 record with the Soppensee SO86-14 record on the basis of stratigraphic markers such as the LST as well as features of the pollen curves (e.g. peaks of *Juniperus, Betula, Pinus, Corylus*). The sedimentation between these markers was assumed to be constant.

The annually laminated sediments of Soppensee provide a preliminary, ^{14}C-independent possibility of estimating the absolute duration of the plateaux of constant age. The plateau at 9500 yr B.P. yields at least 428 ±16 ($^{+88}/_{-58}$) varves, whereas for the 10,000 yr B.P. plateau a time-span of 356 ±22 ($^{+59}/_{-67}$) varves has been inferred. These estimates are in agreement with dendrochronological results of Becker & Kromer (1990) who suggest a duration of ca. 400 years for the Preboreal plateau and at least 250 (probably even up to 400 years) for the Younger Dryas/Preboreal plateau.

Furthermore, the absolute duration of the late-glacial biozones and chronozones can now be estimated on the basis of the floating Soppensee varve chronology and can be compared with their duration in radiocarbon years (estimates in brackets, see Tab.1 and Lotter 1988; Ammann 1989; Ammann & Lotter 1989). The laminated part of the Bϕlling biozone (Ib/c) and chronozone (Bϕ) consists of 335 ±22 ($^{+73}/_{-61}$) varves. If the varve chronology is extrapolated to the base of the Bϕlling a duration of about 500-600 varves (ca. 600 ^{14}C yr) would result for this biozone, whereas about 700-800 varves would result for the chronozone (Bϕ: 13,000-12,000 yr B.P.).

The Allerϕd biozone (II) encompasses 587 ±22 ($^{+95}/_{-91}$) varves (ca. 1300 ^{14}C yr), whereas only 378 ±20 ($^{+70}/_{-65}$) varves have been inferred for the chronozone (AL: 12,000-11,000 yr B.P.). The sediment between the deposition of the LST and the reaction of the vegetation to the climatic impact of the Younger Dryas yields 213 ±6 ($^{+25}/_{-27}$) varves. This time-span agrees with our expectations and is very similar to the results of Zolitschka (1990).

The Younger Dryas biozone (III) contains 684 ±22 ($^{+100}/_{-92}$) varves (ca. 700 ^{14}C yr), whereas an estimate for the duration of the chronozone (DR3: 11,000-10,000 yr B.P.) is strongly inhibited by the problems associated with the 10,000 yr B.P. plateau. However, if the Younger Dryas chronozone is correlated with the sediment portion between the deposition of the LST (commonly dated at 11,000 yr B.P.) and the last major shift in the oxygen-isotope record, a duration of 897 ±26 ($^{+125}/_{-119}$) varves would result.

The Preboreal biozone (IV) comprises 287 ±18 ($^{+45}/_{-49}$) varves (probably ca. 500 ^{14}C yr, but situated between two plateaux!). The duration of the Preboreal chronozone (PB: 10,000-9000 yr B.P.) cannot be estimated with the available data, as there are not enough reliable ^{14}C dates from Rotsee for the end of this chronozone.

These preliminary estimates suggest significant differences between the duration of the late-glacial zones in radiocarbon years and in sidereal years: the Allerϕd and the Preboreal are substantially shorter than commonly estimated by ^{14}C ages (e.g. Welten 1982, Lang 1985a). These age differences may be partly due to the frequent confusion between bio- and chronozonation, as discussed above.

5.1. Possible sources of errors

A varve chronology may always include uncertainties of changing magnitude. The counting error can most easily be estimated through multiple varve counting by one or several investigators and can be expressed as standard deviation from a mean count. Nevertheless, systematic errors originating from intra-annual layers and/or missing varves can occur (Simola & Tolonen 1981; Saarnisto 1986). In the case of Soppensee, however, it is not likely that the phenomenon of missing varves occurs in such regularity that e.g. every second layer would not be present due to erosion or lack of deposition. An alternative hypothesis is that whole portions of the late-glacial sediment are missing. In fact, the sediment record of Soppensee is characterized by two homogenous sediment portions below and above the LST (see Fig.3). Both homogenites are the result of single events within the lake basin (e.g. slumps or turbidites). It is likely that both events have eroded part of the uppermost unconsolidated sediment. However, judging from the pollen curves a major hiatus can be excluded. As the core was taken from the deepest part of the basin the eroded sediment portion should be present at least in a homogenized form.

Consequently, only the analysis and the comparison of several cores within the same lake basin can help to exclude thoroughly systematic errors such as missing or doubled varves in order to establish a reliable varve chronology for a site.

6. Time-scales: ^{14}C *versus* varves

The frequent confusion of the bio- and chronozonation system cannot account for all the discrepancies between the time-scales. As discovered by other absolute dating methods there is an inconsistency between the sidereal and the radiocarbon time-scale. On the basis of the Swedish glacial varve chronology Björck *et al.* (1987, 1988) suggested that the varve chronology is older than the radiocarbon time-scale by 500-600 calendar years for the period 10,500-10,200 yr B.P. Another independent estimate from the Greenland Dye 3 ice-core suggests a discrepancy of 720 calendar yr towards older ages for the last shift in the oxygen-isotope record (Hammer *et al.* 1986). Based on dendrochronological results an offset of 1280 calendar yr towards older ages has been reported by Becker & Kromer (1986) at the onset of the Holocene.

In order to compare the results of our preliminary Soppensee varve chronology with these results from other absolute dating techniques we have matched the floating varve chronology (Fig.4) at the transition from biozone III to IV (i.e. from Younger Dryas to Preboreal). In Central Europe this transition is generally synchronous with the last major $\delta^{18}O$ shift (Eicher & Siegenthaler 1976; Lotter & Zbinden 1989). We may therefore synchronize the transition III/IV with the last major shift in the oxygen-isotope record in Dye 3, dated at 10,720 ±150 ice-core yr B.P. (Hammer *et al.* 1986). The Soppensee varve chronology would then, as a consequence, be set off ca. 700 yr towards older ages at the end of the Younger Dryas (III, Fig.4, curve B). This age-difference gradually diminishes with increasing age and the two chronologies would eventually meet before 12,000 yr B.P.

If we assume the age of 11,280 cal yr B.P. (Becker & Kromer 1986) for the transition III/IV, which is also the transition from the Late-Glacial to the Holocene, the varve chronology would be moved even more towards older ages (Fig.4, curve C), thus producing an age difference of ca. 1300 yr at the onset of the Holocene. The curves would in this case meet around 13,000 yr B.P. This result is in the range of the conclusions of Björck *et al.* (1988). The Soppensee varve chronology, under these assumptions, is not only in good agreement with the Swedish varve chronology but would also indicate higher ^{14}C concentrations between 10,000 and 12,000 yr B.P., as shown in Fig.5, which is based on the data of Fig.4 but expressed as $\Delta^{14}C$ values.

As indicated in Fig.4 our preliminary varve chronology shows uncertainties that are a consequence of the varve-counting errors. Additional uncertainties have been introduced by matching the stratigraphy of Rotsee with the one of Soppensee, as well as by the discrepancies between the absolute ages derived from ice-cores and tree rings at the transition Pleistocene/Holocene, used as anchor points for the floating varve chronology. In order to reduce these problems, new investigations have been initiated. They include the elaboration of a varve chronology and the ^{14}C dating of terrestrial plant macrofossils on the same core and should reach back far enough into the Holocene to allow a precise matching with the existing dendrochronologically established ^{14}C calibration curve.

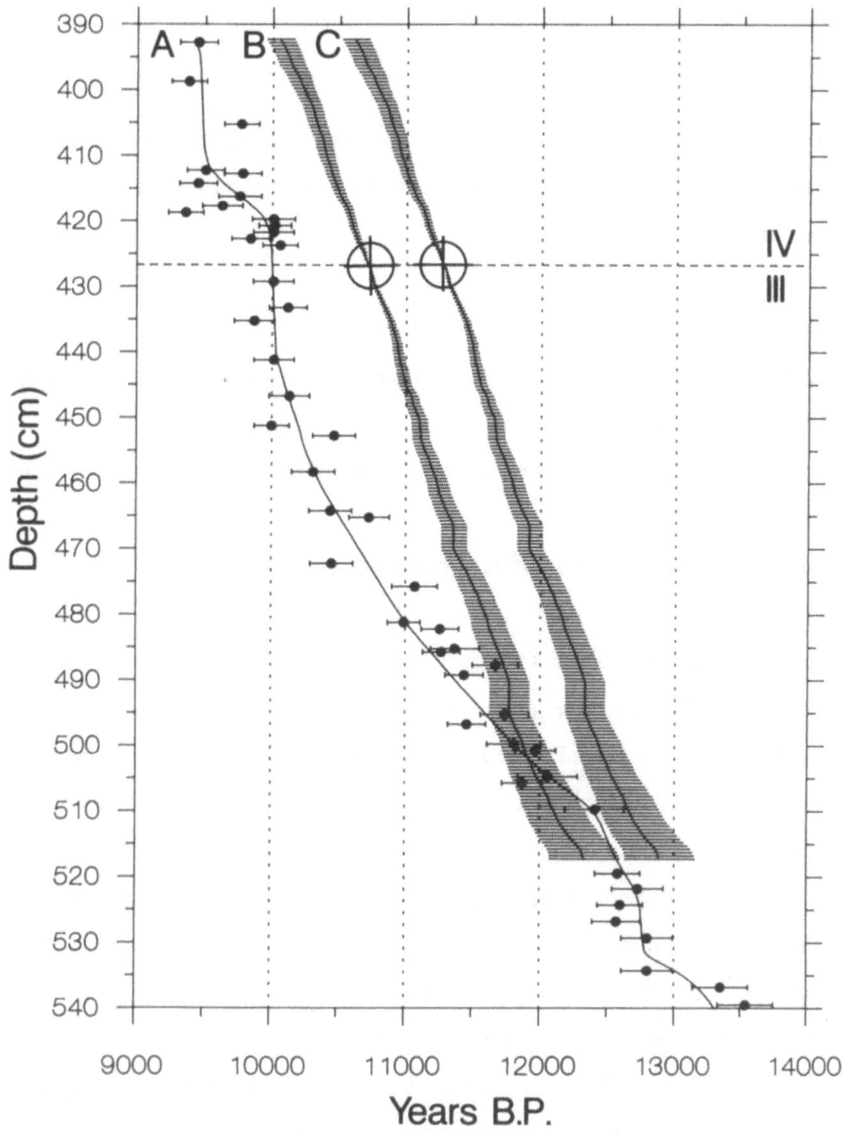

Fig.4 Preliminary comparison of radiocarbon and varve chronology. The depth scale refers to Soppensee core SO86-14.

A: Smoothed curve of the radiocarbon chronology from Rotsee core RL-300 (see Fig.2), matched into the depth scale of Soppensee core SO86-14 according to biostratigraphy.

B: floating Soppensee varve chronology (see Fig.3) fixed at 10,720 cal yr B.P. (Hammer *et al.* 1986) for the transition from biozone III to IV (see text).

C: floating Soppensee varve chronology (see Fig.3) fixed at 11,280 cal yr B.P. (Becker & Kromer 1986) for the transition from biozone III to IV, i.e. from the late-glacial to the Holocene.

Curves B and C include the mean varve counts as well as the accumulated sum of the maximum and minimum deviations from these mean varve counts.

7. $^{14}C/^{12}C$ ratios during the late-glacial period

The reconstruction of the atmospheric $^{14}C/^{12}C$ ratio during the late-glacial period is not only crucial for absolute dating as discussed above but it also offers fundamental insights into the various changes of the earth system.

Possible causes for changes in the $^{14}C/^{12}C$ ratio may be:

(1) Variations of the ^{14}C production rate as induced e.g. by changes in the solar activity (Maunder minimum).

(2) Changes in the exchange rate between the different carbon reservoirs (atmosphere, ocean, biosphere).

(3) Changes in the size of the different carbon reservoirs.

Changes of the atmospheric CO_2 content have in fact been observed in polar ice-cores (Barnola *et al.* 1987), and there are clear indications of a reduced oceanic deep-water production during the late-glacial period (Broecker *et al.* 1985). The recent results by Bard *et al.* (1990) on U/Th dated corals clearly show the importance of a detailed radiocarbon chronology.

The only way to reconstruct the atmospheric $^{14}C/^{12}C$ ratio is to use absolutely dated organic material from terrestrial plants (wood, seeds and fruits, leaves). The last 10,000 years have already been successfully reconstructed using tree rings; further analyses to expand this chronology are still in progress. As a consequence of the development of high-precision ^{14}C dating by AMS (Bonani *et al.* 1987), annually laminated sediments became another potential archive that will provide detailed data necessary not only for an extension of the calibration curve but also for a better understanding of the climatic changes at the end of the last glacial.

The tentative $\Delta^{14}C$ values for the late-glacial (Fig.5) were estimated by using radiocarbon data from the non-laminated Rotsee sediment (Zbinden *et al.* 1989). The absolute time-scale was established by first correlating the radiocarbon-dated pollen zones of Rotsee with the pollen zones of Soppensee. Then the duration of each pollen zone was estimated by varve counts on the sediment of Soppensee. Finally, we present two scenarios: the first one (Fig.5A) assumes an absolute age of 10,700 cal. yr B.P. for the transition Younger Dryas/Preboreal according to Hammer *et al.* (1986), whereas the second one (Fig.5B) assigns this transition to 11,300 cal yr B.P. according to Becker & Kromer (1986). The resulting $\Delta^{14}C$ values (Fig.5) before 11,000 cal yr B.P. for curve A and before 11,500 cal yr B.P. for curve B are similar to the values published by Zbinden *et al.* (1989). These authors, however, used the AMS radiocarbon chronology from Lobsigensee and Rotsee and the Swedish varve chronology as a tentative absolute time-scale.

Our results are in contrast to the data points of Bard *et al.* (1990) which suggest clearly higher $\Delta^{14}C$ values for the same period. Yet additional investigations using absolute dating methods are needed to explain the discrepancies between these results from different systems (e.g. lacustrine sediments, marine corals).

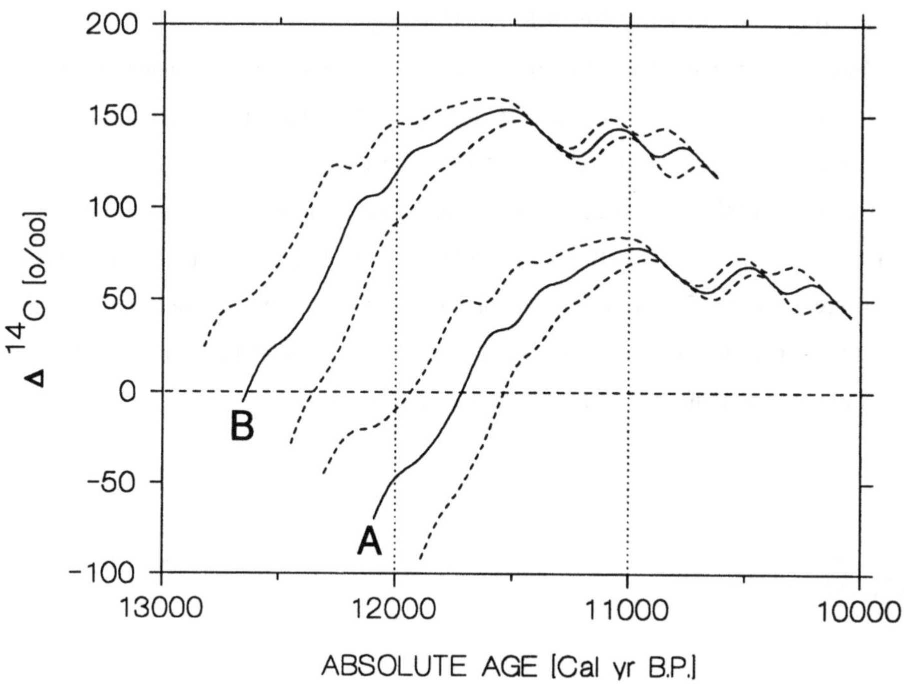

Fig.5 Tentative late-glacial $\Delta^{14}C$ values estimated on the basis of the radiocarbon ages from Rotsee (curve A in Fig.4) and the varve ages (curves B and C in Fig.4). The solid lines represent the smoothed $\Delta^{14}C$ reconstructions based on the mean varve counts (Fig.4) whereas the dashed lines indicate the range (see Fig.4).

A: Matching of the varve chronology according to curve B in Fig.4 (Hammer *et al.* 1986).

B: Matching of the varve chronology according to curve C in Fig.4 (Becker & Kromer 1986).

8. Conclusions and outlook

(1) Varved sediments provide an excellent tool to assess an absolute time-scale as a basis for high-resolution studies of climatic and environmental change.

(2) We emphasize the importance of biozones as a useful basis for palaeoenvironmental history and as important units of climatic significance. Moreover, in order to prevent further confusion in comparing stratigraphies we strongly suggest to state explicitly whether ambiguous terms such as e.g. the "Younger Dryas" have a bio- or a chronostratigraphic connotation (see Ammann & Lotter 1989).

(3) If ages and/or time-spans are estimated it is important to specify the time-scale used: absolute ages (i.e. sidereal, calendar, calibrated yr B.P.), or varve ages, or radiocarbon ages (i.e. conventional ^{14}C yr B.P.), or U/Th ages.

(4) More detailed chronological studies, combining biostratigraphical investigations, varve counts, and high-resolution AMS ^{14}C-dating on the same core, which are at present in progress at Soppensee and at other sites, are needed to avoid correlation problems.

(5) It is indispensable to first correlate the varve chronologies from different cores within the same basin to get a well established local chronology and then to correlate between different basins. This procedure is already a tradition in dendrochronology and the Swedish glacial varve chronology and will help to achieve a reliable regional or even continental absolute time-scale for the Late-Glacial.

9. Acknowledgements

We should like to thank the following persons for their help and support: M.Andrée, H.Bolliger, J.Hofmann, M.Kummer, J.Merkt, H.E.Wright Jr., H.Zbinden, A.Zwyssig. This study was financially supported by the Swiss National Science Foundation (Project Nr 21-26567.89).

10. References

Ammann B (1989) Late-Quaternary Palynology at Lobsigensee. Regional vegetation history and local lake development. Diss Bot 137: 1-157

Ammann B, Lotter AF (1989) Late-Glacial radiocarbon- and palynostratigraphy on the Swiss Plateau. Boreas 18: 109-126

Andrée M, Oeschger H, Siegenthaler U, Riesen T, Moell M, Ammann B, Tobolski K (1986) ^{14}C dating of plant macrofossils in lake sediment. Radiocarbon 28: 411-416

Bard E, Hamelin B, Fairbanks RG, Zindler A (1990) Calibration of the ^{14}C timescale over the past 30,000 years using mass spectrometric U-Th ages from Barbados corals. Nature 345: 405-410

Barnola JM, Raynaud D, Karotkevich YS, Lorius C (1987) Vostok ice core provides 160,000-year record of atmospheric CO_2. Nature 329: 408-414

Becker B, Kromer B (1986) Extension of the Holocene dendrochronology by Preboreal pine series, 8800 to 10,100 BP. Radiocarbon 28: 961-967

Becker B, Kromer B (1990) Dendrochronology and radiocarbon calibration of the early Holocene. In: Barton N, Roe D, Roberts A (eds) The Late Glacial of NW Europe. Oxford: in press

Birks HJB, Gordon AD (1985) Numerical Methods in Quaternary Pollen Analysis. Academic Press, London.

Björck S (1984) Bio- and chronostratigraphic significance of the Older Dryas chronozone on the basis of new radiocarbon dates. GFF 106: 81-91

Björck S, Sandgren P, Holmqvist B (1987) A magnetostratigraphic comparison between ^{14}C years and varve years during the Late Weichselian, indicating significant differences between the time-scales. Journal Quaternary Science 2: 122-140.

Björck S, Berglund, BE, Digerfeldt G (1988) New aspects of the deglaciation chronology of south Sweden. Geographia Polonica 55: 37-49

Bonani G, Beer J, Hofmann H, Synal HA, Suter M, Wölfli W, Peleiderer C, Kromer B, Junghans C, Munnich KO (1987) Fractionation, precision and accuracy in C14 and C13 measurements. Nucl Instr Meth B29: 87-90

Broecker WS, Peteet D, Rind D (1985) Does the ocean-atmosphere system have more than one stable mode of operation? Nature 315: 21-26

Cato I (1987) On the definite connection of the Swedish time scale with the present. Sveriges Geologiska Undersökning Ca 68:1-55

Clark JS, Merkt J, Müller H (1989) Post-glacial fire, vegetation and human history on the northern alpine forelands, south-western Germany. J Ecol 77: 897-925.

Cushing EJ (1967) Late-Wisconsin pollen stratigraphy and the glacial sequence in Minnesota. In: Cushing EJ, Wright HE (eds) Quaternary Paleoecology. Yale University Press, New Haven.

De Geer G (1912) A geochronology of the last 12000 years. XI. Internat Geol Congr CR 1: 241-253

De Vries H (1958) Variations in concentration of radiocarbon with time and location on earth. Kon Ned Akad Wet Proc Ser B 61: 94-102

Eicher U (1987) Die spätglazialen sowie frühpostglazialen Klimaverhältnisse im Bereich der Alpen: Sauerstoffisotopenkurven kalkhaltiger Sedimente. Geographica Helvetica 42: 99-104

Eicher U, Siegenthaler U (1976) Palynological and oxygen isotope investigations on Late-Glacial sediment cores from Swiss lakes. Boreas 5: 109-117

Fairbridge RW (1983) The Pleistocene-Holocene boundary. Quaternary Science Reviews 1: 215-244

Firbas F (1949) Spät- und nacheiszeitliche Waldgeschichte Mitteleuropas nördlich der Alpen. Fischer, Jena. 480 pp

Firbas F (1954) Die Synchronisierung der mitteleuropäischen Pollendiagramme. Danm Geol Unders II/80: 12-21

Gaillard M-J (1985) Late-Galcial and Holocene environments of some ancient lakes in the western Swiss Plateau. In: Lang G (ed.) Swiss lake and mire environments during the last 15000 years. Diss Bot 87: 273-336

Geyh M, Schleicher H (1990) Absolute age determination. Springer Verlag, Heidelberg.

Hammer CU, Clausen HB, Tauber H (1986) Ice-core dating of the Pleistocene/Holocene boundary applied to a calibration of the ^{14}C time scale. Radiocarbon 28: 284-291

Hartz N, Milthers K (1901) Det senglaciale ler i Alleröd teglvaerksgrav. Medd Dansk Geol För 2/8: 31-60

Hedberg HD (1972a) Introduction to an International Guide to Stratigraphic Classification, Terminology, and Usage. Boreas 1: 199-211

Hedberg HD (1972b) Summary of an International Guide to Stratigraphic Classification, Terminology, and Usage. Boreas 1: 213-239

Iversen J (1942) En pollenanalytisk tidsfaestelse af ferskvandslagene ved Nørre Lyngby. Medd Dansk Geol For 10: 130-151

Jessen K (1935) Archaeological dating in the history of north Jutland's vegetation. Acta Archaeologica 5: 185-214

Jessen K (1938) Some west Baltic pollen diagrams. Quartär 1: 124-139

Lang G (1985a) Palynologic research in Switzerland 1925-1985. In: Lang G (ed.) Swiss lake and mire environment during the last 15000 years. Diss Bot 87: 11-82.

Lang G (1985b) Palynologic and stratigraphic investigations of Swiss lake and mire deposits. In: Lang G (ed) Swiss lake and mire environments during the last 15000 years. Diss Bot 87: 107-114.

Lotter A (1985) Amsoldingersee - Late-Glacial and Holocene environments of a lake at the southern edge of the Swiss Plateau. In: Lang G (ed) Swiss lake and mire environments during the last 15000 years. Diss Bot 87: 185-208

Lotter A (1988) Paläoökologische und paläolimnologische Studie des Rotsees bei Luzern. Pollen-, grossrest-, diatomeen- und sedimentanalytische Untersuchungen. Diss Bot 124: 1-187

Lotter AF (1989) Evidence of annual layering in Holocene sediments of Soppensee, Switzerland. Aq Sci 51: 19-30

Lotter AF (1991a) Absolute dating of the late-glacial period in Switzerland using annually laminated sediments. Quat Res 35: 321-330

Lotter AF (1991b) How long was the Younger Dryas? Preliminary evidence from annually laminated sediments of Soppensee (Switzerland). Hydrobiol 214: 52-57

Lotter AF, Zbinden H (1989) Late-Glacial palynology, oxygen isotope record, and radiocarbon stratigraphy from Rotsee (Lucerne), central Swiss Plateau. Eclogae geol Helv 82: 191-202

Lotter AF, Ammann B, Sturm M (1991) Rates of change and chronological problems during the late-glacial period. Climate Dynamics 6: in press

Mangerud J, Andersen ST, Berglund BE, Donner J (1974) Quaternary stratigraphy of Norden, a proposal for terminology and classification. Boreas 3: 109-128

Merkt J (1971) Zuverlässige Auszählungen von Jahresschichten in Seesedimenten mit Hilfe von Gross-Dünnschliffen. Arch Hydrobiol 69: 145-154

North American Commission on Stratigraphic Nomenclature (1983) North American Stratigraphic Code. Bull Am Ass Petr Geol 67: 841-875

Oeschger H, Welten M, Eicher U, Möll M, Riesen T, Siegenthaler U, Wegmüller S (1980) ^{14}C and other parameters during the Younger Dryas cold phase. Radiocarbon 22: 299-310

Pazdur A, Pazdur MF, Wicik B, Wieckowski K (1987) Radiocarbon chronology of annually laminated sediments from Gosciaz lake. Bull Pol Acad Sci 35: 139-145

Rösch M (1985) Nussbaumer Seen - spät- und postglaziale Umweltsveränderungen einer Seengruppe im östlichen Schweizer Mittelland. In: Lang G (ed) Swiss lake and mire environments during the last 15000 years. Diss Bot 87: 337-379

Saarnisto M (1986) Annually laminated sediments. In: Berglund BE (ed) Handbook of Holocene Palaeoecology and Palaeohydrology. J Wiley & Sons, Chichester. pp. 343-370

Siegenthaler U, Eicher U, Oeschger H, Dansgaard W (1984) Lake sediments as continental $\delta^{18}O$ records from the Glacial Post Glacial transition. Ann Glaciology 5: 149-152

Simola H, Tolonen K (1981) Diurnal laminations in the varved sediment of Lake Lovojärvi, south Finland. Boreas 10: 19-26

Strömberg B (1985) Revision of the lateglacial Swedish varve chronology. Boreas 14: 101-105

Stuiver M, Polach HA (1977) Reporting of ^{14}C-data. Radiocarbon 19: 355-363

Stuiver M, Kromer B, Becker B, Ferguson CW (1986) Radiocarbon age calibration back to 13,300 years BP and the ^{14}C age matching of the German oak and US bristlecone pine chronologies. Radiocarbon 28: 969-979

Suess HE (1970) The three causes of the secular C14 variations, their amplitudes and time constants. In: Olsson IU (ed) Radiocarbon variations and absolute chronology. Wiley, New York. pp. 595-605

Troels-Smith J (1955) Characterization of unconsolidated sediments. Danm Geol Unders IV, 3/10: 38-73

Van den Bogaard P, Schmincke HU (1985) Laacher See Tephra: a widespread isochronous late Quaternary tephra layer in central and northern Europe. Geological Society of America Bulletin 96: 1554-1571

Watson RA, Wright HE (1980) The end of the Pleistocene: a general critique of chronostratigraphic classification. Boreas 9: 153-163

Welten M (1944) Pollenanalytische, stratigraphische und geochronologische Untersuchungen aus dem Faulenseemoos bei Spiez. Veröff Geobot Inst Rübel 21: 1-201

Welten M (1979) Late Glacial and Late Weichselian (Spätglazial und Spät-Würm): a comment. Boreas 8: 396

Welten M (1982) Vegetationsgeschichtliche Untersuchungen in den westlichen Schweizer Alpen: Bern-Wallis. Denkschr Schweiz Naturf Ges 95: 1-104

Wright HE (1981) Holocene chronostratigraphy for the United States and southern Canada. Striae 16: 53-55

Zbinden H, Andrée M, Oeschger H, Ammann B, Lotter A, Bonani G, Wölfli W (1989) Atmospheric radiocarbon at the end of the last glacial: an estimate based on AMS radiocarbon dates on terrestrial macrofossils from lake sediments. Radiocarbon 31: 795-804

Zolitschka B (1989) Jahreszeitlich geschichtete Seesedimente aus dem Holzmaar und dem Meerfelder Maar. Z dt geol Ges 140: 25-33

Zolitschka B (1990) Spätquartäre jahreszeitlich geschichtete Seesedimente ausgewählter Eifelmaare. Documenta Naturae 60: 1-226

Zoller H (1987) Zur Geschichte der Vegetation im Spätglazial und Holozän der Schweiz. Mitt Naturf Ges Luzern 29: 123-149

THE LATE GLACIAL-HOLOCENE TRANSITION IN CENTRAL EUROPE DERIVED FROM ISOTOPE STUDIES OF LAMINATED SEDIMENTS FROM LAKE GOŚCIĄŻ (POLAND)

K. Rozanski[1)], T. Goslar[2)], M. Dulinski[3)], T. Kuc[3)], M.F. Pazdur[2)] and A. Walanus[2)]

1. INTRODUCTION

Lake Gościąż is located in the Plock Basin (central Poland), on a terrace plain developed during the ice sheet recession and transformed by dunes (Fig.1). The lake is the largest and deepest one in a complex of four lakes which drain into the Vistula river (Fig.2). Lake Gościąż has a surface area of $0.45 km^2$, a mean depth of 4.6m (maximum depth = 25.8m), and a water volume of $4.7 x 10^5 m^3$. Hydrogeological details of the lake were summarized by Wicik and Więckowski (1988).

A remarkable feature of the sediments from the Gościąż Lake is the distinct microlamination (couplets of dark and light layers) of the 19m thick sediment column (Goslar et al. 1989) In 1987, an interdisciplinary study of the sediments from this lake was initiated within the framework of the Gościąż Lake Project (Ralska-Jasiewiczowa et al. 1987). So far, several cores have been obtained from different parts of the lake. Preliminary results of studies carried out to date on these sediments were published in a special issue (Vol.5) of the Polish journal "Geochronometry" in 1989.

1) International Atomic Energy Agency, Isotope Hydrology Section, P.O. Box 100, A-1400 Vienna, Austria
2) Radiocarbon Laboratory, Silesian Technical University, 44100 Gliwice, Poland
3) Institute of Physics and Nuclear Techniques, Academy of Mining and Metallurgy, 30-059 Krakow, Poland.

NATO ASI Series, Vol. I 2
The Last Deglaciation: Absolute and Radiocarbon Chronologies
Edited by E. Bard and W. S. Broecker

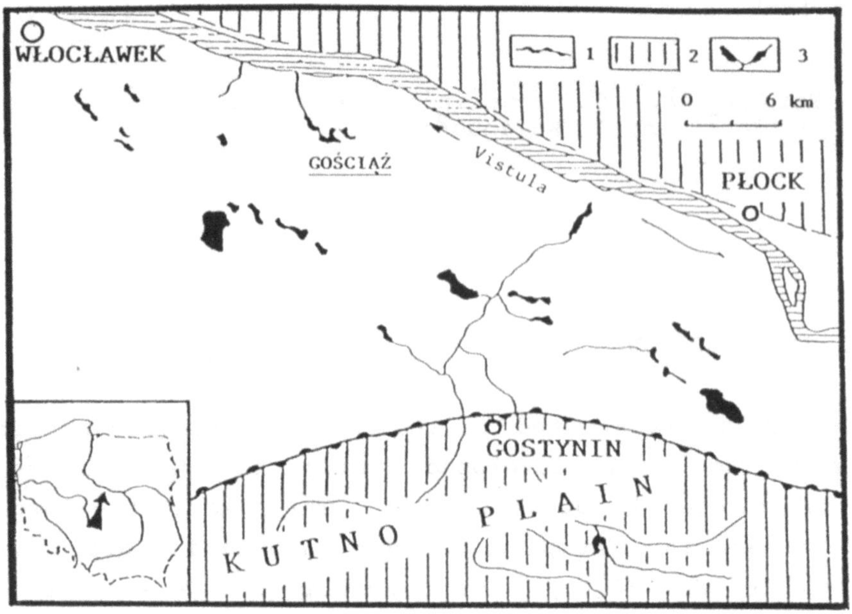

Fig.1 Location of Lake Gościąż (according to Wicik and Więckowski, 1988). 1 - maximum southward extension of the Vistulian ice-sheet, 2 - morainic hills, 3 - lake and river systems.

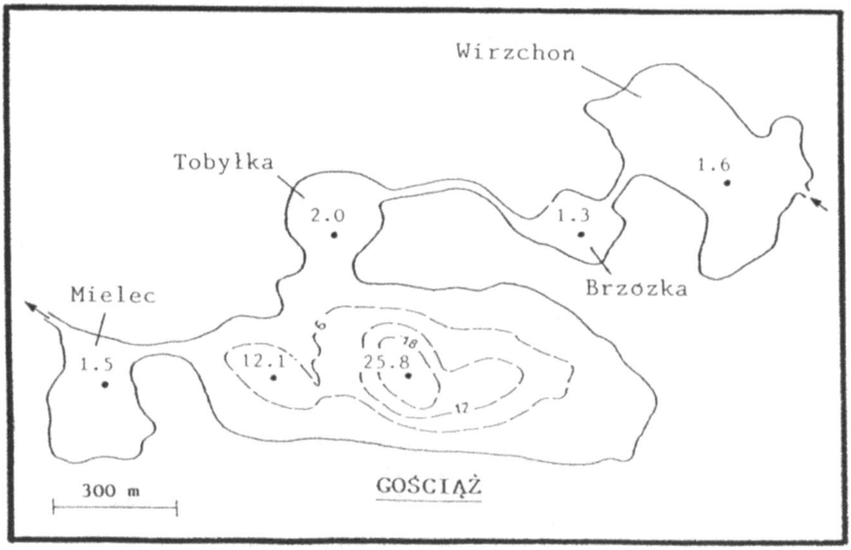

Fig. 2 Bathymetric map of the Gościąż lake complex consisting of four lakes drained by the Ruda creek (according to Ralska-Jasiewiczowa et al., 1987).

2. CHRONOLOGY OF THE SEDIMENT COLUMN

The potential importance of Lake Gościąż for establishing an absolute chronology of the Late Glacial and Holocene was already recognized when the first complete core from the deepest part of the lake was obtained. Several direct counts of the number of varves in this core, labelled as GO, ended up with approximately 12500 couplets. The uncertainty of this number was assessed to be around 500 couplets, with the possibility of missing some varves rather than counting too many. Conventional C-14 dating of bulk organic and inorganic fractions of the GO core (a total of 21 samples covering the whole sediment column) suggested the annual character of microlaminations (Pazdur et al., 1987).

In 1987, two other sediment sequences, labelled G1 and G2, were cored from the deepest part of the lake. Isotope analysis, reported in this paper, was carried out on the material originating from these two cores.

Recently, Goslar et al. (1989) presented a floating chronology from the lower part of the sediment column, covering 9681 varve years. It was based on measurements of varve thickness from cores G1 and G2 and additional correlations with core GO. Due to some breaks in lamination, the chronology is presented as four separate segments (I to IV), starting at the base of the laminated sediments from cores G1 and G2 (Fig.3).

The most important break in the lamination of both cores occurred between segments I and II, caused by a 0.5m thick sandy layer. The duration of this gap will be determined using core G5, taken in 1990 from the western part of the lake. This core was correlated with G1 and G2 and did not contain the sandy layer below segment II. However, the correlation of segment I is still uncertain. The uppermost segment of the sediment column (approximately 2900 years) is still not included in the chronology due to correlation problems among different cores, caused by less distinct and disturbed laminations. Therefore, the error estimates indicated in Fig.3 are relatively large compared to the remaining parts of the sediment column where the overall error should not exceed 50 varves. Hopefully, the ongoing calibration work will reduce the existing uncertainty of the total number of varves in the Lake Gościąż varve chronology.

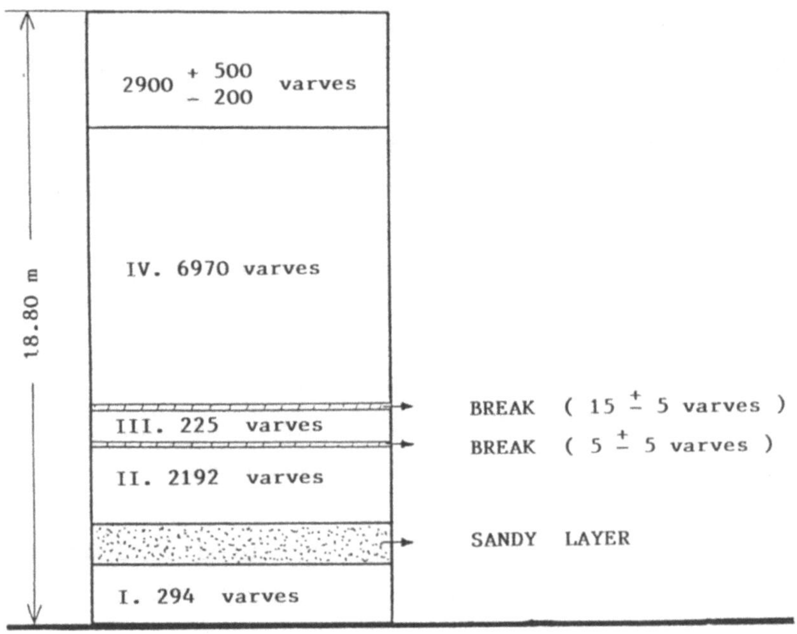

Fig.3 Simplified chronology of the sediment column from Lake Gościąż based on correlation of two complete cores (G1 and G2).

3. ISOTOPE DATA

The $^{13}C/^{12}C$ and $^{18}O/^{16}O$ isotope ratios of bulk inorganic carbon were initially measured in the lower part of core G1 only. Also, $CaCO_3$ content was analyzed. The sampling was continuous (each sample covering a time interval between 5 and 10 varve years) for segment I and the first three hundred couplets of segment II. Starting from laminae No. 680 (segment II), 10-year intervals were sampled every forty years.

Recently, the continuous high-resolution sampling was extended to the lamina No. 1380 using material originating from core G2. Three to four samples were analyzed for every varve decade. The composite profiles of $\delta^{18}O$, $\delta^{13}C$ and $CaCO_3$ content, combining the results obtained from both cores, are presented in Fig.4, covering the first three thousand years of sedimentation in the lake. Results obtained recently from core G2 are plotted as decadal average values.

Stable isotope analyses were carried out using standard techniques. No special treatment of samples except drying ($110^{\circ}C$, ~10 hours) was used. Details of sample preparation and measurement procedures are reported by Rozanski and Wcislo (1988).

Recently, AMS ^{14}C analyses were carried out on two macrofossils (fragments of pine needles) extracted from 10-year intervals of the core G1 (laminae 401-410 and 1260-1270, segment II). Obtained radiocarbon ages are indicated in Fig.4.

4. DISCUSSION

Oxygen isotopic composition of authigenic lake carbonates is controlled by the ^{18}O content of the lake water (e.g. Stuiver, 1970; Fritz, 1983; Siegenthaler and Eicher, 1986). For open lakes with fast water turnover, the isotopic composition of lake water corresponds to that of precipitation over the lake basin, which in turn is temperature dependent. Thus, the ^{18}O content of the deposited carbonate may be used as proxy indicator of cold and mild climatic episodes in the sedimentary history of a lake.

Preliminary results of isotope studies suggest that the water turnover in the Gościąż Lake is rather fast, with a mean

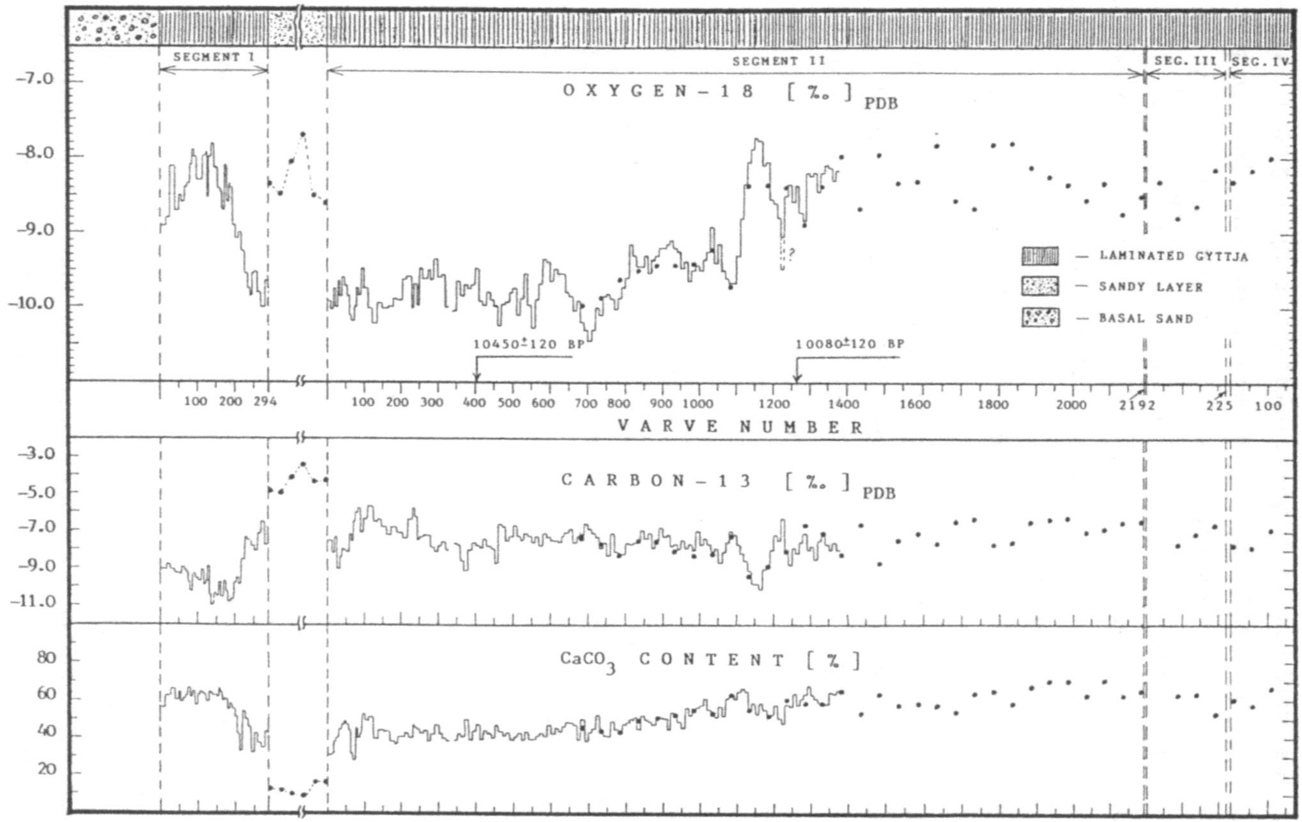

Fig.4 Composite profiles of $\delta^{18}O$, $\delta^{13}C$ and $CaCO_3$ content for the lower part of the sediment column from Lake Gościąż, based on analyses of cores G1 and G2 (see text for details).

residence time of water in the lake probably not exceeding several years (Kuc, unpublished data). There is also no evidence of substantial evaporative enrichment of the lake water. Therefore, the $\delta^{18}O$ profile shown in Fig.4 may be interpreted in terms of climate changes.

Younger Dryas - Preboreal transition

The character of the $\delta^{18}O$ curve leaves little doubt as to the position of the YD/PB boundary in the sediments of Lake Gościąż. It is located between laminae No. 1100 and 1150 (segment II), i.e. within the time interval of approximately 50 years. Fragments of pine needles extracted from laminae 1260-1270 reveal an AMS radiocarbon age of 10080±120 years BP, suggesting reasonably good agreement between the abrupt rise of $\delta^{18}O$ profile and the accepted age of the YD/PB boundary (10000 conventional radiocarbon years BP - Mangerud et al. 1974). More AMS dates from the period of transition are needed to specify this event more exactly. Accepting the chronology presented in Fig.3, YD/PB transition is absolutely dated to 11200^{+500}_{-200} varve years.

The transition from Younger Dryas to Preboreal, as derived from the $\delta^{18}O$ profile, reveals an oscillatory behaviour. An abrupt increase of $\delta^{18}O$ by about 2‰ within fifty years points to a rapid warming, followed by an almost equally rapid cooling and a resulting drop of the isotope signal to less than half of the preceding increase. If the very low $\delta^{18}O$ value, recorded for one single sample (laminae 1220-1222), is considered in the calculation of the decadal mean, this cooling seems to be even more dramatic (broken line in Fig.4). This, however, needs to be confirmed by additional measurements. The following climate seems to improve gradually, with superimposed minor $\delta^{18}O$ oscillations, reaching a nearly constant level after around 400 years. This "overshoot-type" behaviour of the $\delta^{18}O$ profile during the

YD/PB transition reveals striking similarities to the analogous profiles measured from Greenland ice cores, especially those drilled in eastern and southern Greenland (Johnsen et al. - this volume). Presumably this reflects common forcing parameters for the climate system in Europe and in the North Atlantic region.

Alleröd - Younger Dryas transition

The $\delta^{18}O$ profile shown in Fig.3 suggests that sedimentation in the lake started only some 200 years before the beginning of the Younger Dryas, located around lamina 200 (segment I). The transition from Alleröd to Younger Dryas was completed within approximately 100 years, suggesting an abrupt change of climate comparable to the YD/PB transition. The amplitude of the $\delta^{18}O$ shift is very similar to that which occurred at the YD/PB boundary. These facts seem to confirm the recent suggestion that the Younger Dryas cold event was generated by dramatic changes in circulation patterns of the North Atlantic (Broecker and Denton, 1989). During the AL/YD transition other recorded parameters of the lake also reveal drastic changes (increase of $\delta^{13}C$ from -11 to -7‰ , a drop in $CaCO_3$ content from 60 to 30%). Such changes, although visible, are less distinctive at the YD/PB boundary.

Duration and structure of Younger Dryas

The major problem in the precise determination of the duration of this climatic episode from the ^{18}O data is caused by a sandy layer occurring between segments I and II of the cores G1 and G2 (Fig.3). Such a layer has also been recognized in all of the other cores except the recently taken core G5. The data of Fig.4 point to an entirely different character of this sedimentary event in the history of the lake. A sharp increase of $\delta^{18}O$ and $\delta^{13}C$ content, accompanied by a decrease of $CaCO_3$ concentration, suggests an invasion of allochthonus carbonates. The origin of this sedimentary event is unclear. It has been suggested that solifluction caused by melting of dead ice blocks underlying the bottom of the relatively young lake could result in the formation of such a layer (Starkel, 1989). Also, eolian activity has been suggested as a source. The abrupt character of this event seems to be confirmed by the fact that the lowermost sequence of varves in core G1 (275 couplets) is shorter than the same sequence in core G2 (294 couplets); the missing part of core G1 could be destroyed by the moving sand. A more detailed sedimentological study of the sandy layer would be needed in order to test the above mentioned hypothesis.

As indicated above, the recently obtained core G5 does not contain the disturbing sandy layer between segments I and II of the sediment column. Due to the still uncertain correlation between cores G1, G2 and G5 in the segment I, one can only state that the duration of Younger Dryas was not shorter than about 1200 years. A more precise estimation will be possible when ^{18}O analyses from this part of core G5 become available.

It should be mentioned here that the above estimated minimum duration of Younger Dryas is substantially longer than other available estimates ranging from 400 to 1000 years. Ongoing intensive studies of laminated sediments from several European lakes, including Lake Gościąż, should clarify this discrepancy.

The $\delta^{18}O$ profile shown in Fig.4 reveals some minor fluctuations within the Younger Dryas, suggesting a certain variability of climate during this climatic episode. During the first 600 years (segment II), only relatively small, irregular oscillations of $\delta^{18}O$ around the mean value of -9.9‰ are observed. The typical time scale of these fluctuations is between 50 and 100 years. Around laminae 700 the $\delta^{18}O$ curve reaches the lowest point within the Younger Dryas (-10.9‰), probably reflecting the most severe climatic conditions. From this point on, the climate seems to improve gradually over the next 250 years. The Younger Dryas ends up with another minor fluctuation of the $\delta^{18}O$ curve, with a characteristic small minimum preceding the YD/PB boundary. Such a minimum is also seen in the Renland and Dye 3 ice cores (Johnsen et al., this volume).

5. CONCLUDING REMARKS

The isotope data presented in Fig.4 reveal two additional, interesting features which were not discussed so far, but should be mentioned:

- The apparent shift of $\delta^{18}O$ at the AL/YD and YD/PB boundaries amounts to about 2‰. This would correspond to a change in δD of precipitation feeding the lake by approximately 16‰. B. Becker (this volume) reports a shift in δD of a similar magnitude for the YD/PB transition, derived from analyses of deuterium content from cellulose of German pine tree-ring series representing this time interval. Old European groundwaters recharged during the last glaciation also reveal a similar shift in deuterium content, when compared with modern infiltration waters (Rozanski, 1985). This independent evidence supports a climatic interpretation of the $\delta^{18}O$ record derived from the basal part of the sediments from Lake Gościąż.

- A comparison of $\delta^{18}O$ and $\delta^{13}C$ isotope profiles shown in Fig.4 reveals a distinct anticorrelation among these two parameters during periods of major climatic changes (AL/YD, YD/PB). This negative correlation is rather surprising in view of numerous published examples demonstrating rather an opposite trend (e.g. Talbot, 1990).

Nevertheless, it can be understood in the light of presumable changes in the vegetation cover of the catchment area of the lake. A deterioration of climate will likely result in the reduction of the vegetation cover and slowing down the microbial activity of the soil which in turn should result in lower partial pressures of soil CO_2. Lower pCO_2 leads to higher ^{13}C content in the dissolved carbonates, formed under closed or open system conditions.

Isotope investigations of the basal part of the sediment column from Lake Gościąż will be continued. They will concentrate on the new core G5. It is expected to obtain a detailed uninterrupted sequence of $\delta^{18}O$, covering the termination of the last deglaciation. This should help in clarifying some open questions with respect to the onset and duration of Younger Dryas.

ACKNOWLEDGEMENTS

This study was partly supported by grants from the Central Research Project CPBP 03.13, the Ministry of Education (GM-605/RMF1/90), and the International Atomic Energy Agency (Research Contracts No. 5467 and 5179). We gratefully acknowledge the assistance of J.C. Duplessy, M. Arnold and E. Bard (Gif-sur-Yvette Tandetron Facility) in obtaining AMS ^{14}C analyses of two macrofossil samples.

REFERENCES

Broecker WS, Denton GH (1989) The role of ocean-atmosphere reorganizations in glacial cycles. Geochim et Cosmochim Acta 53:2465-2501

Fritz P (1983) Palaeoclimatic studies using fresh water deposits and fossil groundwater in central and northern Canada. In: Palaeoclimates and palaeowaters: a collection of environmental isotope studies. IAEA, Vienna p.157

Goslar T, Pazdur MF, Walanus A (1989) Chronology of the lower part of annually laminated sediments of the Gosciaz Lake. Geochronometry 5:11-20

Mangerud J, Andersen ST, Berglund BE and Donner JJ (1974) Quaternary stratigraphy of Norden, a proposal for terminology and classification. Boreas 3: 109-128

Pazdur MF, Awsiuk R, Goslar T, Pazdur A, Walanus A, Wicik B, Wieckowski K (1987) Calibrated radiocarbon chronology of annually laminated sediments of the Gosciaz Lake. Geochronometry 4:69-83

Ralska-Jasiewiczowa M, Wicik B, Wieckowski K (1987) Lake Gosciaz - a site of annually laminated sediments covering 12000 years. Bull Pol Acad Sci Ser Sci Terre 35:127-137
Rozanski K (1985) Deuterium and oxygen-18 in European groundwaters - links to atmospheric circulation in the past. Chem Geol 52:349-363
Rozanski K, Wcislo D (1988) Oxygen and carbon isotope studies of carbonate sediment cores from some Polish lakes. Freiberger Forschungshefte C420: 79-92
Siegenthaler U, Eicher U (1986) Stable oxygen and carbon isotope analyses. In: Berglund BE (ed) Handbook of holocene palaeoecology and palaeohydrology. Wiley, Chichester, p.407-422
Starkel L (1989) Comment on the paper by Rozanski K, Goslar T, Walanus A, Pazdur MF Comparison of variability of stable isotopes O-18 and C-13 and varve thickness in basal part of laminated sediments of the Gosciaz Lake, Geochronometry 5:72-73
Stuiver M (1970) Oxygen and carbon isotope ratios of fresh water carbonates as climatic indicators. J Geophys Res 75:5247-5257
Talbot MR (1990) A review of the palaeohydrological interpretation of carbon and oxygen isotope ratios in primary lacustrine carbonates. Chem Geol 80:261-279
Wicik B, Wieckowski K (1988) Lakes in the Middle Vistula valley - site Gosciaz. In: Starkel L, Rudkowski J, Ralska-Jasiewiczova M (eds) Late Glacial and holocene environmental changes - Vistula Basin 1988, AGH, Krakow, p 133.

YOUNGER DRYAS OSCILLATION - VARVE DATED MICROSTRATIGRAPHIC, PALYNOLOGICAL AND PALAEOMAGNETIC RECORDS FROM LAKE HOLZMAAR, GERMANY

B. Zolitschka, B. Haverkamp* and J.F.W. Negendank

Department of Geology, University of Trier, D-5500 Trier, Germany

INTRODUCTION

Late Quaternary lakes often contain continuous sedimentary records of palaeoenvironmental and palaeoclimatic changes covering thousands of years. These sediments enable geologists and palaeolimnologists to study lacustrine depositional systems with a high degree of time resolution. The major problem is to obtain a precise timescale. In some lakes these difficulties were solved using annual laminations. Organic varves make available a year-by-year analysis. Rates of change of environmental and climatic conditions are determinable.

The dominating factor controlling sedimentation in lacustrine depositional environments is climate, causing various biotic responses in the lake and in the drainage basin.

The Pleistocene/Holocene transition as determined by Mangerud et al. (1974) in the chronostratigraphic sense to 10,000 conventional radiocarbon years BP is of minor use for palaeoenvironmental investigations, because chronozones cannot be correlated to events in the lithostratigraphic or biostratigraphic record. More appropriate are palaeotemperature reconstructions obtained from $\delta^{18}O$ measurements. They demonstrate a severe temperature setback due to the climatic deterioration of Younger Dryas prior to the early Holocene temperature rise (Eicher & Siegenthaler

*Department of Geophysics, University of Münster, Corrensstr. 24, D-4400 Münster, Germany

NATO ASI Series, Vol. I 2
The Last Deglaciation: Absolute and Radiocarbon Chronologies
Edited by E. Bard and W. S. Broecker

1976, Eicher 1987). This dramatic change is of major interest, because palaeoclimatic and palaeoecological conditions changed rapidly in the course of several centuries or even decades strongly influencing processes
of sedimentation, the surrounding vegetation and therefore also pollenstratigraphy. But still there are no conclusive dates for the duration of Younger Dryas and the beginning of the Holocene.
Three sources of age determination are available: first of all dendrochronology, just recently crossing the chronostratigraphic Holocene/Pleistocene boundary of 10,000 radiocarbon years BP on its way back to the Pleistocene (Becker & Kromer 1986); secondly, varve chronology, improving steadily since early this century (Saarnisto 1986); and, finally, radiocarbon chronology, not based on calendar years but on "Libby years", calculated from the rate of decay of ^{14}C atoms. Radiocarbon dates have to be calibrated to give an equivalent of calendar years (e.g. Stuiver et al. 1991).
This paper presents varve ages of the lithologically, palaeoecologically and palynologically determined Pleistocene/Holocene transition and for the duration of Younger Dryas bio- and lithozones using annually laminated lake sediments from the Eifel area, Germany. Furthermore, palaeoecological variations within Younger Dryas are discussed, based on palynological, palaeomagnetic and thin section investigations.

MAAR LAKES OF THE EIFEL

A maar is a typical geological structure of the West-Eifel Volcanic Field, which is a part of the Rhenish Massif, Germany. More than 50 maars have been determined geophysically. Only eight of the youngest maars are still occupied by lakes. The ages of the volcanic explosions forming these maars are unknown. According to the recovered Weichselian lacustrine sediments from Lake Holzmaar and Lake Meerfelder Maar, these maars are apparently older than 20,000 years.

Five maar lakes were investigated in the course of several palaeolimnological studies (Tab. 1). Only in the eutrophic, relatively shallow lakes with a large catchment area annually laminated sediments have been deposited. High autochthonous productivity leads to several algal blooms in the course of one year which accumulate under anoxic conditions without any bioturbation. Therefore varves have been preserved in this type of environment.

Table 1: Location, morphology, limnology and sediments of investigated Eifel maar lakes: Meerfelder Maar (MFM), Holzmaar (HZM), Schalkenmehrener Maar (SMM), Gemündener Maar (GMM) and Weinfelder Maar (WFM).

lake	MFM	HZM	SMM	GMM	WFM
location	6°45'E 50° 6'N	6°53'E 50° 7'N	6°51'E 50° 8'N	6°50'E 50°11'N	6°51'E 50°11'N
elevation (m a.s.l.)	336.5	425.1	420.5	406.6	484.0
max. depth (m)	18	20	21	39	52
water surface ($10^3 m^2$)	248	58	219	75	159
catchment area ($10^3 m^2$)	5760	2000	1080	430	190
trophic state	eutr.	eutr.	eutr.	oligotr.	oligotr.
recovered sediments (m)	46	32	11	8	8
varved Holocene	yes	yes	yes	no	no

METHODS

The sediments were recovered with a modified Livingstone piston corer (Usinger corer) from a raft just above the deepest part of each lake.
Lake Holzmaar was selected to be the key site, because its laminated deposits are in extraordinary good condition. Two overlapping core series, each consisting of several 2 m long and 80 mm wide core sections, were compiled to one continuous sediment profile according to macroscopic marker horizons, which have been confirmed microscopically in the course of thin section investigations.
For palynological investigations this compiled profile was subsampled continuously from the isochrone of Laacher See Tephra to the sediment/water interface, each subsample containing 2 cm of sediment. Similar subsamples were used for

determination of loss on ignition, expressed as weight loss after ignition at 550°C in percent of dry weight (dried at 105°C) according to Bengtsson & Enell (1986). For microstratigraphic investigations the continuous sediment was subsampled for preparation of large, 10 cm long thin sections (Merkt 1971). Palaeomagnetic studies were carried out continuously on both core series, each subsample containing 2 cm of sediment (Haverkamp 1991).

RESULTS

Sediment sequences from five investigated maar lakes (Tab. 1) turned out to consist of finely laminated lacustrine deposits (Negendank et al. 1990). All of them contain the pyroclastic layer of Laacher See Tephra. Soon after this isochrone a period of minerogenic deposits occurred in every lake inserted inbetween biogenic sediments. These clastic sediments are related to Younger Dryas climatic oscillation, as demonstrated by the pollen record.

Dating of the Sedimentary Record

Thin sections have to be prepared for microscopic investigation, because annual laminations are not visible macroscopically. Using this method, varve chronologies were established for continuous sedimentary records from Lake Holzmaar and Lake Meerfelder Maar (Zolitschka 1990). Results from both lakes are comparable. High resolution palynological and palaeomagnetic investigations have been carried out on the sediments of Lake Holzmaar only. Therefore the following data refer to this site.

The annual character was proved in detail for Holocene and Late-Glacial laminated sediments from Lake Holzmaar and Lake Meerfelder Maar (Zolitschka 1989, 1991). Laminations consist of a spring/summer layer of planktonic diatoms and chrysophytes and of an autumn/winter layer of minerogenic and organogenic detritus. Organic varves occur from the sediment/water interface down to the early Late-Glacial litho-

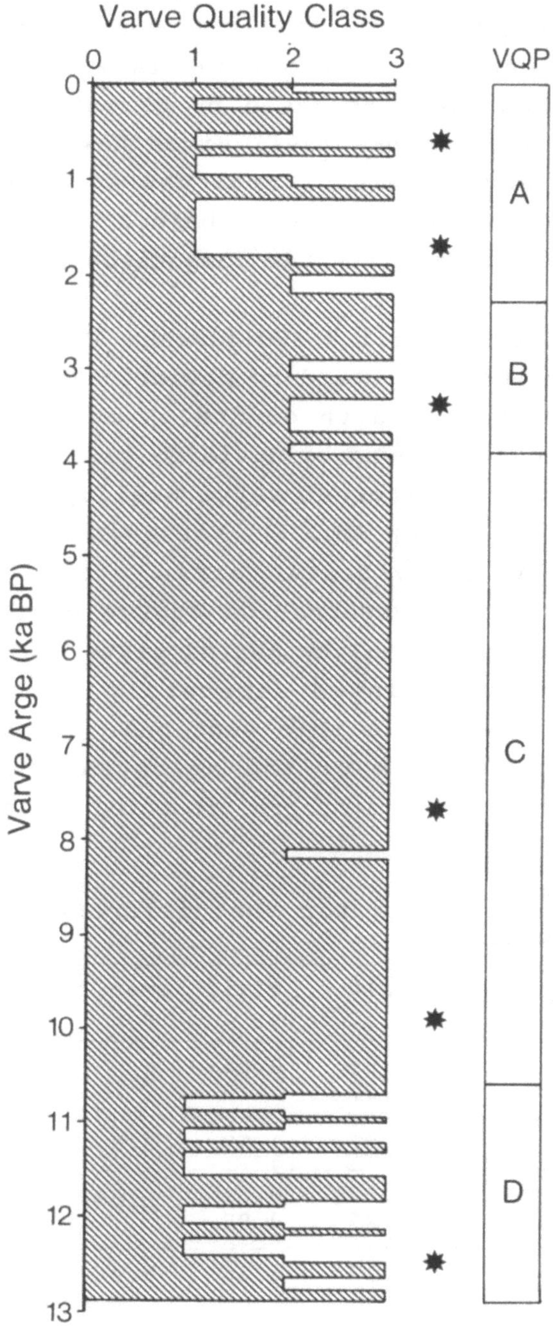

Fig. 1: Varve quality classes from Lake Holzmaar sediment cores HZM-B/C (0 = no varves; 1 = indistinct laminations, only varve estimations; 2 = medium varve preservation; 3 = best varves) with indicated varve quality periods (= VQP) and location of thin sections used for multiple counts (*).

logical boundary from organic to minerogenic deposits.
Counts produced a total of 12,881 varve years for the Holocene and the organic deposits of the Late-Glacial. The number of counted varves has been increased by 87 varve years from a total of 12,794 (Zolitschka 1991) since new cores from 1990 enabled to get better results from the topmost sediments. Varve ages are given in varve years before the reference year 1950 (= v.y. BP) to simplify the comparison with radiocarbon dating.
The dating accuracy of ±1.1 % (Zolitschka 1991) was obtained according to the relative standard deviation (±1σ) of the sum of all of the multiple varve counts. A more precise error estimation has been achieved by dividing the varve chronology into four periods of different varve quality (varve quality periods = VQP: A, B, C, D; Fig. 1). Varve quality periods were distinguished by microstratigraphic evaluation of the distinctness of varve preservation (varve quality classes; Fig. 1). For each varve quality period varve counts and error estimations were calculated seperately at least on two successive, representative thin sections. Six varve counts were carried out on each of the selected thin sections. Results of error calculations are presented in Tab. 2.

Tab. 2: Error estimations for the different varve quality periods from Lake Holzmaar.

varve quality periods and duration (v.y. BP)	±1σ	±1σ%	mean %
A: 0 - 2300	2.59 2.07	3.84 4.17	4.0
B: 2300 - 3900	10.67	2.57	2.6
C: 3900 - 10600	1.97 2.26	0.61 0.82	0.7
D: 10600 - 12881	7.37	5.88	5.9

The standard deviation (±1σ) and the relative standard deviation (±%) declare 68.26 % of the variability of the data.

The column "mean %" gives the arithmetic mean of the relative standard deviation for varve quality periods with more than one data set.

For the boundaries of major lithozones and biozones and for the isochrone of Laacher See Tephra ±1σ errors of varve counts are listed below:

local lithozone boundaries

H5/H4: 10,420 ±180 v.y. BP

H4/H3: 10,630 ±183 v.y. BP

H3/H2: 11,080 ±209 v.y. BP

Firbas pollen zone boundaries

IV/III: 10,630 ±183 v.y. BP

III/II: 11,080 ±209 v.y. BP

Laacher See Tephra

LST: 11,323 ±224 v.y. BP

These errors have been estimated according to the sum of the absolute error of the varve quality period following the age in question.

Microstratigraphic Record

Microscopic investigation of sediment thin sections provided the most accurate counting of annual laminations. Additionally, information about sedimentation rates, sedimentological changes and to some extent even about palaeobiological changes were achieved.

The sedimentary record from Lake Holzmaar has been subdivided into 8 major local lithozones (H1 - H8). H2 characterizes the organic deposits of the Late-Glacial with Laacher See Tephra close to its top. Minerogenic sediments of H3 are supposed to be the imprint of Younger Dryas climatic oscillation. H4 is a transitional lithozone at the very beginning of Holocene organic deposits, followed by H5 lasting for 1620 varve years.

During lithozone H2 the yellowish brown biogenic deposits (Fig. 2) were composed of organic detritus and planktonic diatoms. Precipitated calcites and ostracods occurred. The beginning of H3 is characterized by a change to brown sediments with high amounts of silt and clay, while calcites, ostracods and planktonic diatom layers of ceased. Chrysophytes

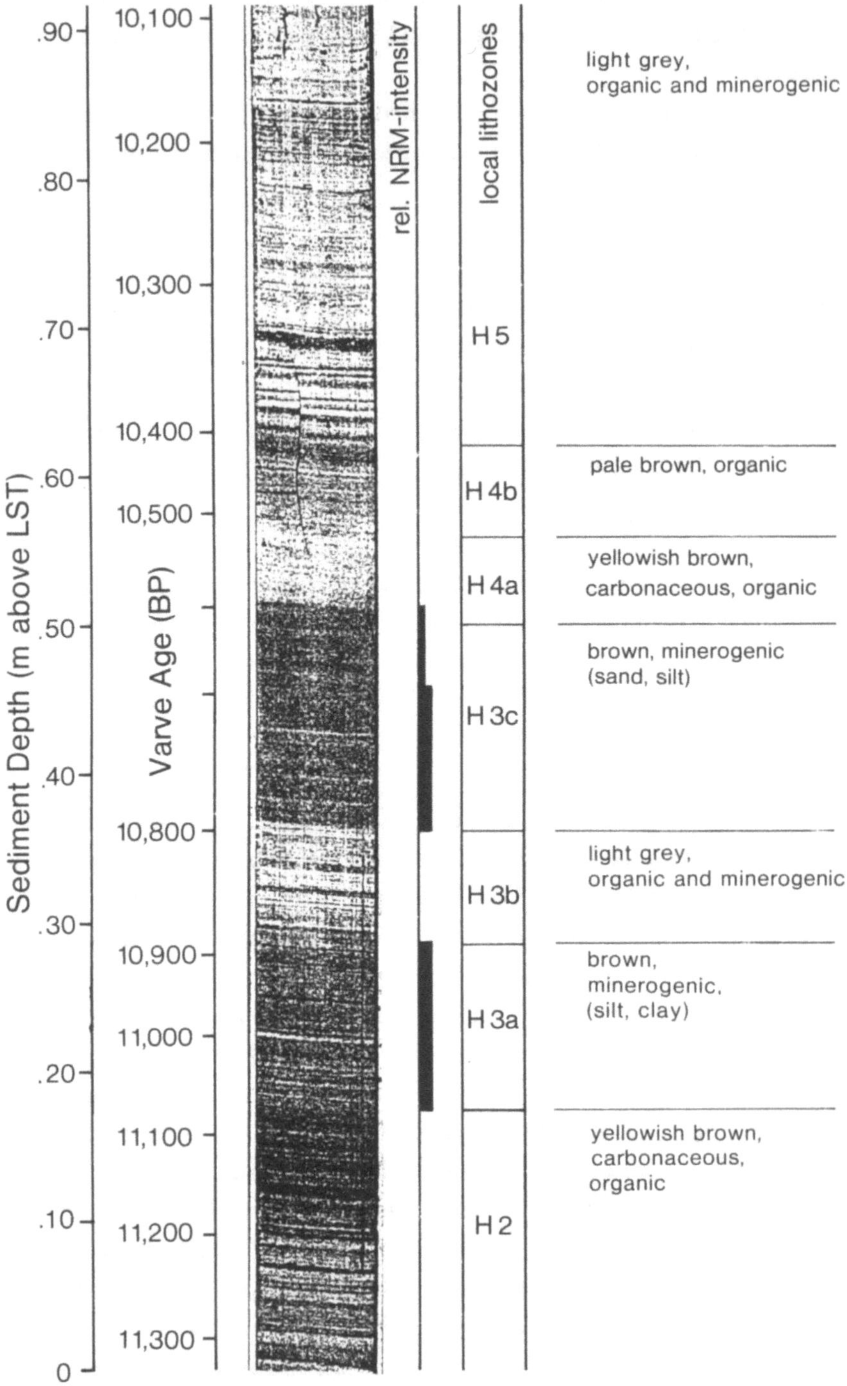

Fig. 2: Macroscopic view of the sediment from Lake Holzmaar (core HZM-C9), including relative NRM-intensity, local lithozones and a brief sediment description.

became abundant. Three subzones have been distinguished for this lithozone (Fig. 2). After the first maximum of clastic deposition (H3a: 11,080 - 10,900 v.y. BP) silt and clay diminished, changing the colour to a light grey (H3b: 10,900 - 10,800 v.y. BP). About 10,800 v.y. BP a rise of silt and sand occurred (H3c: 10,800 - 10,630 v.y. BP). Accordingly, the colour of the sediment turned back to brown (Fig. 2). This second maximum of minerogenic detritus lasted for about 50 years, until anorganic components dropped to lower values (10,750 v.y. BP) and finally disappeared (10,630 v.y. BP). The first increase of autochthonous biogenic productivity (layers of planktonic diatoms) happened during the final four decades of lithozone H3c (since 10.675 v.y. BP). In the course of only 45 years they were abundant again, consisting of 250 μm thick spring laminae, while chrysophytes diminished. Organic detritus and calcite reappeared as well, changing the sediment colour to yellowish brown of lithozone H4 (10,630 - 10,420 v.y. BP). H4 is subdivided into two subzones: during H4a precipitated calcites are present. They vanished around 10,530 v.y. BP, causing a colour change to pale brown of lithozone H4b (Fig. 2). Organic sediments prevailed during H4b for another 110 years, until the sediment changed again to grey of lithozone H5 (10,420 - 8,800 v.y. BP), indicating a slight increase of sand, silt and clay.

Lithozone H3, corresponding to Younger Dryas climatic deterioration, lasted for about 450 varve years. An even shorter duration (405 varve years) of the impact of this palaeoclimatic event is obtained, if the increase of planktonic diatoms is regarded as first signal of the Postglacial.

The outlined changes of sedimentation are recognizable using sedimentation rates as well (Fig. 3). But these data are misleading without any supplementary information about sediment composition, because sedimentation rates result from many different processes in the lake's water body and in its catchment area. Minimum sedimentation rates during H3a are due to missing organic deposits, making up more than 80 % of the deposits during H2. About 180 years later (H3b) a slight increase of biogenic productivity caused a rise in sedimentation rates. During the beginning of H3c highest sedimentation rates are due to coarse grained clastic deposition. Around 10,730 v.y. BP a decline of sedimentation rate occurred. During H4 minerogenic sediments diminished and organic deposits prevailed for 210 varve years. About 10,420 v.y. BP clastic deposition increased in addition to biogenic deposits, causing a rise in sedimentation rates. These sediments dominate until the middle of the Boreal biozone (lithozone H5).

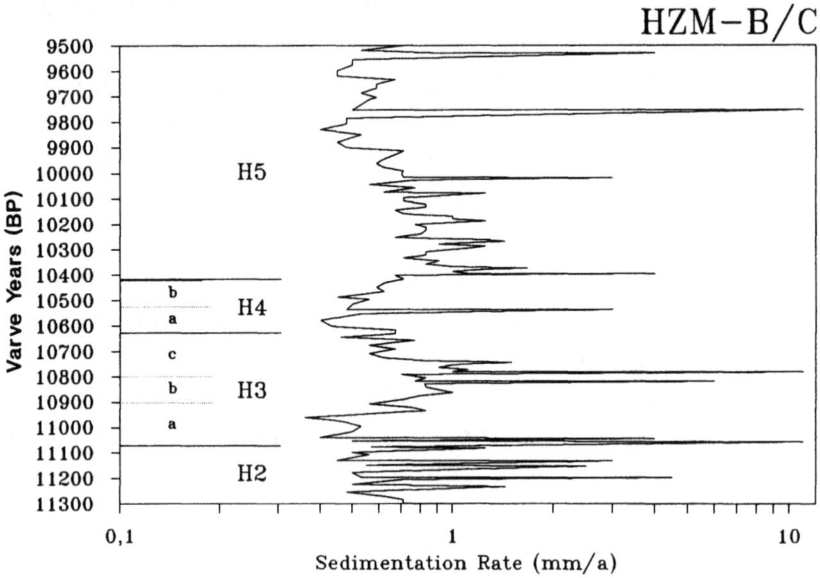

Fig. 3: Sedimentation rates of varve dated sediments with indicated local lithozones, Lake Holzmaar.

Palynological Record

Climatic deterioration of Younger Dryas is recognizable from Lake Holzmaar pollen record. The transition from Firbas pollen zone II (Alleröd) to III (Younger Dryas; regional pollen zones according to Firbas 1949, 1954) is characterized by a marked increase of nonarboreal pollen contents around 17 cm above LST (11.080 v.y. BP). Pollen of Pinus and Betula declined (Fig. 4). Heliophilous herbs (Cyperaceae, Poaceae,

Artemisia, Rumex, Chenopodiaceae) and shrubs (Salix, Juniperus, Empetrum) expanded, whereas thermophilous herbs (Urtica, Filipendula) diminished, causing a tundra-like vegetation (cf. Usinger 1982, 1984).

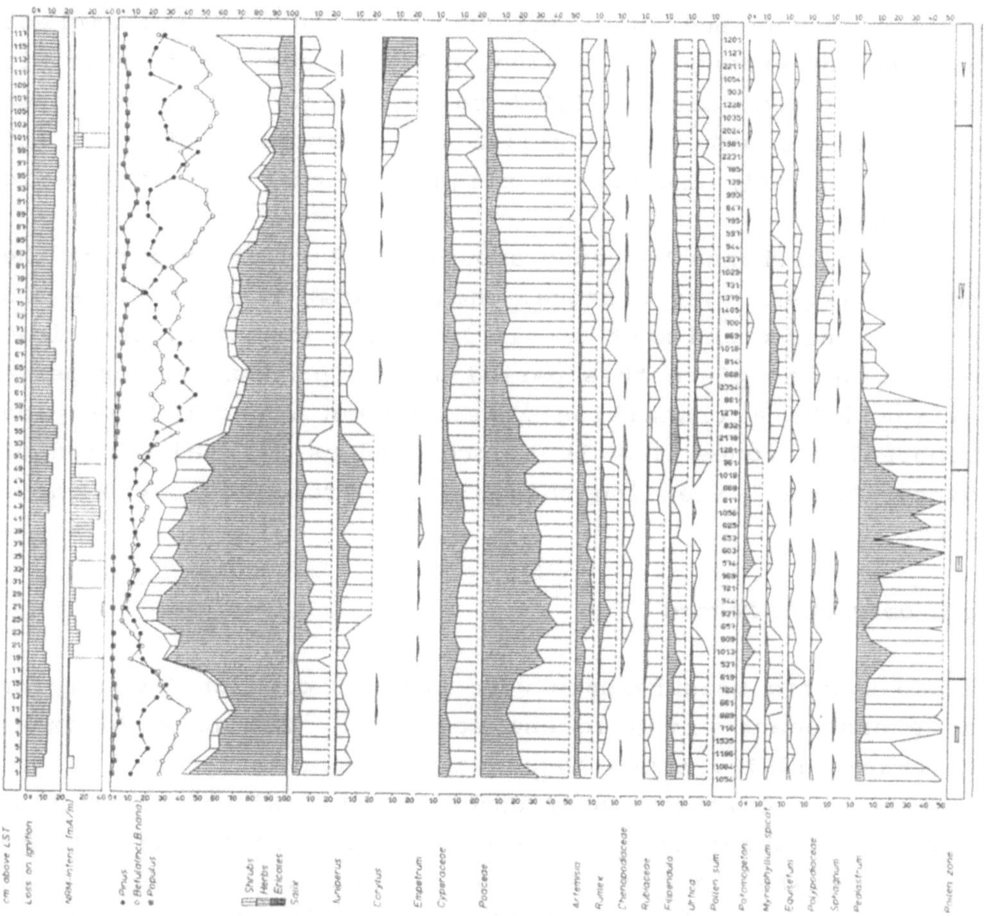

Fig. 4: Pollen stratigraphy from Lake Holzmaar since Laacher See Tephra (LST) until the beginning of Boreal, including loss on ignition and palaeomagnetic NRM-intensity data, (Usinger & Wolf, unpublished, modified).

Younger Dryas pollen zone ended with a pronounced rise of tree pollen values and a decrease of Juniperus, Poaceae, Artemisia and Rumex. The beginning of Preboreal biozone (Firbas pollen zone IV) at 50 cm above LST (10.630 v.y. BP) was fixed to a

marked rise of Betula and the increase of Filipendula and Urtica, indicating warmer climatic conditions (Usinger 1982).
Younger Dryas pollen zone lasted from 11,080 until 10,630 v.y. BP with a duration of 450 varve years. This biozone is not uniform, but can be subdivided at least into two different periods. At the beginning (11,080 - 10,900 v.y. BP) highest contents of nonarboreal pollen, Salix, Poaceae and Artemisia were recognized. Later (10,900 to 10,630 v.y. BP) a decrease of Poaceae and nonarboreal pollen and an increase of Pinus and Betula occurred, indicating slightly warmer palaeoclimatic conditions.
A remarkable change happened around 10,690 v.y. BP during the end of Younger Dryas biozone, when a first rise of Pinus and Betula occurred and most of all Juniperus expanded. Our first opinion was to place this period with a pronounced Juniperus maximum into a transition zone between Younger Dryas and Preboreal biozones. Now it is proposed as a final period of Younger Dryas, extending this biozone from 370 to 450 varve years.

Palaeomagnetic Record

Palaeomagnetic measurements were carried out continuously on the Pleistocene and Holocene sediments from Lake Holzmaar (Haverkamp 1991). Among these data, measurements of natural remanent magnetization (NRM) emphasize the palaeoenvironmental effect of Younger Dryas. Due to a sudden change from pine-birch forests to tundra vegetation erosion increased, resulting in a high content of minerogenic particles in the sediment. This is the reason for increased values of NRM-intensity during lithozone H3, occurring in two distinct maxima (H3a and H3c of Figs. 2, 5).
Fig. 5 demonstrates a slightly similar trend of NRM-intensity and nonarboreal pollen contents. The marked gap between the two maxima of NRM-intensity (lithozone H3b) is hardly represented in the pollen record.

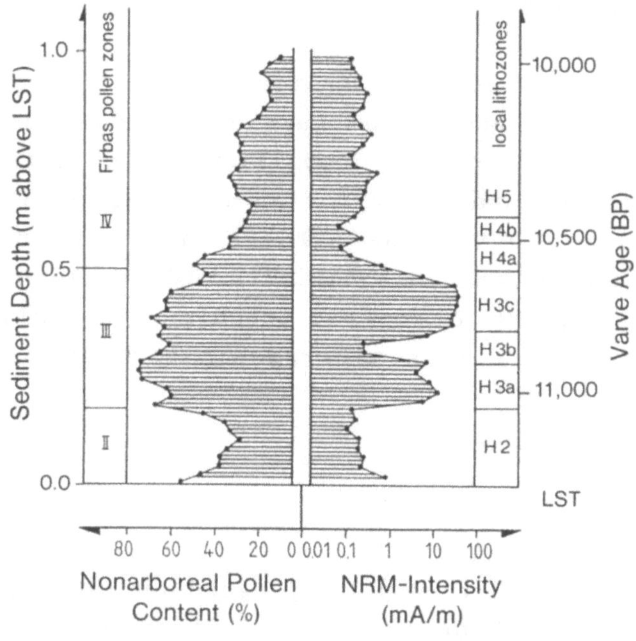

Fig. 5: Nonarboreal pollen contents and palaeomagnetic NRM-intensity including Firbas pollen zones and local lithozones, Lake Holzmaar.

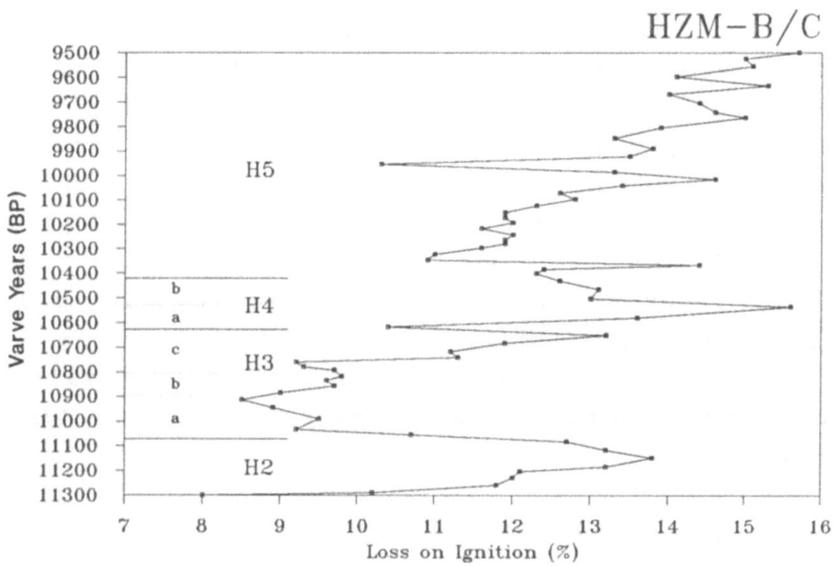

Fig. 6: Loss on ignition of varve dated sediments with indicated local lithozones, Lake Holzmaar.

Supplementary Data

Determinations of loss on ignition (Fig. 6) were obtained from varve dated sediments from Lake Holzmaar, demonstrating the severe change in organic deposition during lithozone H3. Low values were measured only for an interval of 330 varve years (11,080 - 10,750 v.y. BP). There are no coincidences with palynological or palaeomagnetic records explaining this short duration. Only the microstratigraphic record gives some hints for a change to a less minerogenic sediment composition at that time.

IMPLICATIONS

Sediment description (Fig. 2), based on microscopic analysis of varve composition allows a refinement of lithozone H3, which has been confirmed by palaeomagnetic investigations. According to palynology, the final part of lithozone H2 (Fig. 2) corresponds to the Alleröd biozone. The beginning of Younger Dryas biozone and the onset of lithozone H3 coincide, indicating no lag phase between aquatic and terrestrial vegetation. Problems still arise from defining the end of Younger Dryas from available microstratigraphic and palynological data. What appears to be the Pleistocene/Holocene biozone boundary at 10.630 v.y. BP has an equivalent in lithozonation (lithozone boundary H3/H4). Again there is no obvious lag phase. But in fact planktonic diatom populations recovered from Younger Dryas cooling already around 10.675 v.y. BP. This would imply, that there is a time lag of approximately 45 varve years between lacustrine and terrestrial ecosystems in responding to the early Holocene climatic warming.

Climatic deterioration of Younger Dryas caused a distinct break of Late-Glacial to Holocene plant succession and soil development in the catchment area of Lake Holzmaar. Alleröd pine-birch forests changed to an open tundra-like vegetation, indicating a cold and dry palaeoclimate. Due to these climatic conditions autochthonous biogenic production diminished, as minimum values in loss on ignition (Fig. 6) and absent planktonic diatoms suggest. Only chrysophytes attained high abundance. They are sensitive indicators of changes in the nutrient status and occur in large quantities in oligotrophic, coolwater lakes (Smol 1988). Presumably Lake Holzmaar's nutrient conditions turned from at least mesotrophic during Alleröd to oligotrophic during Younger Dryas and back to mesotrophic/eutrophic at the onset of the Preboreal.

Younger Dryas climatic conditions promoted erosion in the catchment area, either by wind or by rain storms. Increased rates of erosion caused high amounts of minerogenic deposition and therefore maxima of NRM-intensity (Fig. 5). Dominating silt and clay grain sizes of early Younger Dryas minerogenic deposits (lithozone H3a) give rise to the suspicion, that soil erosion in the catchment basin was the reason for this phenomenon.

The drop of NRM-intensity during lithozone H3b (Fig. 5) originates from a decrease of clastic deposition, probably due to climatic changes. Arid conditions might give an explanation: Provided that temperature was still as low as during the beginning of Younger Dryas, decreasing precipitation stopped surface runoff and connected silt and clay sedimentation, resulting in an apparent rise of organic deposition. At the same time the lake level was lowered. This might be interpretated as a time lag of nearly two centuries between the drop of temperature and the drop of precipitation at the beginning of lithozone H3, leading to more continental climatic conditions in the course of Younger Dryas.

The second NRM-intensity maximum has no equivalent in the pollen record. Palynological data suggest a weak but steady climatic improvement towards the Preboreal. But between the two NRM-intensity maxima (lithozones H3a and H3c) there is a qualitative difference: In contrast to H3a, clastic deposits of H3c have a coarser, sandy grain size. Therefore these minerogenic particles are probably of local origin, eroded from lake shores as the result of a low lake level (cf. Digerfeldt 1972).

Loss on ignition supports this theory, because it rises already since lithozone H3b without any palynological indication of a climatic amelioration. This might result from erosion of lake shore sediments, dried out after the lake level has fallen. A palynological evidence for erosion of litoral deposits is given by abundance of Pediastrum during this period (Fig. 4).

The Pleistocene/Holocene transition as determined lithologically shows an increase of lacustrine productivity: Chrysophytes vanished in the plankton and were replaced by thick diatom layers. A further indication of warmer summer climates and increased photosynthesis is precipitation of calcites, starting at the beginning of lithozone H4. Calcium probably derived from weathering of a thin carbonaceous loess veil of Pleniglacial age. Therefore it soon was depleted and precipitation of calcite ended after 100 years. The increase of loss on ignition also reflects this rise in lacustrine productivity at the onset of lithozone H4.

Lithozone boundary H4/H5 is reflected in the loss of ignition data (Fig. 6) by a marked decrease, due to slightly more minerogenic deposits causing an increase of sedimentation rates (Fig. 3). This weak rise in clastic deposition probably results from increased rainfall in the catchment basin. Possibly, there is another time lag between the rise of temperature and precipitation similar to the one at the beginning of lithozone H3, when temperature and precipitation dropped. This time the lag phase comprises somewhat more than 200 varve years.

DISCUSSION

The varve dated, high resolution sedimentary record from Lake Holzmaar is still unique in Central Europe. Generally this record agrees with palynological studies on sediments from eight dry maars (Straka 1975) and with palynological (Usinger 1982, 1984), palaeomagnetic (Haverkamp 1984, 1991) and microstratigraphic investigations (Zolitschka 1990) on sediments from Lake Meerfelder Maar. But due to the high resolution time series presented data give many additional information.

Three subzones of Younger Dryas, assumed by palaeomagnetic and microstratigraphic records from Lake Holzmaar sediments, might be related to the following palaeoclimatic development: The cooling-off at the beginning of Younger Dryas (lithozone H3a) was not related to dry climatic conditions, which became apparent during lithozone H3b and H3c, resulting in a lowering of the lake level. Lake shore erosion during these subzones was caused by decreasing precipitation. First indications of an improvement of palaeoenvironmental conditions after Younger Dryas climatic deterioration are marked by an increase of planktonic diatom abundance 45 varve years before the beginning of lithozone H4. At the lithozone boundary H3/H4 deposits changed from minerogenic to organogenic, precipitation of autochthonous calcites started and terrestrial vegetation responded (pollen zone boundary III/IV).

The use of varve chronology in comparison with other dating methods is still problematic. Varve dated sediments from Lake Holzmaar and Lake Meerfelder Maar match each other. The Pleistocene/Holocene boundary as determined palynologically and lithologically was dated at Lake Holzmaar to 10,630 v.y. BP. A similar age was obtained by Greenland ice-core dating: 10,720 years (Hammer et al. 1986). The Swedish varve chronology produced an age of 10,750 varve years for the end of Younger Dryas (Björck et al. 1987). The termination of Younger Dryas, derived from ^{14}C calibration using a mixture of dendroyears and varve years, was determined to 11,300 calibrated conventional radiocarbon years BP (Becker & Kromer 1986, Stuiver et al. 1991). These ages for Younger Dryas termination imply, that 120 up to 700 varves might be missing in the Holocene record from Lake Holzmaar. But there are neither problems with the connection of varve chronology to the present nor with any erosion features, that might explain the differences. Missing varves or short, non-laminated, homogeneous periods, so far interpreted as turbidites, may give rise to some errors as well as the occurance of indistinct and ambiguous varves.

Duration of Younger Dryas was proposed by Swedish varve chronology to 260 - 400 varve years (Björck et al. 1987), in Greenland ice-cores to 450 years (Hammer et al. 1986). This is in the same order as Lake Holzmaar varve counts (450 varve years for Younger Dryas biozone). Varve counts on Swiss lake sediments end up with a Younger Dryas biozone of 680 varve years duration. But the Preboreal biozone has a duration of only 290 varve years in this record (Lotter 1991), which is in contrast to the Lake Holzmaar record with a Preboreal biozone comprising about 770 varve years. These problems probably are due to the long distance between investigated sites and also to different geographical settings making a comparison, based on biozones, impossible. These data also raise the question, if the duration of biozones is similar all over Europe. Until now there is no answer, because comparisons between sites and methods are still at the very beginning.

The necessity to calibrate radiocarbon years to calendar years represents a major problem to compare varve dated records with ^{14}C-dated records. During Late-Glacial this becomes even more complicated, for severe environmental changes caused fluctuations of atmospheric ^{14}C concentrations, related to detected periods of constant radiocarbon ages in sedimentary records, so-called ^{14}C-plateaux (Ammann 1989).
To overcome these problems varve chronological, dendrochronological and U-Th ^{14}C-age calibrations are in use. But for example U/Th calibration of ^{14}C-dates obtained a duration for Younger Dryas of 1,200 to 1,300 U/Th years (Bard et al. 1990b), which is in contrast to those data obtained from varved records.
To demonstrate the effect of calibration methods on the varve and radiocarbon dated isochrone of Laacher See Tephra (LST), the radiocarbon age for this isochrone of 11,000 conventional ^{14}C years BP (Bogaard & Schmincke 1985) was calibrated according to the listed methods (Tab. 3). Even if a 2σ error is considered for the data from Lake Holzmaar (±450 varve years), there are still more than 500 varve years missing to come close to the age obtained by "tree-ring" and U/Th calibrations of the radiocarbon time scale.

Table 3: Comparison of the varve age of LST from Lake Holzmaar with different attempts to calibrate the radiocarbon age of LST (11,000 conventional radiocarbon years BP).

11,320 v.y. BP	Lake Holzmaar organic varves	
11,400 cal. BP	Swedish glacial varves	(Björck et al. 1988)
12,250 cal. BP	adjusted lake-sediment and tree-ring records	(Stuiver et al. 1991)
13,300 cal. BP	U-Th from Barbados corals	(Bard et al. 1990a)

CONCLUSION

The continuous depositional record from Lake Holzmaar provides palaeoenvironmental data for reconstruction and dating of Younger Dryas litho- and biozones.

Varve ages from Lake Holzmaar differ from latest efforts to calibrate radiocarbon years. Therefore it is an urgent need to obtain a set of AMS-^{14}C dates on terrestrial plant remains from Lake Holzmaar sediments, just as to replicate varve counting on additional sediment cores from the same lake and from lakes Meerfelder Maar and Schalkenmehrener Maar to minimize errors that may arise from inter-lake variability of laminations and from local disturbances.

Finally, the prospect to obtain information from other sites with annually laminated sediment sequences to confirm or to refuse the results presented seem to be promising. Recently, high resolution time series from lacustrine archives were reported from Soppensee, Switzerland (Lotter 1989, 1991) and Lake Gosciaz, Poland (Goslar et al. 1989).

ACKNOWLEDGEMENTS

With appreciation we acknowledge the support of H. Usinger and A. Wolf, making available unpublished pollen data. Thanks are also due to E. Bard, A. Lotter and an anonymous reviewer for valuable comments on this paper.

We wish to thank Deutsche Forschungsgemeinschaft and European Community (EC-project "Geomaar") for financial support.

REFERENCES

Ammann B (1989) Periods of rapid environmental change around 12,500 and 10,000 years BP, as recorded in Swiss lake deposits. J Paleolimnol 1: 269-277

Bard E, Hamelin B, Fairbanks RG, Zindler A (1990a) Calibration of the 14C timescale over the past 30,000 years using mass spectrometric U-Th ages from Barbados corals. Nature 345: 405-410

Bard E, Hamelin B, Fairbanks RG, Zindler A, Mathieu G, Arnold M (1990b) U/Th and 14C ages of corals from Barbados and their use for calibrating the 14C time scale beyond 9000 years B.P. Nuc Instr Meth in Phys Res B52: 461-468

Becker B, Kromer B (1986) Extension of the Holocene dendrochronology by the Preboreal pine series, 8800 to 10,100 BP. Radiocarbon 28 (2B): 961-967

Bengtsson L, Enell M (1986) Chemical analysis. In: Berglund B (ed) Handbook of Holocene palaeoecology and palaeohydrology, 423-451, Chichester

Björck S, Sandgren P, Holmquist B (1987) A magnetostratigraphic comparison between 14C years and varve years during the late Weichselian, indicating significant differences between the time scales. J Quat Sci 2: 133-140

Björck S, Berglund BE, Digerfeldt G (1988) New aspects on the deglaciation chronology of south Sweden. Geogr Pol 55: 37-49

Bogaard P, Schmincke H-U (1985) Laacher See Tephra: a widespread isochronous late Quaternary tephra layer in central and northern Europe. Geol Soc Am Bull 96: 1554-1571

Digerfeldt G (1972) The Post-Glacial development of Lake Trummen. Fol Limnol Scand 16: 1-104

Eicher U (1987) Die spätglazialen sowie frühpostglazialen Klimaverhältnisse im Bereich der Alpen: Sauerstoffisotopenkurven kalkhaltiger Sedimente. Geogr Helv 42/2: 99-104

Eicher U, Siegenthaler U (1976) Palynological and oxygen isotope investigations on Late Glacial sediment cores from Swiss lakes. Boreas 5: 109-117

Firbas F (1949) Spät- und nacheiszeitliche Waldgeschichte Mitteleuropas nördlich der Alpen. p 480, Jena

Firbas F (1954) Die Synchronisierung der mitteleuropäischen Pollendiagramme. Danm Geol Unders II 80: 12-21

Goslar T, Pazdur MF, Walanus A (1989) Chronology of lower part of annually laminated sediments of the Gosciaz Lake. Zeszyty Naukowe Politech. Slaskiej, Seria: Matematika-Fizyka 57: 11-20

Hammer CU, Clausen HB, Tauber H (1986) Ice core dating of the Pleistocene/Holocene boundary applied to a calibration of the C-14 time scale. Radiocarbon 28: 284-291

Haverkamp B (1984) Paläomagnetische Messungen an den Sedimentkernen der Bohrung B4 im Meerfelder Maar. Cour Forsch Inst Senckenberg 65: 21-28

Haverkamp B (1991) Paläomagnetische Untersuchungen an spätquartären Maarsedimenten zur Paläosäkularvariation im Gebiet der Westeifel während der letzten 20-25000 Jahre. Ph D thesis Univ Münster (unpubl), p 233

Lotter A (1989) Evidence of annual layering in Holocene sediments of Soppensee, Switzerland. Aq Sci 51: 19-30

Lotter A (1991) Absolute dating of the Late-Glacial period in Switzerland using annually laminated sediments. Quat Res 35: 321-330
Mangerud J, Andersen ST, Berglund BE, Donner JJ (1974) Quaternary stratigraphy of Norden, a proposal for terminology and classification. Boreas 3: 109-128
Merkt J (1971) Zuverlässige Auszählungen von Jahresschichten in Seesedimenten mit Hilfe von Groß-Dünnschliffen. Arch Hydrobiol 69: 145-154
Negendank JFW, Brauer A, Zolitschka B (1990) Die Eifelmaare als erdgeschichtliche Fallen und Quellen zur Rekonstruktion des Paläoenvironments. Mainzer Geowiss Mitt 19: 235-262
Saarnisto M (1986) Annually laminated lake sediments. In: Berglund B (ed) Handbook of Holocene palaeoecology and palaeohydrology, 343-370, Chichester
Smol JP (1988) Chrysophycean microfossils in palaeolimnological studies. Palaeogeogr Palaeoclimatol Palaeoecol 62: 287-297
Straka H (1975) Die spätquartäre Vegetationsgeschichte der Vulkaneifel. Beitr Landespfl Rheinl-Pfalz Beih 3: 1-163
Stuiver M, Braziunas TF, Becker B, Kromer B (1991) Climatic, solar, oceanic, and geomagnetic influences on Late-Glacial and Holocene atmospheric 14C/12C change. Quat Res 35: 1-24
Usinger H (1982) Pollenanalytische Untersuchungen an spätglazialen und präborealen Sedimenten aus dem Meerfelder Maar (Eifel). Flora 172: 373-409
Usinger H (1984) Pollenanalytische Untersuchungen zum Alter des Meerfelder Maares und zur Vegetationsentwicklung in der Westeifel während der ausklingenden Eiszeit. Cour Forsch Inst Senckenberg 65: 49-66
Zolitschka B (1989) Jahreszeitlich geschichtete Seesedimente aus dem Holzmaar und dem Meerfelder Maar (Eifel). Z Dt Geol Ges 140: 25-33
Zolitschka B (1990) Spätquartäre jahreszeitlich geschichtete Seesedimente ausgewählter Eifelmaare. Doc Nat 60: 1-226
Zolitschka B (1991) Absolute dating of late Quaternary lacustrine sediments by high resolution varve chronology. Hydrobiol 214: 59-61

$^{230}Th/^{234}U$ AND ^{14}C AGES OBTAINED BY MASS SPECTROMETRY ON CORALS FROM BARBADOS (WEST INDIES), ISABELA (GALAPAGOS) AND MURUROA (FRENCH POLYNESIA).

E. Bard[1,2,3], R. G. Fairbanks [2], M. Arnold[1] and B. Hamelin [2,3]

1/ Centre des Faibles Radioactivités, CNRS-CEA
91198 Gif-sur-Yvette, France
2/ Lamont-Doherty Geological Observatory
Palisades, New-York 10964, USA
3/ Géosciences de l'Environnement,
Université Aix-Marseille III
13397 Marseille, France

abstract:

We present in this paper an update of the isotopic data obtained on scleratinian corals for the comparison of ^{14}C and $^{230}Th/^{234}U$ ages. A special emphasis of the paper is the comparison of those radiochronological results with the tree ring data base which now extends from 0 to ca 11,400 cal yr BP (Kromer & Becker 1991, this volume). In particular it is shown that for the eight youngest samples of the Barbados collection there is an excellent agreement between the calendar ages and the $^{230}Th/^{234}U$ age determinations. This brings further support for the accuracy and precision of measurement of time by means of the $^{230}Th/^{234}U$ thermal ionization mass spectrometric technique.

Introduction:

In 1988, Fairbanks conducted a drilling expedition off the south coast of Barbados to recover submerged corals contemporaneous of the last deglaciation. The core recovery was excellent and more than thirty different samples were dated by conventional β counting techniques (Fairbanks 1989). In the meantime we developed at Lamont the thermal ionization mass spectrometry (TIMS) technique in order to obtain precise U-Th ages and to compare them with the radiocarbon estimates measured on the very same samples. One of the biggest surprises was that the discrepancy between ^{14}C and Th-U ages increases through time to reach about 3000-3500 years at about 15,000 ^{14}C-yr BP (Bard et al. 1990a). Because the three youngest samples gave U-Th ages in agreement with their calibrated ^{14}C ages, we preliminarily concluded that the TIMS U-Th determinations were not only precise but also accurate and that this ^{14}C versus Th-U data set could be considered as a tool for a first order ^{14}C calibration.

Since that time we pursued this effort by replicating the ^{14}C measurements by means of accelerator mass spectrometry (AMS) on very small samples which enables us to use strong acid leaching to get rid of surface contaminants (see Bard et al. 1990b for the technical procedure and the report of the first third of those measurements). In order to establish the

NATO ASI Series, Vol. I 2
The Last Deglaciation: Absolute and Radiocarbon Chronologies
Edited by E. Bard and W. S. Broecker

global uniformity of the observed phenomenon, we also measured ages in new samples from very different environnements: the atoll of Mururoa and the island of Isabela in the tropical Pacific ocean.

Results obtained by AMS and TIMS on Barbados corals:

During three recent runs of the Gif-sur-Yvette Tandetron (June 1990, April 1991 and July 1991) we obtained about sixty new ^{14}C determinations on about twenty samples already dated by β-counting (Fairbanks 1989) and TIMS (Bard et al. 1990a). The high number of replicates enables us to calculate weighted mean AMS ^{14}C ages with an improved precision (better than ± 200 at 2σ for the last 15 kyr). The new results obtained by AMS agree to first order with the β-counting estimates. However, we further confirm our previous observation (Bard et al. 1990b) that for ages beyond the Holocene the AMS determinations are slightly older than the β-counting estimates. The samples prepared for AMS have been strongly leached with acid (loss of about 30 to 40% of the total weight) whereas the samples measured by β-counting were not (except the four samples measured at Geochron which also seem to be systematically older than the other β-counting estimates). Consequently it is possible that a minor (<1% modern C) contamination is responsible for this discrepancy. To illustrate how the calculation of $\Delta^{14}C$ is sensitive for old ages it is informative to cite the deepest *A. palmata* sample (RGF 9-27-5) which has now been dated three times by AMS in Gif (weighted mean of 16,360 ± 230 2σ ^{14}C yr BP), one time by AMS in Tucson (16,145 ± 260 2σ ^{14}C yr BP) and two times by TIMS at Lamont (weighted mean of 19,000 ± 70 2σ). The $\Delta^{14}C$ of this sample is thus about 300 ‰ instead of the 400-500 ‰ initially reported for that period using the β-counting ages obtained without leaching procedures.

Figure 1 shows the results obtained by AMS and β-counting versus the ages already obtained by TIMS. The maximal difference between the ^{14}C and U-Th chronometers is reached at about 16,000 ^{14}C yr BP and is about 2500 years if one takes into account the true half-life of ^{14}C (5730 years, Godwin 1962). Figure 2 shows a blow-up of the intercomparison which clearly illustrates the good agreement between our reconstruction and the dendrochronological determinations given by Kromer & Becker (1991, this volume). It should also be noted that the agreement would be even better if, as suggested by the data from varved sediments of Lake of the Cloud (these ones were directly ^{14}C dated, cf Stuiver 1970), there was a 150 cal yr gap between the main dendrochronology and the German pine extension which are for the moment only ^{14}C matched with a minimal calendar difference (Figure 3). The gap between the German oak and German pine was chosen by Kromer & Becker in order to produce a symetric $\Delta^{14}C$ wiggle between 9600 and 10,000 cal yr. Nevertheless one could also argue that the wiggle could have had the same slope as in the German pine section. This would lead again to the

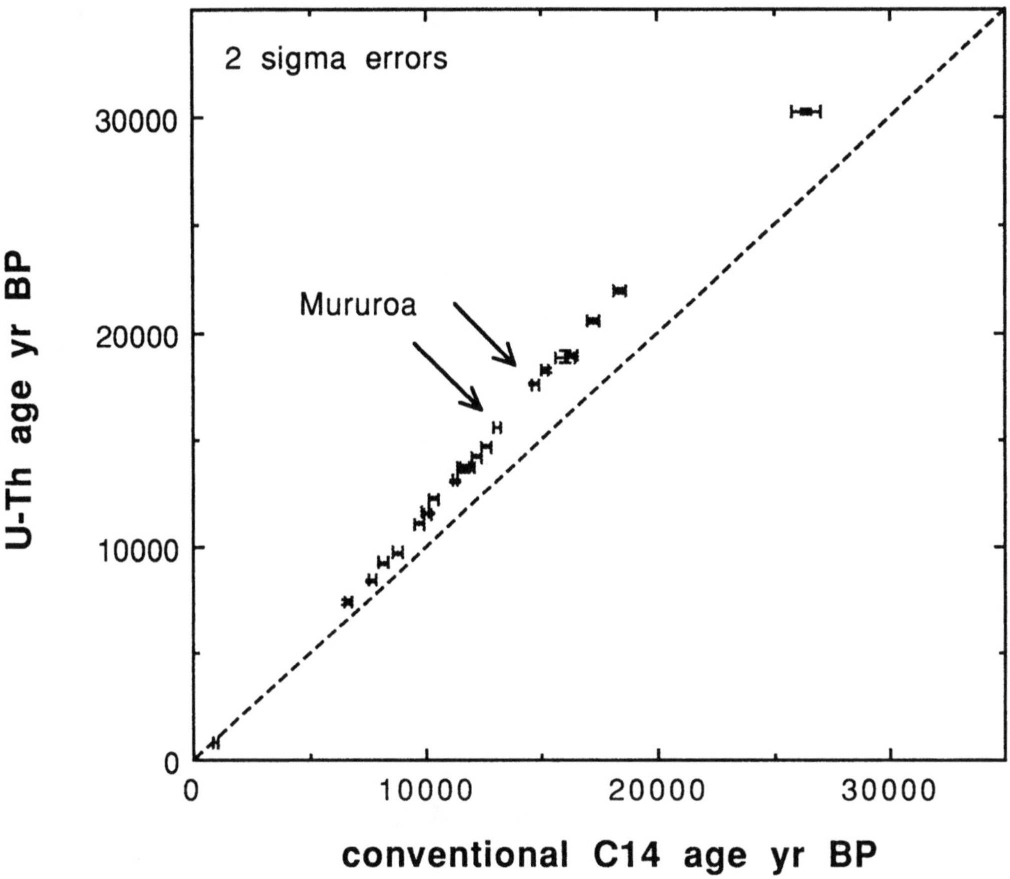

Figure 1: U-Th ages obtained by TIMS plotted versus AMS ^{14}C ages obtained on corals from Barbados and Mururoa. ^{14}C ages are conventional ages in yr BP with errors given at 2 σ.

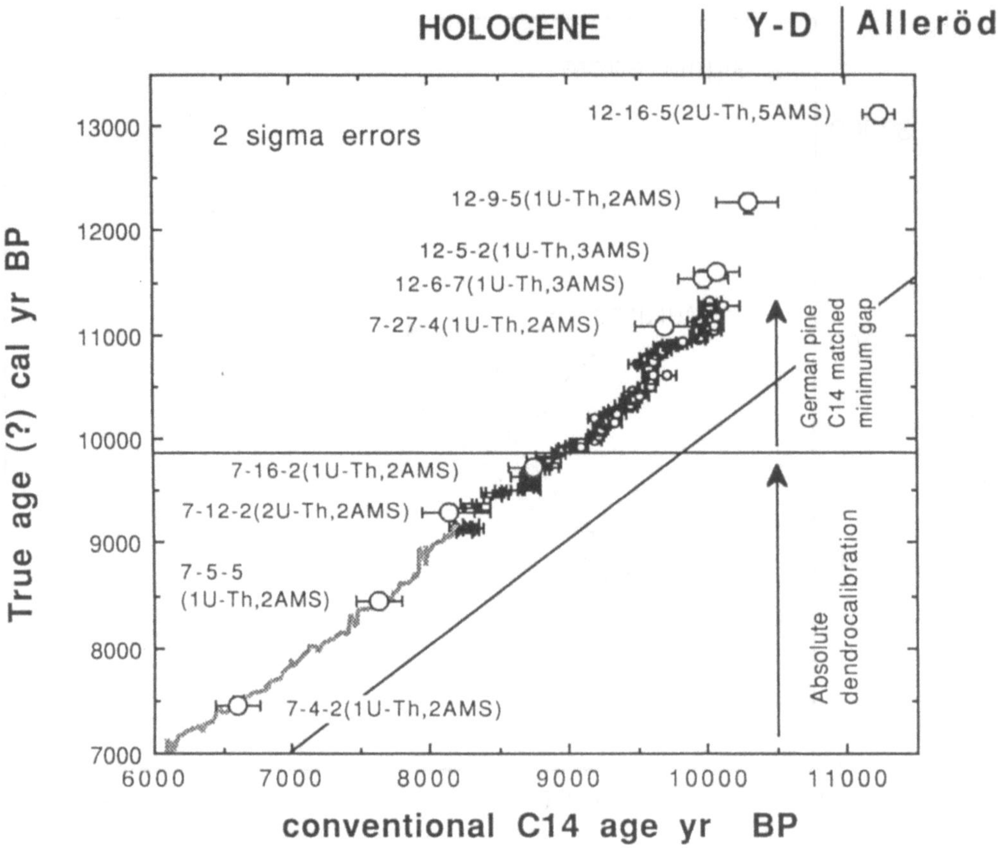

Figure 2: Blow-up of the diagram Figure 1 also showing the tree ring data base. Until 9000 cal yr the solid curve is the 20 year running average of the dendrochronology. Between 9100 and 9800 cal yr it is comprised of the German oak data (small symbols, cf Kromer & Becker 1991, this volume). Beyond 9800 cal yr the additional data are from the German pine sections (open symbols, cf Kromer & Becker 1991, this volume). All errors are given at 2 σ.

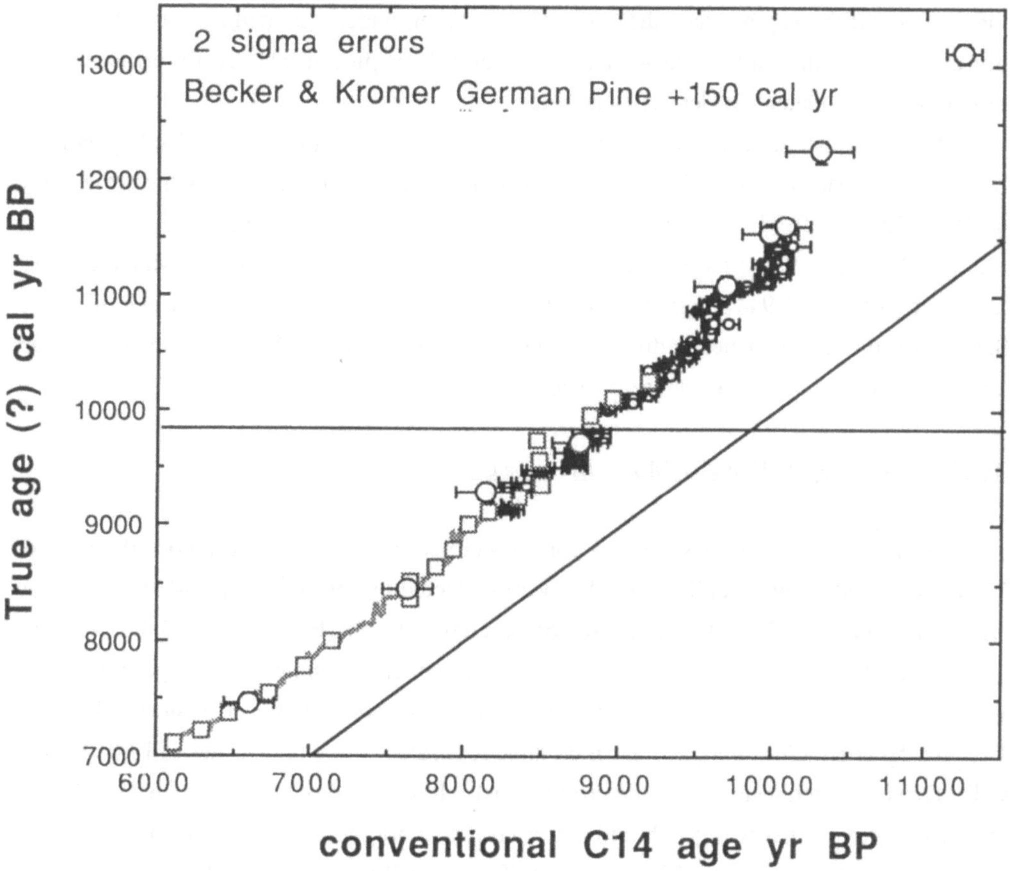

Figure 3: Same as Figure 2 except that the gap between the German oak and German pine has been augmented by 150 cal yr. This procedure is needed to optimize the agreement between the main dendrochronology, the Lake of the Cloud ^{14}C data (squares, cf Stuiver 1970) and the German pine serie.

conclusion that the gap between the two sections is a bit underestimated by about 150 yr. In any case we should wait for new discoveries of oak or pine trunks by Bernd Becker to definitely dendro-match the German pine series.

Another important check was of the reservoir age correction (400 yr) obtained by dating a very recent Barbados coral drilled on the flat part of the reef at about 7m below the present day sea level (RGF B52). The U-Th age is 773 ± 10 yr BP (2σ) (weighted mean on two replicates) while the ^{14}C age is 955 ± 85 BP (2σ) (weighted mean on four determinations) which lead to a calibrated age of 777-944 in agreement with the U-Th estimate. The reservoir age off shore Barbados is thus not statistically different from our adopted value of 400 year which is a mean for low and middle latitude open ocean waters (Bard 1988).

Results obtained by AMS and TIMS on Mururoa corals:

We have measured ages for two coral specimen collected from offshore the atoll of Mururoa, French Polynesia. The samples were collected by means of a land-based deviated drill (≈30°) at about 170 meters below the present day sea level. For sample 313 the U-Th age is 17,595 ± 70 yr BP (2σ) (weighted mean on three replicates) while the ^{14}C age is 14,730 ± 160 yr BP (2σ) (weighted mean on five determinations). For sample 315 the U-Th age is 15,580 ± 50 yr BP (2σ) (weighted mean on three replicates) while the ^{14}C age is 13,060 ± 140 yr BP (2σ) (weighted mean on five determinations). These new results show again a significant difference between the two geochronometers corresponding to $\Delta^{14}C$ values of 340 ± 30 ‰ and 295 ± 25 ‰ at about 18,000 and 16,000 Th-yr BP respectively. These values are in excellent concordance with the redeterminations of the Barbados ^{14}C ages by means of AMS (cf Figure 1). This new comparative data set of U-series and radiocarbon ages demonstrates that the large discrepancies first evidenced in Barbados are not the results of local alteration processes.

Results obtained by TIMS on Galapagos corals:

In order to further study the reliability of the TIMS age estimates on very young samples, we dated samples from a *Pavona clavus* colony which was sampled on Isabela island in the Galapagos. This recent coral head was killed in 1954 as a result of tectonic uplift from the shelf of Urvina Bay. By means of X-raying sections of this colony it was possible to study annual growth bands and to give a high resolution chronology for this colony (Shen et al. 1991). We dated two samples by TIMS: sample Gal 1 which has a sclerochronological age of 283-285 cal yr BP and gave a U-Th age of 268 ± 6 yr BP (2σ) (weighted mean on two replicates), and sample Gal 2 dated at 296-298 cal yr BP with a U-Th age of 275 ± 10 yr BP (2σ) (weighted mean on two replicates). The results obtained by TIMS are thus very close to the true ages but not in complete agreement if one takes into account the statistical errors.

It should be noted that these measurements were our first attempts to date very young corals on a rather old solid source mass spectrometer (VG-MM30) equipped with an analog Daly detector. For those analyses the $^{230}Th+$ ion currents were thus very close to the background and it could have been preferable to make additional base line corrections at mass 229.5 and 230.5 instead of our simpler procedure of measuring a common base line at mass 227.5. However, by applying such a procedure we would have significantly lost ^{230}Th measurement time which would have deteriorated the precision of the determinations. Consequently, we feel that those preliminary measurements constitute a trade-off between accuracy and precision. In any case, the optimal solution will be to replicate those determinations in 1992 by means of a new generation mass spectrometer equipped with an ion counting device able to decrease significantly the detection limit for small ion currents (by about 2 orders of magnitude).

Another explanation for the 10-20 yr discrepancy could also come from the sclerochronological estimates themselves. It could be either due to problems in correlating the different coral sections which had to be corrected for small hiatuses or due to the presence of additionnal stress bands in response to exceptionally cold years (see for example Druffel & Linick 1978, who showed evidence for three stress bands in a thirty year section of *Montastrea annularis* from Florida Keys).

Conclusions:

The new AMS and TIMS results presented in this paper confirm the precision and accuracy of the U-Th geochronometer applied to scleratinian corals. Is is shown again that the ^{14}C ages are systematically younger than true ages during most of the radiocarbon application range (except for the brief period between 500 and 2500 cal yr BP). More data will be needed before giving duration estimates for the different ^{14}C chronozones which occured during the last deglaciation. However based on the available data it seems inescapable that the calendar duration of the Holocene chronozone (0-10,000 ^{14}C yr BP) ranges from 11,200 to 11,500 cal yr and that the Younger Dryas chronozone (10,000-11,000 ^{14}C yr BP) exceeded one millenium. Assuming that the ^{2}H record measured by Becker et al. (1991) on the German pine sections is representative of the European climate one would estimate an age of about 9700 ^{14}C yr BP for the midpoint of the Younger-Dryas/Preboreal ^{2}H transition. Consequently the U-Th age of the midpoint of this climatic transition can be dated at about 11,100 yr by means of the U-Th versus ^{14}C age comparison (close to sample RGF 7-27-4).

acknowledgements: We thank D. Buigues and C. Caristan for providing the samples from Mururoa and G. Shen for the samples from Galapagos.

references:

Bard E, (1988). Correction of accelerator mass spectrometry ^{14}C ages measured in planktonic foraminifera: Paleoceanographic implications. *Paleoceanography* 3, 635-645.

Bard E, Hamelin B, Fairbanks RG, Zindler A (1990a). Calibration of ^{14}C timescale over the past 30,000 years using mass spectrometric U-Th ages from Barbados corals. *Nature* 345, 405-410

Bard E, Hamelin B, Fairbanks RG, Zindler A, Arnold M, Mathieu G (1990b). U/Th and ^{14}C ages of corals from Barbados and their use for calibrating the ^{14}C timescale beyond 9000 years BP. *Nuclear Inst. and Methods in Physics Research* B52, 461-468

Becker B, Kromer B, Trimborn P (1991). A stable isotope tree-ring time scale of the Late Glacial/Holocene boundary. *Nature* in press.

Druffel E, Linick T (1978) Radiocarbon in annual coral rings of Florida. *Geophys. Res. Let.* 5, 913-916.

Fairbanks RG (1989). A 17,000-year glacio-eustatic sea level record: influence of glacial melting rates on the Younger Dryas event and deep ocean circulation. *Nature* 342, 637-647

Godwin (1962). Half-life of radiocarbon. *Nature* 195, 984

Kromer B, Becker B (1991). Tree-ring ^{14}C calibration at 10.000 BP. This volume.

Shen G.T., Campbell T.M., Dunbar R.B., Wellington G.M., Colgan M.W., Glynn P.W. (1991). Paleochemistry of manganese in corals from the Galapagos Islands. *Coral Reefs* 10, 91-100

Stuiver M (1970) Evidence for the variation of atmospheric ^{14}C content in the late Quaternary. in *Late Cenozoic Glacial Ages* (ed. Turekian K.K.) (Yale Univ. Press, New Haven), 57-70.

PART II:

COSMONUCLIDE PRODUCTION CHANGES DURING THE PAST

EXPECTED SECULAR VARIATIONS IN THE GLOBAL TERRESTRIAL PRODUCTION RATE OF RADIOCARBON

Devendra Lal
Scripps Institution of Oceanography
Geological Research Division
La Jolla, CA 92093-0220 (U.S.A.)

ABSTRACT

I review our present day knowledge on the production of natural radiocarbon on the earth. Most of the radiocarbon is produced in the earth's atmosphere by the ^{14}N (n, p) ^{14}C reaction at thermal neutron energies. The remainder (< 4%) is produced in high cosmic ray spallations of oxygen and argon in the atmosphere, and of other target nuclei in the lithosphere and in the hydrosphere. The *atmospheric radiocarbon production* rate depends on the incident primary cosmic ray flux at the top of the atmosphere, which is appreciably modulated by both the geomagnetic field strength and solar activity.

Because of the present uncertainties in the determination of the absolute atmospheric neutron source function, absolute radiocarbon production rates are known to only within ± 20%. Relative variations in ^{14}C production rates for any postulated geomagnetic field strength and solar activity can be estimated fairly accurately, within ± 20%. The approach followed allows estimation of radiocarbon production rates during prolonged periods of very low solar activity, e.g. the Maunder Minimum period. Estimated global mean production rates of ^{14}C for different geomagnetic field intensities and solar activity levels are presented. For completeness, similar calculations are also presented for the atmospheric cosmogenic nuclide, ^{10}Be.

1. Introduction

Studies of the atmospheric $^{14}C/^{12}C$ ratios in tree rings (cf. Stuiver *et al.* 1991) and in U-Th dated coral samples (Bard *et al.* 1990) showed that this ratio varied appreciably in the past, by up to + 50% at ca. 20 k yr B.P. There are four possible causes of this variation associated with *changes in*: (i) the cosmic ray (C.R.) flux incident on the heliosphere (e.g. due to a nearby supernova, (ii) the solar activity leading to variable modulation of the C.R. flux on the earth, (iii) the geomagnetic field intensity and finally (iv) the carbon cycle parameters induced by climate changes. It is important to correctly gauge the magnitude of changes expected due to the

NATO ASI Series, Vol. I 2
The Last Deglaciation: Absolute and Radiocarbon Chronologies
Edited by E. Bard and W. S. Broecker

first three causes, namely the changes in the cosmic ray intensity so that climatic effects can be assessed properly (cf. Lal, 1985). ^{14}C is well mixed in the atmosphere within time scales short compared to the time of circulation of ^{14}C between the atmosphere and the oceans, and therefore it should suffice to obtain the global average production rate of ^{14}C. However, in order to obtain this value, it becomes necessary to ascertain the latitudinal variation in ^{14}C production rate for each of the three cases separately, since the C.R. flux variation is a different function of the latitude for the three cases, (i), (ii) and (iii). In this paper, I review expected magnitude of changes in the global ^{14}C production rate, Q(^{14}C), for cases (ii) and (iii) only; more explicitly, the mean Q(^{14}C) values for different geomagnetic field intensities as a function of changes in the solar activity. [For discussions on the expected magnitude of Q(^{14}C) changes with time due to a nearby supernova, reference is made to Lingenfelter (1969) and Kocharov (1990).] The temporal changes in global Q(^{14}C) values, presented here are primarily based on the estimates of Castagnoli and Lal (1980) and Lal (1988a). For completeness, I have also included the corresponding changes in global Q(^{10}Be) values. The fall-out of ^{10}Be depends sensitively on meteorological factors (Lal, 1987).

2. Radiocarbon production rates in the atmosphere

Subsequent to the earlier calculations of Anderson and Libby (1951), the global production rates of ^{14}C were revised by several authors with the availability of new cosmic ray measurements. Reference is made to Lingenfelter for a review up to 1963 and his own detailed calculations (Lingenfelter, 1963).

As seen from Table 1, radiocarbon production in the atmosphere occurs mainly by the thermal neutron capture reaction with atmospheric nitrogen. Other reactions contribute to less than 4% of the global production of radiocarbon in the atmosphere.

The production rates of radiocarbon in the atmosphere during solar minimum period in 1953-1954 and solar maximum period in 1957-1958 were calculated by Lingenfelter (1963), using multigroup neutron diffusion theory in conjunction with the experimentally measured altitude-latitude dependence in the production rate of nuclear disintegrations and slow neutrons in the atmosphere. This procedure allowed Lingenfelter (op. cit.) to estimate accurately the relative neutron leakage fluxes from the upper atmosphere, as well as the relative slow neutron fluxes in the atmosphere. Absolute ^{14}C production rates were then obtained by determining the effective areas of neutron counters. Lingenfelter determined the normalization factor to convert neutron counting rates to ^{14}C production rates for several investigators and arrived at a mean normalization factor of 9.0 ± 1.7 for all the measurements. Thus, the absolute ^{14}C production rates have an uncertainty of about ± 20%; no improvements on the absolute value have since been made.

Assuming an inverse relationship between the mean sunspot number and mean ^{14}C production rate, he then obtained for the mean global production rate of ^{14}C, Q_T during period T with mean sunspot number S_T, the following relation (Lingenfelter, 1963):

$$\overline{Q}_T = 2.61 - 0.53\,(\overline{S}_T - 9.1)/178.4 \quad (1)$$

$$= 2.58 - 2.97 \times 10^{-3}\,\overline{S}_T \quad (2)$$

Equation (1) yields a mean value of 2.5 (± 0.5) ^{14}C atoms/cm^2·sec for the global production rate during the past ca. century (S_T = 47.7). The error refers to the uncertainty in the absolute value of $Q(^{14}C)$. In a later paper, Lingenfelter and Ramaty (1970) recalculated the solar modulation in ^{14}C production rate using the solar modulation function, ζ used by Lockwood and Webber (1967) to describe temporal changes in the primary cosmic ray flux and energy spectrum. These authors also renormalized the neutron production data and obtained a 7% lower value for the absolute production rate of neutron than determined earlier by Lingenfelter (1963). Based on these calculations, Lingenfelter and Ramaty (1970) obtained a mean global ^{14}C production rate of 2.2 (± 0.4) ^{14}C atoms/cm^2 · sec, where the error refers to the uncertainty in the absolute value of $Q(^{14}C)$. The temporal variations in $Q(^{14}C)$ associated with changes in solar activity are < ± 10% for the maximum variation in the modulation parameter, ζ.

Subsequent to the work of Lingenfelter and Ramaty (1970), Korff and Mendell (1980) published new measurements of cosmic ray neutron fluxes up to 1976, and gave the following relation between annual mean sunspot number, R_Z and global mean $Q(^{14}C)$:

$$Q(^{14}C) = 2.60 - 5.53 \times 10^{-3}\,R_Z \quad (3)$$

with the global mean ^{14}C production rate during the solar cycle #20 (1965-1976) being 2.25 ^{14}C atoms/cm^2·sec. It should be noted that whereas the time averaged mean production rate obtained by Korff and Mendell (1980) agrees with that given by Lingenfelter (1963), their coefficient of variation (with sunspot number) is about 2× as large as obtained by Lingenfelter (1963) and Lingenfelter and Ramaty (1970).

Equations (1) and (3) take into account only the ^{14}C produced by galactic cosmic ray particles. Contributions to ^{14}C due to sporadic acceleration of cosmic rays in solar flares have to be taken into account separately on an event by event basis. The expected ^{14}C production during a flare for a specified number-energy spectrum can be calculated (Lingenfelter and Flamm, 1964 and Lingenfelter and Ramaty, 1970). Castagnoli and Lal (op. cit.) recalculated ^{14}C production by solar flare protons and commented that whereas solar flare protons can lead to appreciable ^{14}C production in the atmosphere at high latitudes in some flare events (with large fluence and hard energy spectrum), they do not constitute any appreciable addition to the mean global $Q(^{14}C)$ due to galactic cosmic radiation even on decadal time scale. Lingenfelter and Ramaty (1970) however had pointed out earlier that geomagnetic storms following large solar flares can considerably reduce the geomagnetic cut-off rigidities. The effect of solar flares on $Q(^{14}C)$ has to be therefore considered carefully.

3. In situ production of ^{14}C in the lithosphere, hydrosphere and the cryosphere

Most of the cosmic ray energy is dissipated in nuclear interactions in the earth's atmosphere. The secondary flux at the surface of the earth consists of neutrons, muons, protons, photons and neutrinos. Nuclear disintegration rates in the atmosphere are lower by factors of about 200-1000 at sea level depending on the latitude (Lal and Peters, 1967). For relative rates of nuclear disintegrations at sea-level caused by different particles, reference is made to a recent review by Lal (1988a).

As the cosmic ray energy is further absorbed in interactions with matter underground, nuclear interactions due to radiogenic neutrons become important (Lal, 1988b; Andrews *et al.* 1989).

Approximate *in situ* cosmogenic production rates in different terrestrial materials are given in Table 2 (Lal, 1991), including the production of ^{14}C due to capture of radiogenic neutrons by ^{17}O, ^{14}N, ^{13}C and ^{11}B.

The integrated cosmogenic column production of ^{14}C in the lithosphere/hydrosphere/ cryosphere is seen to be small compared to its total atmospheric production, ca. 2.5 ^{14}C atoms/cm$^2\cdot$sec, amounting to $< 10^{-2}$ %.

4. Dependence of Q(^{14}C) on geomagnetic field intensity and solar activity

Calculations have been presented earlier for the global Q(^{14}C) values as a function of geomagnetic field (Elsasser *et al.* 1956; Lal and Venkatavaradan, 1970). The relative geomagnetic field variation with the dipole field, M, approximately follows the relation:

$$Q_M(^{14}C) / Q_{M_o}(^{14}C) = \text{constant} * (M/M_o)^{-0.5} \quad (4)$$

as was first noted by Elsasser *et al.* (op. cit.); M_o is the reference field intensity. Relation (4) is correct to a first approximation only for values of M ~1 (Lingenfelter and Ramaty, 1970; Lal and Venkatavaradan, 1970). In fact, Q(^{14}C) depends both on the dipole field intensity and the solar activity. In the following, we will first discuss the question of dependence of Q(^{14}C) on solar activity.

The cosmic ray flux, within the heliosphere, is modulated due to scattering of cosmic rays on the irregularities in the interplanetary magnetic fields convected outwards by conductive solar wind plasma. To a first approximation, the higher the solar activity the lower the cosmic ray flux. However, the use of just annual sunspot numbers or Aa index to parametrise solar activity can lead to appreciable errors in estimation of cosmic ray flux. Firstly, marked (variable) phase lags occur between solar activity and cosmic ray flux change at the Earth. Secondly, cosmic ray flux changes appreciably from cycle to cycle near low sunspot numbers. It may be noted here that the geomagnetic indices, e.g. Aa appear to be more linear than the sunspot numbers. For a detailed discussion of the cosmic ray modulation data, reference is made to Castagnoli and Lal (1980).

Of particular interest is the question how cosmic ray fluxes changed in the past when solar activity was low for several decades. The interplanetary magnetic irregularities would then clear off and one may expect much higher cosmic ray fluxes. Such a period has been identified in the historical record, during 1645-1715 AD; Eddy called this the Maunder Minimum epoch after Maunder who was the first to point to this period of prolonged solar low activity (Eddy, 1976). Oscillations in tree ring $^{14}C/^{12}C$ ratios, with a period of ca. 200 yr are identified with Maunder Minimum type epochs (Damon and Linick, 1986; Stuiver *et al.*, 1991).

For estimating expected changes in cosmic ray production of ^{14}C on the earth, the best approach at present seems to be to consider the modulation of cosmic ray flux in terms of a modulation potential. A reasonable theoretical framework exists for this, in terms of the Fokker-Planck equation for diffusion of cosmic ray particles within the heliosphere (see references in

Castagnoli and Lal, 1980). Garcia-Munoz *et al.* (1975), Garcia-Munoz (1977) and Garcia-Munoz and Simpson (priv. comm.) have fitted the observed cosmic ray energy spectra for different epochs with that expected theoretically, and deduced the free parameters of the modulation equation. This parametrization allows characterizing the modulation intensity during different periods by a single parameter, ϕ (MV). The parameter ϕ is directly correlated with solar activity; higher the activity, higher the value of ϕ). For the modulation equation used and the measured value of ϕ, reference is made to Castagnoli and Lal (1980) and Lal (1988a). The observed values of ϕ vary between ~ 400-1250 MV during 1965-1979 (Lal, 1988a). The period of lowest modulation was observed during 1974 and 1976-77 (Webber and Lezniak, 1974; Badhwar and Golden (1979). During periods of Maunder Minimum, Castagnoli and Lal (1980) have argued that the value of ϕ may have been as low as 100 MV. In any case, the lowest possible value of ϕ is 0, corresponding to the interstellar cosmic ray fluxes.

We present in Table 3 the estimated average global production rates of ^{14}C in the atmosphere for ϕ = 100, 450 and 1000 MV. These estimates are based on the procedure of Castagnoli and Lal (1980); they differ slightly from the earlier estimates especially for low values of ϕ, because they include modulation of cosmic ray flux at energies above 5 GeV.

For completeness, we have given in Table 4, the estimated changes in global production rates of ^{10}Be (Lal, 1988a), for which long time series are available in polar ice. Of particular interest are the recent high resolution data obtained by Beer *et al.* (1990).

In Tables 3 and 4 we have given the estimated oscillations in Q(^{14}C) and Q(^{10}Be) for the corresponding fluctuations in $100 < \phi < 1000$ MV.

The estimated % oscillation in the global average production rates of ^{14}C around the mean value $(\Delta Q/\bar{Q}) \times 100$, are summarized below for the present day dipole field for different excursions in ϕ.

$$0 < \phi < 1000 \text{ MV} \pm 34\%$$

$$100 < \phi\ 1000 \text{ MV} \pm 29\%$$

$$450 < \phi\ 1000 \text{ MV} \pm 16\%.$$

It is of some usefulness to compare the magnitude of oscillations in Q(^{14}C) in Table 3 with the two earlier estimates, relations 1 and 3 (Lingenfelter, 1963 and Korff and Mendell, 1980). The variations in solar activity during a typical cycle correspond to $450 < \phi < 1000$ MV, and to sunspot numbers, $5 < R_Z < 150$. Relations (1) and (3) yield oscillations of ± 7.7% and ±

13.4% respectively. Our estimated magnitude of variation for the same range in ϕ is $\pm$ 16%, higher than the estimate of Korff and Mendell (1980) by 20%.

The more interesting aspect of the ϕ model is that it allows realistic estimates of changes in the cosmic ray flux and energy spectra during prolonged periods of low solar activity. As was mentioned earlier, the lowest expected ϕ value is ~ 100 MV, and the extreme case is $\phi = 0$, corresponding to the interstellar cosmic ray flux being incident outside the earth's magnetosphere. The variations in Q(^{14}C) for these modulation states are given in Table 3.

The mean global production rate of ^{14}C during the Holocene, based on its inventory on the earth has been estimated to be 2.4 $\pm$ 0.2 atoms/cm^2 · sec (Damon and Sternberg, 1989). Note that the error bar on this estimate is smaller than the present uncertainty in the cosmic ray production rate estimate of ^{14}C ($\pm$ 20%). The mean value is in good agreement with that expected during the Holocene, for $0.5 \leq M/M_o \leq 1$ and $450 < \phi < 1000$ MV (Table 3).

Conclusions

We have reviewed the published estimates of production of ^{14}C in the atmosphere and in the surface materials of the earth. Special emphasis is given on the estimates of oscillations in the mean global production rates of ^{14}C during different periods of solar activity. The estimates (Castagnoli and Lal, 1980; Lal, 1988a) are presented as a function of earth's dipole field intensity, (0-10)$\times$ present field, using the cosmic ray flux modulation theory developed by Urch and Gleason (1972) and Axford (1972). Earlier calculations of the expected variations in Q(^{14}C) with solar activity by Lingenfelter (1963), Lingenfelter and Ramaty (1970) and Korff and Mendell (1980) are reviewed. Calculations similar to those by Castagnoli and Lal (1980), using ϕ as the modulation parameter, were made by O'Brien (1979). His absolute global ^{14}C production rate estimates as well as the altitude-latitude curves are at considerable variance with those of Lingenfelter (1963). This is not unexpected since his calculations are based on the nucleonic cascade theory. Furthermore, his estimates are based on limited cosmic ray data; see Lal (1988a.) The results of O'Brien are therefore not reviewed here.

The ϕ modulation model allows realistic estimates of relative oscillations in the global Q(^{14}C), during periods of prolonged low solar activity. Evidence for the occurrence of such periods is based on $^{14}C/^{12}C$ ratios in tree rings (cf. Stuiver *et al.*, 1991). There is also a strong evidence for such an epoch in the historical records of sunspots (Eddy, 1976); the period 1645-1715 AD was characterized by few sunspots and was called the Maunder Minimum epoch by Eddy. The highest recorded solar activity on the other hand was during 1958, with $\phi \approx 1250$ MV. The global $Q(^{14}C)$ was then depressed by about 14% compared to the typical solar high ($\phi = 1000$ MV).

Table 3 is a summary of the estimated variations in $Q(^{14}C)$ for periods of different ϕ and earth's dipole field intensity. Typical solar cycles correspond to $450 < \phi < 1000$ MV. The lowest value of ϕ is estimated to be ~ 100 MV; the highest observed to date is 1250 MV. The minimum possible value of ϕ is of course 0. For completeness, we have also given the corresponding estimated oscillations in global average production rates of ^{10}Be.

Acknowledgements

I thank Dr. P. E. Damon and Dr. R. E. Lingenfelter for several helpful comments and suggestions.

Table 1. Production reactions for ^{14}C in the atmosphere†

Nuclear reaction	Threshold energy (MeV)	Relative production rate in the atmosphere
<u>Neutron</u> @		
$N^{14}(n, p)\ C^{14}$	+ 0.626	1
$O^{16}(n, He^3)\ C^{14}$	– 14.6	$< 3 \times 10^{-2}$
$O^{17}(n, \alpha)\ C^{14}$	+ 1.82	2.3×10^{-5}
$N^{15}(n, d)\ C^{14}$	– 7.98	$< 4 \times 10^{-5}$
$C^{13}(n, \gamma)\ C^{14}$	+ 8.17	1.1×10^{-9}
<u>Charged particle</u> @		
(p, ^4He)		
spallation of O, Ar	–(15-30)	$< 10^{-2}$

† Based on Lingenfelter (1963).
@ Incident particle.

Table 2. Approximate *in situ* cosmogenic radiocarbon production rates at geomagnetic latitude ≥ 60° (Lal, 1991)

a) Surface Production Rates

Altitude	**^{14}C atoms/gm·sec**	
	Rock†	**Ice**
5 km	1.8×10^{-5}	1.6×10^{-5}
2 km	3.1×10^{-6}	2.7×10^{-6}
Sea-level	5.6×10^{-7}	4.9×10^{-7}

b) Integrated column production rates

In a rock exposed at:	**^{14}C atoms/cm^2·sec**
i) 5 km	3×10^{-3}
ii) 2 km	5×10^{-4}
iii) sea-level	6×10^{-5}
In the oceans:	1×10^{-4}
Radiogenic production in the crust@	10^{-3}

†The production rates are for SiO_2.

@Note that except for this entry, others are for production of ^{14}C by cosmic ray particles.

Table 3. Global average production rates† of ^{14}C as a function of modulation parameter, ϕ and geomagnetic dipole field, M

Dipole field, M/M$_o$*	Global Production Rate, ^{14}C atoms/cm^2 · sec				
	ϕ (MV)				Mean@ for
	0	100	450	1000	$100 < \phi < 1000$
0	10.35	9.27	5.90	3.69	6.48 ± 2.79 (±43%)
0.25	7.87	7.05	4.93	3.32	5.18 ± 1.87 (±36%)
0.5	5.84	5.23	3.82	2.70	3.96 ± 1.27 (±32%)
1.0	4.03	3.61	2.72	1.99	2.8 ± 0.81 (±29%)
2.0	2.71	2.43	1.83	1.39	1.91 ± 0.52 (±27%)
4.0	1.91	1.71	1.28	0.95	1.33 ± 0.38 (± 28%)
10.0	1.28	1.15	0.83	0.60	0.87 ± 0.28 (± 30%)

† Based on Castagnoli and Lal (1980) and Lal (1988a).

*M_o = present field.

*The value denotes the mean and the oscillation in Q(^{14}C) for the range in ϕ considered. Uncertainties in the absolute production rate of ^{14}C, ± 20% are not included.

Table 4. Global average production rates† of ^{10}Be as a function of modulation parameter, ϕ and geomagnetic dipole field, M

Dipole field, M/M_o*	Global Production Rate, ^{10}Be atoms/cm^2 · sec			
	ϕ (MV)			Mean@ for
	100	450	1000	$100 < \phi < 1000$
0	11.6×10^{-2}	8.13×10^{-2}	5.4×10^{-2}	$(8.5 \pm 3.1) \times 10^{-2}$ (±36%)
0.25	9.32×10^{-2}	6.99×10^{-2}	4.88×10^{-2}	$(7.1 \pm 2.22) \times 10^{-2}$ (±31%)
0.5	7.28×10^{-2}	5.65×10^{-2}	4.15×10^{-2}	$(5.7 \pm 1.57) \times 10^{-2}$ (±27%)
1.0	5.22×10^{-2}	4.32×10^{-2}	3.18×10^{-2}	$(4.35 \pm 1.17) \times 10^{-2}$ (±27%)
2.0	3.42×10^{-2}	2.86×10^{-2}	2.21×10^{-2}	$(2.82 \pm 0.61) \times 10^{-2}$ (±21%)
4.0	2.43×10^{-2}	2.01×10^{-2}	1.55×10^{-2}	$(1.99 \pm 0.44) \times 10^{-2}$ (±22%)
10.0	1.53×10^{-2}	1.26×10^{-2}	0.97×10^{-2}	$(1.25 \pm 0.28) \times 10^{-2}$ (±22%)

† Based on Lal (1988a).

*M_o = present field.

@The value denotes the mean and oscillation in Q(^{10}Be) for the range in ϕ considered. Uncertainties in the absolute production rate of ^{10}Be, ~ ± 20%, are not included.

References

Anderson EC, Libby WF (1951) World-wide distribution of natural radiocarbon. Phys Rev 81:64-69.

Andrews JN, Davis SN, Fabryka-Martin J, Fontes J-Ch, Lehmann BE, Loosli HH, Michelot J-L, Moser H, Smith B, Wolf M (1989) The *in situ* production of radioisotopes in rock matrices with particular reference to the stripa granite. Geochim Cosmochim Acta 53:1803-1815.

Axford WI (1972) The interaction of the solar-wind with the interstellar medium. In: Sonett CP, Coleman Jr. PJ, Wilcox JM (eds) Conf. "Solar Wind" Proc. Washington DC, NASA-SP-308, 609.

Badhwar GD, Golden RL (1979) Solar modulation of protons and alpha particles at rigidities of 4-10 GV/c. In Proc Internatl Cosmic Ray Conf 16th, MG 3-11, 3:249-253.

Bard E, Hamelin B, Fairbanks RG, Zindler A (1990) Calibration of the ^{14}C timescale over the past 30,000 years using mass spectrometric U-Th ages from Barbados corals. Nature 345:405410.

Beer J, Blinov A, Bonani B, Finkel RC, Hofmann HJ, Lehmann B, Oeschger H, Sigg A, Schwander J, Staffelbach T, Stauffer B, Suter M, Wölfli W (1990) Use of ^{10}Be in polar ice to trace the 11-year cycle of solar activity. Nature 347:164-166.

Castagnoli G, Lal D (1980) Solar modulation effects in terrestrial production of carbon-14. Radiocarbon 22:133-158.

Damon PE, Linick TW (1986) Geomagnetic-heliomagnetic modulation of atmospheric radiocarbon production. Radiocarbon 28:266-278.

Damon PE, Sternberg RE (1989) Global production and decay of radiocarbon Radiocarbon 31:697-703.

Eddy JA (1976) The Maunder Minimum. Science 192:1189-1202.

Elsasser W, Ney EP, Winckler JR (1956) Cosmic-ray intensity and geomagnetism. Nature 178:1226-1227.

Garcia-Munoz M, Mason GM, Simpson JA (1975) The anomalous ^{4}He component in the cosmic-ray spectrum at $\leq$ 50 MeV per nucleon during 1972-1974. Astrophys Jour 202:265-275.

Garcia-Munoz M (1977) New aspects of the cosmic-ray modulation in 1974-1975 near solar minimum. Astrophys Jour 213:263-268.

Kocharov GE (1990) Radiocarbon and astrophysical-geophysical phenomenon. In: Taylor RE, Long A and Kra RS (eds) Radiocarbon after four decades. Univ. of Arizona Press, Tucson (in press).

Kocharov GE, Blinov AV, Konstantinov AN, Levchenko VA (1990) Cosmogenic isotopes and geomagnetic field in the past. Proc 21st Int Cosmic Ray Conf 7(SG):116-119.

Korff SA, Mendell RB (1980) Variations in radiocarbon production in the Earth's atmosphere. Radiocarbon 22:159-165.

Lal D (1985) Carbon cycle variations during the past 50,000 years: atmospheric $^{14}C/^{12}C$ ratio as an isotopic indicator. In: Sundquist ET and Broecker WS (eds) American Geophy Union Monograph 32, Washington DC, 221-233.

Lal D (1987) ^{10}Be in polar ice: Data reflect charges in cosmic-ray flux or polar meteorology? Geophy Res Letts 14:785-788.

Lal D (1988a) Theoretically expected variations in the terrestrial cosmic-ray production rates of isotopes. Proc. Fermi Sch. Phys. 95th, Varenna, Ital. Acad. Sci., 216-233.

Lal D (1988b) In-situ produced cosmogenic isotopes in terrestrial rocks. Ann Rev Earth Planet Sci 16:355-388.

Lal D (1991) Cosmogenic *in situ* radiocarbon on the Earth. In: Taylor RE, Long A and Kra RS (eds) Radiocarbon after four decades. Univ of Arizona Press, Tucson (in press).

Lal D, Peters B (1967) Cosmic ray produced radioactivity on the earth. In: Handbook of Physics, Springer, Berlin 46/2, 551-612.

Lal D, Venkatavaradan VS (1970) Analysis of causes of C-14 variations in the atmosphere. Proc 12th Nobel Symposium John Wiley and Sons, Inc, 549-569.

Lingenfelter RE (1963) Production of carbon 14 by cosmic-ray neutrons. Rev Geophs 1:35-55.

Lingenfelter RE (1969) Pulsars and local cosmic-ray prehistory. Nature 224:1182-1186.

Lingenfelter RE, Flamm EJ (1964) Production of carbon-14 by solar protons. Jr Atmos Sci 21:134-140.

Lingenfelter RE, Ramaty R (1970) Astrophysical and geophysical variations in C-14 production. Proc 12th Nobel Symposium, John Wiley and Sons, Inc, 513-537.

Lockwood JA, Webber WR (1967) The 11 yr solar modulation of cosmic rays as deduced from neutron monitor variations and direct measurements at low energies. J Geophys Res 72:5977-5989.

O'Brien K (1979) Secular variation in the production of cosmogenic isotopes in the earth's atmosphere. Jr Geophys Res 84:423-431.

Stuiver M, Braziunas TF, Becker B, Kroner B (1991) Climatic, solar, oceanic and geomagnetic influences on Late-Glacial and Holocene atmospheric $^{14}C/^{12}C$ change. Quat Res 35:1-24.

Urch IH, Gleeson LJ (1972) Galactic cosmic-ray modulation from 1965-1970. Astron Space Sci 17:426-446.

Webber WR, Lezniak JA (1974) The comparative spectra of cosmic ray protons and helium nuclei. Astron Space Sci 30:361.

^{10}Be DEPOSITION AT VOSTOK, ANTARCTICA DURING THE LAST 50,000 YEARS AND ITS RELATIONSHIP TO POSSIBLE COSMOGENIC PRODUCTION VARIATIONS DURING THIS PERIOD

G.M. Raisbeck[1], F. Yiou[1], J. Jouzel[2,3], J.R. Petit[3], N.I. Barkov[4] and E. Bard[5]

1 *Centre de Spectrométrie Nucléaire et de Spectrométrie de Masse, IN2P3-CNRS, Bât. 104-108, 91405 ORSAY Campus, FRANCE*

2 *Laboratoire de Geochimie Isotopique, DSM, CEN-CEA Saclay, 91191 GIF SUR YVETTE Cedex, FRANCE*

3 *Laboratoire de Glaciologie et Géophysique de l'Environnement, B.P. 96, 38402 SAINT MARTIN D'HERES Cedex, FRANCE*

4 *Arctic and Antarctic Research Institute, Fontanka 34, LENINGRAD 191104, USSR*

5 *Laboratoire de Géosciences de l'Environnement, 13397 Marseille Cedex 13, FRANCE.*

ABSTRACT

Using measured concentrations of ^{10}Be in ice cores from Vostok, Antarctica, and assuming that past ice accumulation rates in these cores can be calculated from their stable isotope ratio δ^2H, we give the ^{10}Be deposition rate at this location for the past 50,000 years. The ^{10}Be flux appears to have been relatively constant (± 20 %) during this period, except for a ~1000-2000 year interval 35,000 years ago, when it increased by a factor of ~2. On the basis of these data we argue that it is unlikely that primary cosmic ray or solar modulation induced cosmogenic production rate changes can completely account for the ^{14}C/^{12}C changes in the atmosphere during the late glacial period implied by the coral results of Bard et al. (1990a,b). No similar conclusions regarding possible production variations due to geomagnetic field variations can be made at the present time because it is not known how these will be reflected in ^{10}Be deposition over polar regions. We also show that the discrepancies in the U-Th and ^{14}C ages of corals for the period 10,000-20,000 years ago cannot be related to the ^{10}Be "peak" at ~35,000 B.P. unless the latter is due to some mechanism (solar flares, supernova shock wave, γ-rays) which produces much larger production rate changes in ^{14}C than in ^{10}Be.

NATO ASI Series, Vol. I 2
The Last Deglaciation: Absolute and Radiocarbon Chronologies
Edited by E. Bard and W. S. Broecker

1. INTRODUCTION

We have previously measured ^{10}Be concentration profiles in several Antarctic ice cores (Raisbeck et al. 1981a, 1987, 1990 ; Yiou et al. 1985, Raisbeck and Yiou 1988). Variations in the ^{10}Be concentration with depth in these cores can be ascribed to two causes ; (i) temporal variations in the ^{10}Be production rate (due to changes in primary cosmic ray flux, solar modulation, or geomagnetic field intensity) ; (ii) temporal variation in the ^{10}Be deposition pattern due to climatic effects (precipitation rate, atmospheric circulation). Over periods of relative climatic stability, such as the Holocene, the main source of ^{10}Be variations appears to be solar modulation (Raisbeck et al. 1981a, 1990 ; Raisbeck and Yiou 1988). On the other hand, in going from glacial to interglacial climates, we have observed strong (approximately factor of two) concentration variations which we have tentatively attributed to being mainly due to precipitation rate changes (Raisbeck et al. 1981a, 1987 ; Yiou et al. 1985). This makes it difficult to infer production rate variations between such periods using ^{10}Be concentration data alone.

An independent method of estimating paleo-precipitation rates in the Antarctic is through stable isotope measurements (δ^2H or $\delta^{18}O$). The procedure rests on the hypothesis that precipitation on the Antarctic plateau is essentially limited by the available moisture at the atmospheric inversion layer, which is itself related to the temperature and isotopic composition of water vapour at this layer (Robin 1977). This hypothesis derives from the observation that present day spatial variations of precipitation in Antarctica are well correlated with mean annual surface temperatures. Whether such a hypothesis can correctly account for temporal variations in precipitation at a given location is of course open to question. However, by using such an assumption it has been possible to produce a paleo-precipitation rate (and thus timescale) for the Vostok ice core, which appears to be at least approximately correct over the last climatic cycle (Jouzel et al. 1989).

By adopting this hypothesis it is thus possible to combine ^{10}Be concentrations and inferred paleo-precipitation estimates, to obtain paleo-flux or deposition rates of ^{10}Be over periods which include glacial-interglacial transitions. We argue that these deposition rates give the best estimates that can presently be made from ^{10}Be regarding cosmogenic production changes during such periods. In particular, we examine here what information these ice cores can give regarding ^{10}Be production variations during the past 50,000 years, and how these compare with the variations in ^{14}C production hypothesized by Bard et al. (1990a) as being necessary to explain the difference between ^{14}C and U-Th ages in corals over this period.

2. DATA

The ^{10}Be data used here are part of a new high resolution series of measurements in ice at Vostok, Antarctica currently being prepared for publication. Samples, from both the 3G and 4G cores, were taken approximately every 8 m, corresponding to a time resolution of ~ 300-600 years. In addition we have made essentially continuous measurements at 1 m intervals from 590-630 m depth which contains a previously identified ^{10}Be concentration "peak" (Raisbeck et al. 1987). Details of the sample treatment and measurement procedures, which were similar to those described earlier (Raisbeck et al. 1987) will be given elsewhere. Uncertainties in the ^{10}Be measurements (1 sigma statistics plus a 5% instrumental contribution) are estimated as ~6 %.

Paleo-precipitation rates have been calculated with the δ^2H data presented in Jouzel et al. (1987), using the procedure described in Jouzel et al. (1989).

3. RESULTS

In Fig. 1 we show the ^{10}Be concentrations, calculated precipitation rate, and resulting ^{10}Be deposition rate as a function of time, using the time scale of Jouzel et al. (1989) (see also Jouzel et al., this volume, for a discussion of this timescale). As can be seen, the glacial-interglacial difference in ^{10}Be concentration is roughly mirrored in the calculated precipitation curve, giving rise to an approximately constant ^{10}Be deposition rate between the two periods. On the other hand the concentration peak at ~35,000 B.P. not only remains, but stands out even more clearly in the deposition curve. This supports our earlier interpretation (Raisbeck et al. 1987) that this peak is most likely a reflection of increased production, and not a simple climatic effect.

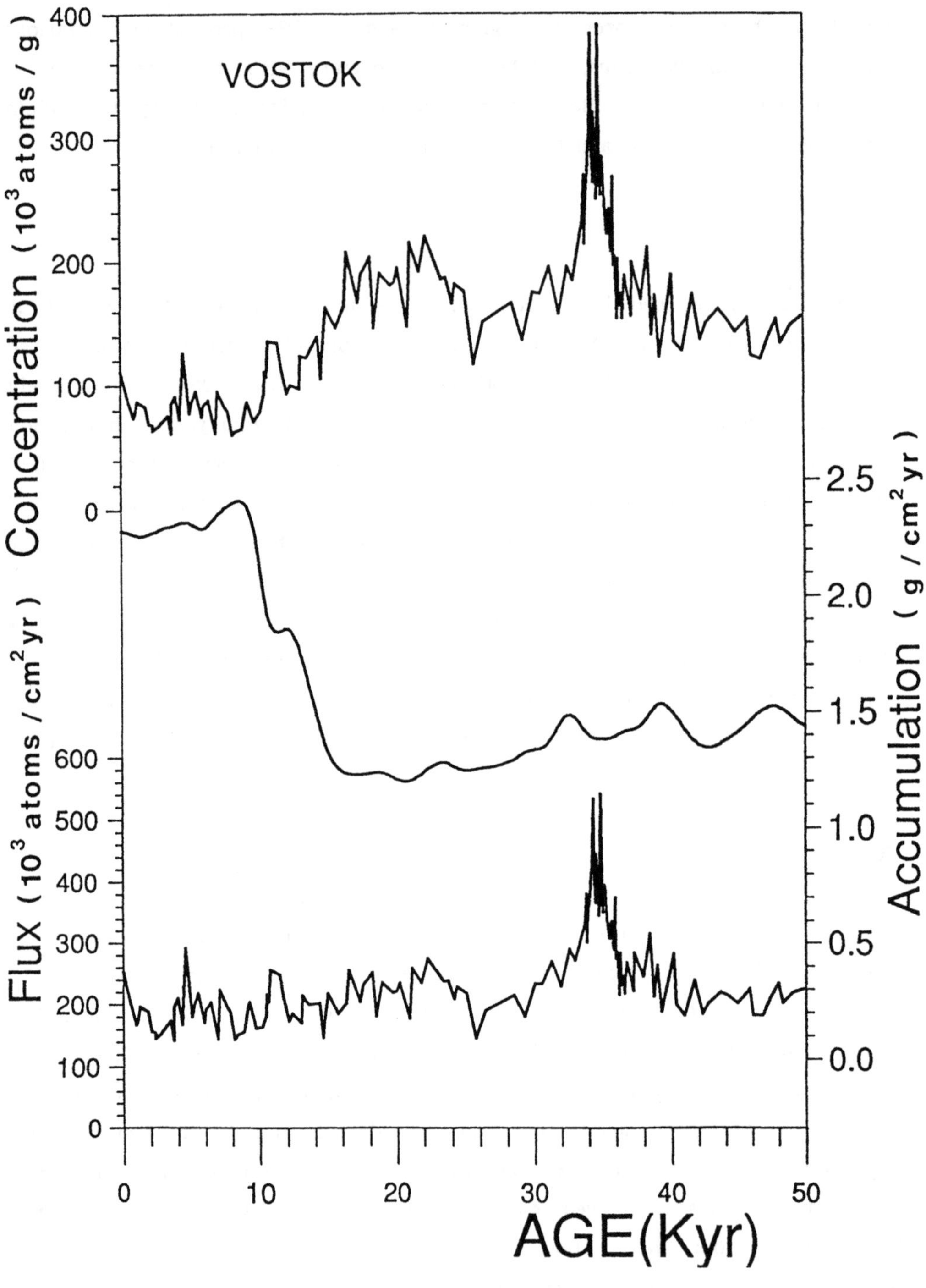

Fig. 1 : Measured ^{10}Be concentration, estimated precipitation rate, and estimated ^{10}Be flux in ice at Vostok, Antarctica, during the past 50,000 years.

4. DISCUSSION

We now examine how this ^{10}Be data might be related to the variation in ^{14}C production postulated by Bard et al. (1990a) as being responsible for the difference between ^{14}C and U-Th ages measured in corals from the late glacial period.

4.1 - ^{10}Be peak at 35,000 B.P.

We first examine whether the discrepancy between the ^{14}C and U-Th ages could be related to the observed ^{10}Be peak at ~35,000 B.P. in the ice core data. To treat this in the most general and model independent way, let us consider all the earth's exchangeable ^{14}C (atmosphere-biosphere-ocean) as a single reservoir. For a constant ^{14}C production rate (P), the total inventory of ^{14}C in this reservoir (Q) is :

$$Q = Pt_m$$

where t_m is the mean life of ^{14}C.

The ice core data suggest that during a period of ~1000-2000 years, represented by the ^{10}Be "peak", cosmogenic production (P^*) might have increased by a factor of ~2(ie $P^* \sim 2P$). The "excess" ^{14}C (ΔQ) produced during a time (Δt) is given by :

$$\Delta Q = (P^*-P)\ t_m\ (1-e^{-\Delta t/t_m})$$

Thus, taking $P^*=2P$, $\Delta t=2000$ years and $t_m=5730/\ln 2$ one finds that the excess production, compared to the equilibrium quantity, is :

$$\Delta Q/Q = 1-e^{-2000\ \ln 2/5730} \sim 20\%$$

A few thousand years (a few times the ocean circulation time) after this event, this excess ^{14}C will be distributed almost uniformly in the three compartments of the reservoir mentioned earlier. By 20,000 B.P. most of this excess will have decayed, leaving a residual enhancement of only ~3%. This is far less than that required to explain the age difference observed by Bard et al. (1990) at this time.

To illustrate this more specifically we show in Fig. 2 the expected $^{14}C/^{12}C$ response assuming the production profile described above and applying the simple 2 box model described in Bard et al. (1990a), consisting of atmosphere-mixed layer and deep ocean reservoirs.

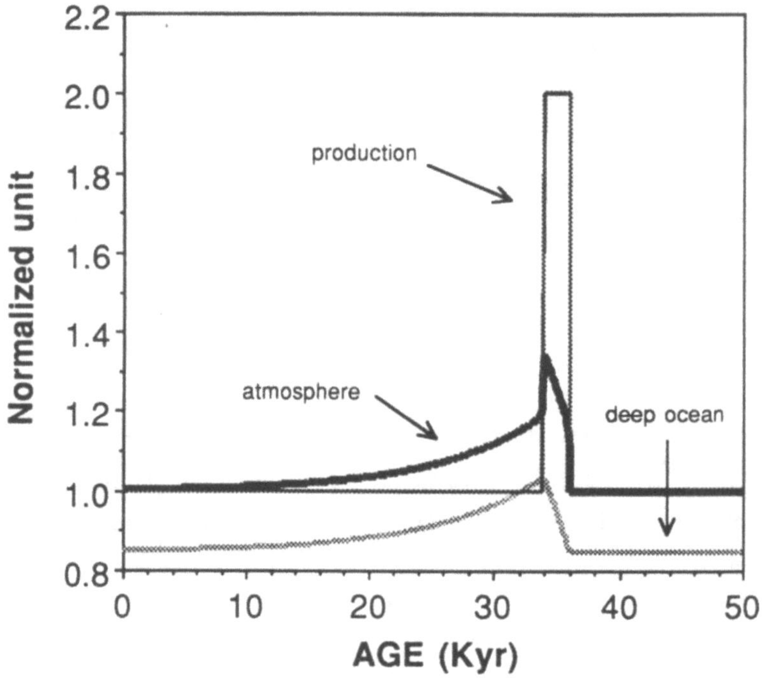

Fig. 2: Calculated response of $^{14}C/^{12}C$ in atmosphere and deep ocean assuming a factor of 2 increase in production during 2000 year interval 35,000 years ago.

4.2. ^{10}Be deposition between 10,000-30,000 B.P.

A careful examination of Fig. 1 suggests that there is some indication of a slightly higher ^{10}Be deposition rate between 15,000 and 25,000 years ago, compared to the Holocene. To consider the possible implications of this we apply the same 2-box model mentioned above. We then make the assumption that the ^{14}C production rate is proportional to the ^{10}Be deposition rate. While we believe such an assumption is quantitatively incorrect (see Raisbeck and Yiou 1988 and discussion below) it is interesting to consider it since a similar assumption has been used previously by others (Siegenthaler and Beer 1988, Beer et al. 1984, 1988, Stuiver et al. 1991). The result of such a calculation is shown in Fig. 3, together with the most recent summary of $^{14}C/^{12}C$ ratios measured in corals by Bard et al. (this volume). As expected from the discussion in the preceeding section, the predicted effect of increased production during the ^{10}Be peak interval has largely disappeared by 20,000 B.P.. On the other hand, due mostly to the increased ^{10}Be deposition from

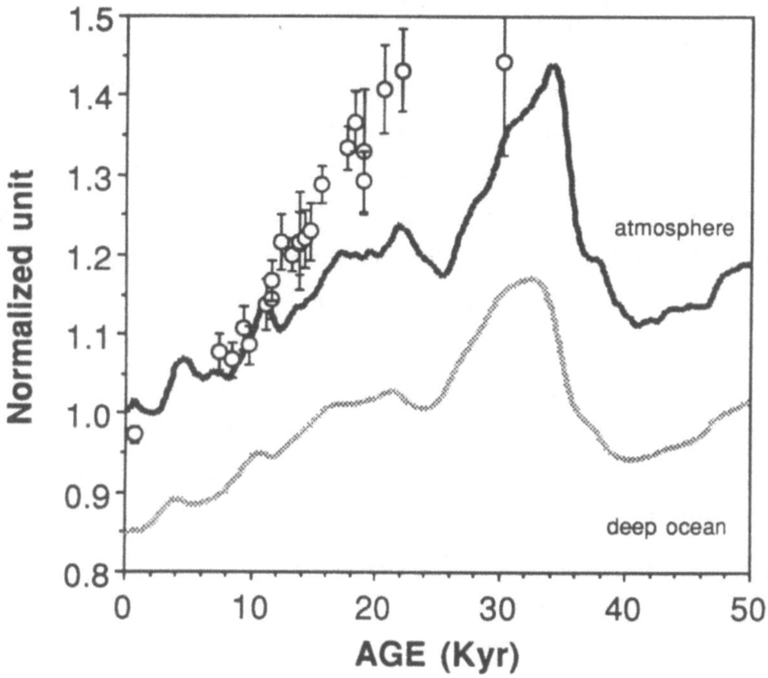

Fig. 3 : Calculated response of $^{14}C/^{12}C$ in atmosphere and deep ocean assuming ^{14}C production is proportional to ^{10}Be flux in Fig. 1. Open circles are most recent estimate of $^{14}C/^{12}C$ inferred from AMS ^{14}C measurements in U/Th dated corals (Bard et al. this volume).

~15,000-25,000 B.P., the model predicts between 10,000 and 20,000 B.P. a $^{14}C/^{12}C$ ratio in the atmosphere (or ocean mixed layer) ~ 15-20 % higher than the present day (pre-industrial) value. While far from negligible, this is substantially less than the enhancement inferred by the $^{14}C/^{12}C$ measurements in corals.

What is the significance of these results ? Even putting aside uncertainties associated with our calculated ice accumulation rates, and the fact that the ^{10}Be profile is not yet continuous, the answer to this question is not clear.

As discussed elsewhere (Raisbeck and Yiou 1985), because of its short atmospheric residence time of ~1 year (Raisbeck et al. 1981b), changes in ^{10}Be production will not always give rise to the same deposition rate variations at all latitudes. In particular, geomagnetic induced production variations, which occur primarily at low geomagnetic latitudes, are expected to be attenuated in high latitude reservoirs. This is the reason we (Raisbeck and Yiou 1988), in contrast to some others (Beer et al. 1984), have remained cautious about interpreting the rather flat ^{10}Be profile in the Holocene period (see also Fig. 1) as being inconsistent with long period $^{14}C/^{12}C$ production variations during this period. The latter are often assumed to result from geomagnetic secular variation. Thus, until a quantitative relationship can be established between geomagnetic intensity variations and ^{10}Be deposition at high latitudes, it will be difficult to infer paleomagnetically induced global production variations from ^{10}Be profiles in ice cores. About the most we can say is that, if the variations in the ^{10}Be deposition curve of Fig. 1 are due to a geomagnetic effect, then the modelled ^{14}C results in Fig. 3 represent a <u>lower</u> <u>limit</u> to the expected global $^{14}C/^{12}C$ changes.

At this point, one may wonder what is the potential implication of the same argument as applied to the preceeding section. However calculations (see for example Lal , this volume) show that the increase in global ^{14}C production, even for a <u>zero</u> magnetic field, is only about a factor of ~ 2. Since this is the enhancement observed for ^{10}Be, and assumed in the calculation of Fig. 2, the conclusions of the preceeding section would not be altered even if the ^{10}Be peak were due to a reduced geomagnetic field.

By contrast, cosmogenic production variations due to solar modulation are expected to be enhanced in high latitude ^{10}Be reservoirs. This is because the low energy cosmic rays most sensitive to modulation are not excluded by the geomagnetic field, and are thus more abundant at high latitudes. This is one of the reasons we predicted that polar ice cores would be favourable reservoirs for retreiving a solar activity related record (Raisbeck and Yiou 1980a). Thus, in this case the modelled ^{14}C variations in Fig. 3 represent an <u>upper</u> <u>limit</u> to those expected for a change in solar modulation.

Therefore, it is only variations in the primary cosmic ray flux, which for example have been suggested by Sonnett et al. (1987) and Kocharov (1990) as the origin of the ^{10}Be peak at ~35,000 B.P., that are expected to give an unbiased response in all reservoirs (but see below for a possible exception). In this case, assuming atmosphere deposition patterns have not changed, the ^{10}Be ice core data should give a reasonably quantitative measure of the global production changes.

In summary then, despite some early optimism (Raisbeck and Yiou, 1980b) it appears that ^{10}Be in polar ice can, at present, give only semi-quantitative information on global ^{14}C production rates. The difficulty is compounded by the fact that, even if reliable correlations between production and deposition could be established for present day atmospheric circulation patterns, there is no assurance that these would be valid for earlier and different climatic periods. For the moment, then, about the only conclusion we can draw is that the ^{10}Be data in Fig. 1 are not inconsistant with the increased ^{14}C production rate inferred by Bard et al. (1990a,b), providing this enhancement was due to a lower paleomagnetic field intensity. On the other hand, the most straightforward interpretation of the ^{10}Be data suggests that decreased solar modulation or increased primary cosmic ray intensity cannot entirely explain the coral results.

4.3 Potential implications of an enhanced $^{14}C/^{10}Be$ production ratio

So far we have implicitely assumed that ^{14}C and ^{10}Be production rates scale with each other. For any solar or geomagnetically modulated modification of the current galactic cosmic ray flux, this is probably a reasonably good assumption (see for example, Lal, this volume). There are however, at least two general cases where this approximation would not be valid : (i) production by particles having a much softer (more steeply falling) energy spectra ; (ii) production by high energy γ-rays. Are there any conditions when such considerations might be relevant, particularly with regard to the ^{10}Be peak at 35,000 B.P. ?

Power law, or other steeply falling, spectra are believed to be quite characteristic of many natural acceleration mechanisms. For example, solar flare particles have such spectra, and the unmodulated interstellar galactic cosmic ray spectrum is also generally believed to be of this form. Because the energy threshold for ^{14}C production is much lower than that for ^{10}Be, particles with such spectra are much more effective at producing ^{14}C than ^{10}Be. Lal (1988) for example estimates that the current solar flare particles have a $^{14}C/^{10}Be$ production ratio at least a factor of five larger than that for galactic cosmic ray production.

Solar flares are estimated to account for ~ 1-10 % of present day cosmogenic ^{14}C production (Lingenfelter and Ramaty 1970, Lal 1988). Since the current solar flare flux is dominated by a few large events each 11 year cycle, it is not difficult to imagine a one or two order of magnitude increase during an unusually active cycle. However, to give the observed ^{10}Be peak, such behavior would have to persist for one or two thousand years, which seems considerably more unlikely.

During the passage of a supernova shock wave, of the type envisaged by Sonett et al. (1987), particles accelerated in the shock are likely to have a power law spectra (Axford 1981) and, depending on the spectral index, could also produce a higher $^{14}C/^{10}Be$ ratio than at present.

High energy (>10 MeV) γ-rays interact with nuclei primarily via the giant dipole resonance, leading mainly to the ejection of a single nucleon. As a result, spallation cross sections for products several mass numbers lighter than the target are generally much smaller than corresponding nucleon induced reactions. Several years ago we measured the ^{10}Be production cross sections for high energy γ-ray induced spallation of ^{14}N and ^{16}O precisely in order to be able to estimate the possible contribution to ^{10}Be from a nearby supernova (Raisbeck et al. 1981c). The cross sections for both reactions were ~10μb about 3 orders of magnitude smaller than for the corresponding nucleon induced reactions. Although we are not aware of any similar experimental cross sections for ^{14}C production, they are expected to be substantially larger than for ^{10}Be. This is because, in addition to direct formation by $^{16}O\ (\gamma,2p)^{14}C$, ^{14}C can also be produced from secondary neutrons (by $^{14}N(n,p)^{14}C$) emitted in the reactions $^{14}N(\gamma,n)^{13}N$ or $^{16}O(\gamma,n)^{15}O$. On the basis of such considerations Konstantinov and Kocharov (1965) have estimated that ~1% of the γ-ray energy absorbed in the atmosphere will produce ^{14}C. Assuming a total γ-ray interaction cross section of ~250mb, this is equivalent to an effective ^{14}C cross section of 2.5mb. Using these estimates the $^{14}C/^{10}Be$ production rate from high energy γ-rays would be ~250, or ~4 times the current cosmogenic ratio.

In view of the duration of the ^{10}Be peak, the type of γ-rays we are thinking of here are not "prompt" γ-rays from a supernova explosion itself (which would give rise to enhanced ^{10}Be deposition over only one or two years), but rather those produced by the interaction of high energy electrons (via bremstrahlung and inverse Compton effect) and protons (via π° generation) with interstellar gas. Such interactions could occur in a number of ways (Morfill and Tenorio-Tagle, 1983). An additional interesting implication of such a production mechanism is that, because γ-rays, unlike charged particles, propagate in straight lines in interstellar space, they would not irradiate the earth's atmosphere homogeneously. Because the atmospheric residence time of ^{10}Be is comparable to the inter-hemispheric mixing time of the atmosphere, ^{10}Be deposition from such a process would also not necessarily be symmetric between the northern and southern hemispheres.

In order to estimate the plausibility of such an event, we note that, using the above ^{10}Be cross section, one requires the deposition of ~0.1 ergs/cm^{2}sec of high energy γ-rays for ~10^3 years to produce the observed ^{10}Be peak. Since the galaxy is relatively transparent to γ-rays, any process which has occured during the last 10^5 years, and lasted ~10^3 years, is likely to be observable elsewhere in the galaxy at the present time. Currently observable γ-ray sources emit energies of $< 10^{36}$ ergs/sec (Morfill and Tenorio-Tagle, 1983). Such a source would need to have occurred at ~10^{18} cm (~1 light year) from the solar system to produce the required γ-ray flux at the earth. Since this is closer than the nearest star, the probability for such an event during the past 10^5 years seems extremely small.

In summary then, while one can at least conceive of scenarios which could give rise to a significantly higher ^{14}C/^{10}Be production ratio than that occurring at present, thus potentially linking the inferred ^{14}C discrepancies during the late glacial with the observed ^{10}Be peak at 35,000 B.P., these scenarios are all highly speculative. In addition, of course, such a model would imply a very characteristic ^{14}C time profile, with the discrepancy between ^{14}C and calender ages exponentially increasing with age until the time of the ^{10}Be peak. Such a pattern should be relatively easy to test. For example, the point of Bard et al. (1990) at 30,000 B.P. may be already incompatible with such a hypothesis, although the error bar is still rather large.

5. CONCLUSIONS

Making the assumption that the relative paleo-precipitation rates in the Antarctic can be estimated from δ^2H profiles in ice cores, we calculate the ^{10}Be deposition rate at Vostok, Antarctica over the past 50,000 years. Except for a very significant (~ factor of 2) peak at ~35,000 B.P., the calculated

^{10}Be deposition rate is approximately constant, thus giving no evidence for any large (>20 %) variation in cosmogenic production during this period. However, because the relationship between production changes due to geomagnetic field variation and the ^{10}Be deposition in polar ice has not yet been established, no quantitative limitations on paleointensity induced ^{14}C variation can be made at present. On the other hand, production variations due to primary cosmic ray flux or solar modulation changes can probably be ruled out as the source of the recently observed discrepancy between ^{14}C and U-Th ages in corals during the late glacial period. It has also been shown that this discrepancy cannot be related to the ^{10}Be "peak" observed in Antarctic ice ~ 35,000 years ago unless this latter were due to some mechanism (solar flare particles, supernova shock waves, γ-rays) having a much larger $^{14}C/^{10}Be$ production ratio than that due to present galactic cosmic ray particles.

REFERENCES

Axford W.I. (1981). Acceleration of cosmic rays by shock waves, in Proceedings of the 17th Int. Cosmic Ray Conf. Vol 12, 155-203

Bard E., Hamelin B., Fairbanks R.G. and Zindler A. (1990a). Calibration of the ^{14}C time scale over the past 30 000 years using mass spectrometric U-Th ages from Barbados Corals. *Nature* **345**, 405-410

Bard E., Hamelin B., Fairbanks R.G., Zindler A., Mathieu G. and Arnold M. (1990b) U/Th and ^{14}C ages of corals from Barbados and their use for calibrating the ^{14}C time scale beyond 9000 B.P. *Nuclear Instrument and Method in Physics Research* **B52**, 461-468

Beer J. et al. (1984). The Camp Century ^{10}Be record : implications for long-term variations of the geomagnetic dipole moment. *Nuclear Instruments and Methods in Physics Research* **85**, 380-384

Beer J., Siegenthaler U., Bonani G., Finkel R.C., Oeschger H., Suter M. and Wolfli W. (1988), Information on past solar activity and geomagnetism from ^{10}Be in the Camp Century ice core *Nature* **331**, 675-679

Jouzel J., Lorius C., Petit J.R., Genthon C., Barkov N.I., Kotlyakov V.M. and Petrov V.N. (1987). Vostok ice core : a continuous isotope temperature record over the last climatic cycle (160 000 years). *Nature* **329**, 403-408

Jouzel J., Raisbeck G., Benoist J.P., Yiou F., Lorius C., Raynaud D., Petit J.R., Barkov N.I., Korotkevitch Y.S. and Kotlyakov V.M. (1989). A comparison of deep Antarctic ice cores and their implications for climate between 65000 and 15000 years ago. *Quat. Res.* **31**, 135-150

Kocharov G.E. (1990). Investigation of astrophysical and geophysical problems by AMS : successes achieved and prospects. *Nucl. Instr. and Meth. in Phys. Res.* **B52**, 583-587

Konstantinov B.P. and Kocharov G.E. (1965). Astrophysical phenomena and radiocarbon, Doklady Akad. Nauk. **165**, 63-64 (translation Sov. Phys. Dolk (1966) **10**, 1043-1044

Lal D. (1988). Theoretically expected variations in the terrestrial cosmic-ray production rates of isotopes, in Cini Castagnoli G. (Ed.), *Solar-terrestrial relationships and the earth environment in the last millennia*, North Holland, 216-245

Lingenfelter R.E. and Ramaty R. (1970). Astrophysical and geophysical variations in ^{14}C production, in Olsson I.U. (Ed.). *Radiocarbon variations and absolute chronology*, Almqvist and Wiksell, Stockholm, 513-537

Morfill G.E. and Tenorio-Tagle G. (1983) Point like gamma-ray sources, *Space Sci. Rev.* 36, 93-143

Raisbeck G.M. and Yiou F. (1980a). ^{10}Be in polar ice cores as a record of solar activity, in *Pepin R.O., Eddy J.A. and Merrill R.B.(Eds.) The Ancient Sun*- Pergamon, N.Y., 185-190

Raisbeck G.M. and Yiou F. (1980b) Temporal variations in cosmogenic ^{10}Be production : implications for radiocarbon dating, *Radiocarbon* 22, 245-249

Raisbeck G.M. and Yiou F. (1985). ^{10}Be in polar ice and atmosphere. *Annals of Glaciology* 7, 138-140

Raisbeck G.M. and Yiou F. (1988). ^{10}Be as a proxy indicator of variations in solar activity and geomagnetic field intensity during the last 10,000 years, in Stephenson F.R. and Wolfendale A.W. (Eds.) *Secular solar and geomagnetic variations in the last 10 000 years* - Kluwer Academic Publishers), 287-296

Raisbeck G.M., Yiou F., Fruneau M., Loiseaux J.M., Lieuvin M., Ravel J.C. and Lorius C. (1981a). Cosmogenic ^{10}Be concentrations in Antarctic ice during the past 30,000 years. *Nature* 292, 825-826

Raisbeck G.M., Yiou F., Fruneau M., Loiseaux J.M., Lieuvin M., Ravel J.C. (1981b). Cosmogenic $^{10}Be/^{7}Be$ as a probe of atmospheric transport processes. *Geophysical Research Letters* 8, 1015-1018

Raisbeck G.M., Austin S. and Yiou F. (1981c). Possible identification of supernova explosions in the past by means of cosmogenic ^{10}Be in geophysical reservoirs, in Proceedings of the 17th Int. Cosmic Ray Conf. Vol 2, 365

Raisbeck G.M., Yiou F., Bourlès D., Lorius C., Jouzel J. and Barkov N.I. (1987). Evidence for two intervals of enhanced ^{10}Be deposition in Antarctic ice during the last glacial period. *Nature* 326, 273-277

Raisbeck G., Yiou F., Jouzel J. and Petit J.R. (1990). ^{10}Be and $\delta^{2}H$ in polar ice cores as a probe of the solar variability's influence on climate. *Philosophical Transactions of the Royal Society London* A330, 463-470

Robin G. de Q. (1977). Ice cores and climatic changes. *Philosophical Transactions of the Royal Society London* B280, 143-168

Siegenthaler U. and Beer J. (1988) Model comparison of ^{14}C and ^{10}Be isotope records. In Stephenson F.R. and Wolfendale A.W. (Eds.) *"Secular solar and geomagnetic variations in the last 10 000 years"* (Kluwer Academic Publishers) 315-328.

Sonett C.P., Morfill G.E. and Jokipii J.R. (1987). Intersteller shock waves and ^{10}Be from ice cores. *Nature* 330, 458-460

Stuiver M., Braziunas T.F., Becker B. and Kromer B. (1991) Climatic, solar, oceanic and geomagnetic influences on late-glacial and Holocene atmospheric $^{14}C/^{12}C$ change. *Quat. Research* 35, 1-24

Yiou F., Raisbeck G.M., Bourlès D., Lorius C. and Barkov N.I. (1985). ^{10}Be in ice at Vostok, Antarctica, during the last climatic cycle. *Nature* 316, 616-617

Lingenfelter R.E. and Ramaty R. (1970). Astrophysical and geophysical variations in ^{14}C production, in Olsson I.U. (Ed.). *Radiocarbon variations and absolute chronology*, Almqvist and Wiksell, Stockholm, 513-537

Morfill G.E. and Tenorio-Tagle G. (1983) Point like gamma-ray sources, *Space Sci. Rev.* 36, 93-143

Raisbeck G.M. and Yiou F. (1980a). ^{10}Be in polar ice cores as a record of solar activity, in *Pepin R.O., Eddy J.A. and Merrill R.B.(Eds.) The Ancient Sun*- Pergamon, N.Y., 185-190

Raisbeck G.M. and Yiou F. (1980b) Temporal variations in cosmogenic ^{10}Be production : implications for radiocarbon dating, *Radiocarbon* 22, 245-249

Raisbeck G.M. and Yiou F. (1985). ^{10}Be in polar ice and atmosphere. *Annals of Glaciology* 7, 138-140

Raisbeck G.M. and Yiou F. (1988). ^{10}Be as a proxy indicator of variations in solar activity and geomagnetic field intensity during the last 10,000 years, in Stephenson F.R. and Wolfendale A.W. (Eds.) *Secular solar and geomagnetic variations in the last 10 000 years* - Kluwer Academic Publishers), 287-296

Raisbeck G.M., Yiou F., Fruneau M., Loiseaux J.M., Lieuvin M., Ravel J.C. and Lorius C. (1981a). Cosmogenic ^{10}Be concentrations in Antarctic ice during the past 30,000 years. *Nature* 292, 825-826

Raisbeck G.M., Yiou F., Fruneau M., Loiseaux J.M., Lieuvin M., Ravel J.C. (1981b). Cosmogenic $^{10}Be/^{7}Be$ as a probe of atmospheric transport processes. *Geophysical Research Letters* 8, 1015-1018

Raisbeck G.M., Austin S. and Yiou F. (1981c). Possible identification of supernova explosions in the past by means of cosmogenic ^{10}Be in geophysical reservoirs, in Proceedings of the 17th Int. Cosmic Ray Conf. Vol 2, 365

Raisbeck G.M., Yiou F., Bourlès D., Lorius C., Jouzel J. and Barkov N.I. (1987). Evidence for two intervals of enhanced ^{10}Be deposition in Antarctic ice during the last glacial period. *Nature* 326, 273-277

Raisbeck G., Yiou F., Jouzel J. and Petit J.R. (1990). ^{10}Be and $\delta^{2}H$ in polar ice cores as a probe of the solar variability's influence on climate. *Philosophical Transactions of the Royal Society London* A330, 463-470

Robin G. de Q. (1977). Ice cores and climatic changes. *Philosophical Transactions of the Royal Society London* B280, 143-168

Siegenthaler U. and Beer J. (1988) Model comparison of ^{14}C and ^{10}Be isotope records. In Stephenson F.R. and Wolfendale A.W. (Eds.) *"Secular solar and geomagnetic variations in the last 10 000 years"* (Kluwer Academic Publishers) 315-328.

Sonett C.P., Morfill G.E. and Jokipii J.R. (1987). Intersteller shock waves and ^{10}Be from ice cores. *Nature* 330, 458-460

Stuiver M., Braziunas T.F., Becker B. and Kromer B. (1991) Climatic, solar, oceanic and geomagnetic influences on late-glacial and Holocene atmospheric $^{14}C/^{12}C$ change. *Quat. Research* 35, 1-24

Yiou F., Raisbeck G.M., Bourlès D., Lorius C. and Barkov N.I. (1985). ^{10}Be in ice at Vostok, Antarctica, during the last climatic cycle. *Nature* 316, 616-617

^{10}BE PEAKS AS TIME MARKERS IN POLAR ICE CORES

J. Beer, S.J. Johnsen[2,3], G. Bonani[4], R.C. Finkel[5], C.C. Langway[6], H. Oeschger[7], B. Stauffer[7], M. Suter[8] & W. Woelfli[4]

Environmental Physics
ETH-Zürich
Institute for Aquatic Sciences and Water Pollution Control
c/o EAWAG
CH-8600 Dübendorf
Switzerland

Abstract

^{10}Be concentration peaks observed in the ice cores from Vostok (around 35 ka B.P. and 60 ka B.P.) and Dome C (around 35 ka B.P.) offer the possibility of synchronizing the southern and the northern hemisphere ice core records provided that these peaks are of global extent. A search in the ^{10}Be ice core records of Byrd station, Antarctica, and Camp Century, Greenland, revealed the following results: The 35 ka B.P. peak is clearly present in the Byrd core. Despite the coarse time resolution and missing data there is a good indication that the peak is also present in the Camp Century record. Comparisons with $\delta^{18}O$ and SO_4 suggest that the peak is caused by a higher production rate of cosmogenic radionuclides rather than by climatic effects. The 60 ka B.P. peak could not be found in either of these two cores. At present it is not known whether this is due to inadequate sampling or to a non-production origin of the 60 ka peak at Vostok.

2) Geophysical Isotope Laboratory, Geophysical Institute, University of Copenhagen, Denmark

3) Science Institute, University of Iceland, IS-107 Reykjavik, Iceland

4) Institute for Intermediate Energy Physics, ETH-Hönggerberg, CH-8093 Zürich, Switzerland

5) Lawrence Livermore National Laboratory, Livermore, CA, 94550, USA

6) Ice Core Laboratory, Dept. of Geology, SUNY at Buffalo, Amherst, NY 14226, USA

7) Physics Institute, University of Bern, CH-3012 Bern, Switzerland

8) Paul Scherrer Institute, CH-5232 Villigen, Switzerland

NATO ASI Series, Vol. I 2
The Last Deglaciation: Absolute and Radiocarbon Chronologies
Edited by E. Bard and W. S. Broecker

Introduction

Ice cores have proven to be the most important archive for study of climate history of the last ~100'000 years. They contain not only all the atmospheric constituents removed by dry or wet deposition but also, enclosed as air bubbles, samples of the atmospheric gases. This unique property of ice to store a record of atmospheric composition is of fundamental importance for understanding the climate forcing due to greenhouse gases. However, to take full advantage of the information stored in polar ice caps a precise time scale is needed. Due to the flow properties of polar ice and to changes in the precipitation rates the establishment of such a time scale is a difficult task. If the accumulation rate is relatively high the seasonal variations of parameters such as $\delta^{18}O$, H_2O_2 or dust content can be used to count single years over the last 8'000 to 11'000 years. At greater depth the annual layer thickness decreases and one has to rely more and more on calculations based on ice flow models (Dansgaard and Johnsen 1969). At sites with low accumulation rates as often occurs in Antarctica annual layers cannot be determined even at the surface. Since all flow models are of necessity based on various assumptions about climate history and ice temperature, the uncertainty in age increases with increasing depth. The determination of absolute ages becomes less and less certain. There remains however the possibility of synchronizing records from different locations by using as time markers well-defined stratigraphic features such as acidity peaks caused by large remote volcanic eruptions or periods when a particular parameter (e.g. $\delta^{18}O$) shows a typical pattern in its variation. If these time markers are global in the sense that they can be detected unambiguously worldwide, they offer the very important opportunity to synchronize the northern and southern hemisphere records. While there are many clear time markers during the last few millennia, some even dated historically (e.g. volcanic eruptions) (Hammer 1984), there are practically no well-defined signals for the period before 10'000 B.P.. Detailed $\delta^{18}O$ records obtained from several deep drilling ice cores from Greenland and Antarctica show quite similar patterns (Dansgaard and Oeschger 1989) (Jouzel et al. 1989). However, the degree of similarity is not high enough to allow a reliable synchronization over the whole of the last glacial period. In addition, there remains the open question of whether, in spite of the similar patterns of the northern and southern hemispheric $\delta^{18}O$ variations, the climatic changes really were simultaneous. What is needed, therefore, are well-defined time markers of global extent within the period 10'000 to 100'000 years B.P.. As discussed below such time markers can be provided by peaks of the ^{10}Be concentration. YiouRaisbeck and Raisbeck (Yiou et al. 1985) discovered in the Vostok ice core two outstanding peaks approximately dated to 35 and 60 ka B.P.. The younger ^{10}Be peak could also be seen in the Dome C core (Raisbeck et al. 1987). These peaks could well serve as ideal time markers if they are indeed of global extent.

^{10}Be peaks

^{10}Be is a cosmogenic radioisotope which is produced by the interaction of cosmic ray particles with the atmosphere. Its production rate therefore depends directly on the cosmic ray flux, which itself is modulated by the magnetic properties of the solar wind and the intensity of the geomagnetic dipole field. After production ^{10}Be becomes attached to aerosols and is removed from the atmosphere after a mean residence time of about one year by wet or dry deposition. The ^{10}Be concentration measured in polar ice is therefore determined by the production rate and the regional or global precipitation rate. An increase in ^{10}Be concentration in ice can result from either a higher production rate, caused by a higher cosmic ray flux, a lower solar activity or a decrease of the geomagnetic dipole moment, or from a lower precipitation rate, caused by glacial period climate conditions. The ^{10}Be peak found in the Vostok core at 35 ka B.P. extends for period of 1000 to 1500 years and shows an increase in ^{10}Be concentration by a factor of 2.5 (Raisbeck et al. 1987). Several hypothesis have been proposed to explain the peaks. While Raisbeck and Yiou (Raisbeck et al. 1987) favor, to some extent, the possibility of an extremely long quiet sun period, Sonett (Sonett et al. 1987) proposed a cosmic ray shock wave and Kocharov (Kocharov 1990) a super nova explosion to account for the higher ^{10}Be concentration. At present the question of cause remains open and more work needs to be done to solve this problem. The ^{10}Be peak at 60 ka B.P. is smaller in amplitude and broader than the one at 35 ka B.P. (Raisbeck, personal communication). If the ^{10}Be peaks are in fact the result of increased production, then the peaks should be global and similar peaks would be expected for other cosmogenic radioisotopes like ^{36}Cl and ^{14}C. If the peaks are the results of concentration effects due to a sudden decrease in the precipitation rate or to an atmospheric focusing process, then similar changes in other atmospheric constituents not produced by cosmic rays should be found in the cores. As a first step to resolve this puzzle we searched for the 35 ka B.P. and 60 ka B.P. peaks in existing ^{10}Be records from other locations in the northern and southern hemispheres.

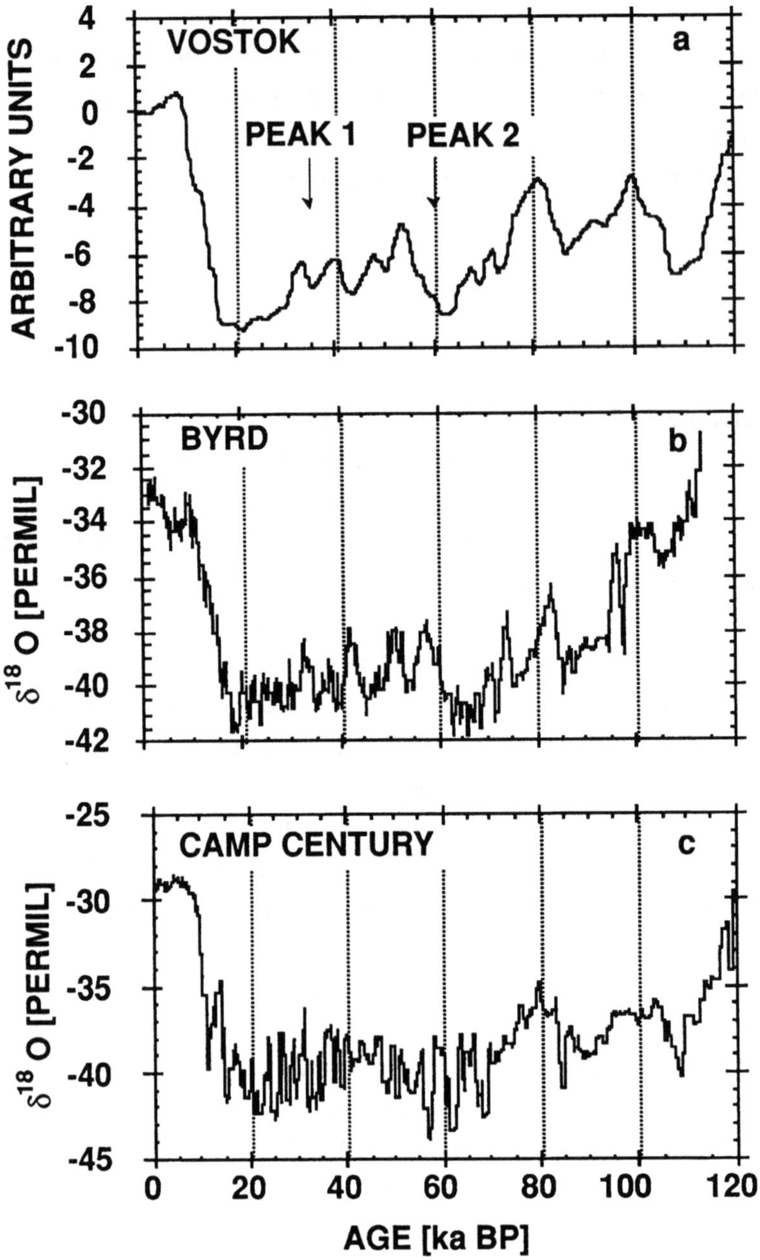

Fig. 1 Comparison of the climate records from Vostok (a), Byrd (b) and Camp Century (c) based on $\delta^{18}O$ measurements. The positions of the ^{10}Be peaks in the Vostok record are marked with arrows. The data of the Vostok record are taken from (Lorius et al. 1989). The time scale for Camp Century is discussed by Johnsen (this volume).

Search for ^{10}Be peaks

Beside Vostok and Dome C there exists a third ^{10}Be record from Antarctica, the one of Byrd which has partly been published earlier (Beer et al. 1987). The Byrd record is quasicontinuous in the sense that only every second core tube has been analyzed. In Fig. 1 the δ^{18}O records of Vostok, Byrd and Camp Century are compared and the expected positions of the ^{10}Be peaks in the Byrd and the Camp Century cores are marked.

The Byrd core is dated by assuming i) uniform vertical strain rate, ii) present accumulation rate of 0.13 m of ice per year, iii) past accumulation rates were corrected by assuming an 8 % change for 1 permil change in δ^{18}O (Clausen 1988) and iv) 10 meter extra melting at the bottom. Since the δ^{18}O records of Vostok and Byrd are quite similar the correlation of the δ^{18}O variations is rather straightforward.

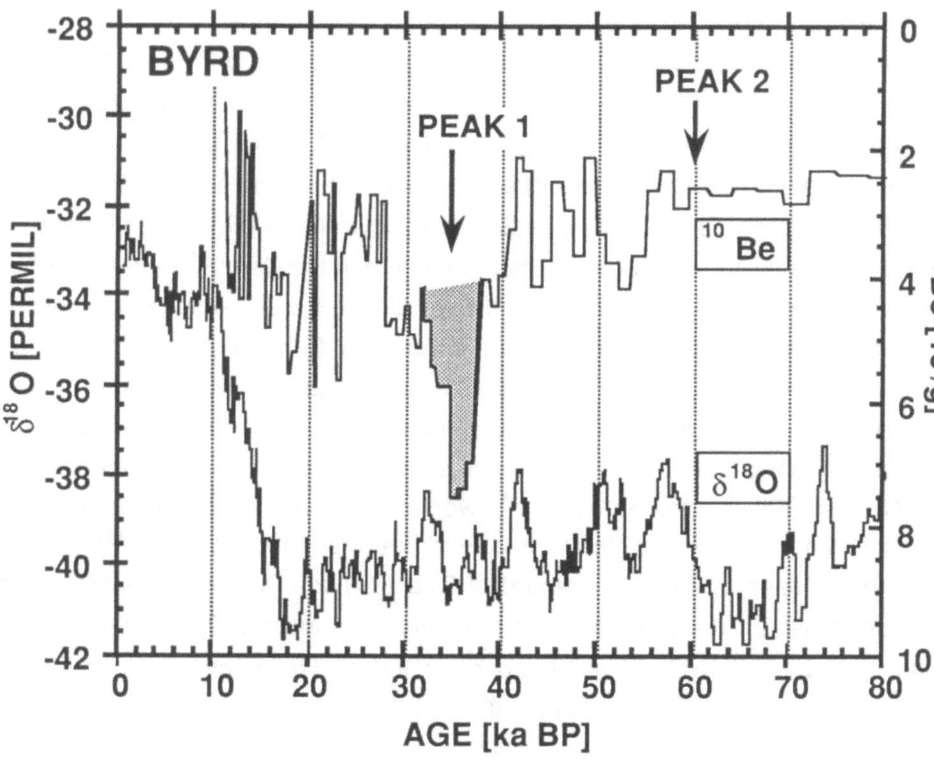

Fig. 2 Comparison of the δ^{18}O and the ^{10}Be records from Byrd. The ^{10}Be peak 1 at 35 ka B.P. is clearly visible but not peak 2 at 60 ka B.P.. Typical errors (1 σ) of the ^{10}Be concentration are 4-7%.

The dating for the Camp Century core is mainly based on the assumption that the high $\delta^{18}O$ values, found close to the bottom, represent the stage 5e or the Eemian interglacial. This is supported by correlation with the Dye 3 core where volcanic ash is found 88 meters above the bottom (Ram and Gayley 1991). This ash is believed to correspond to the Z2 volcanic zone found in deep sea cores at the boundary of stage 3 and 4. The details of the dating are discussed by Johnsen (this volume). The accuracy is ± 5 ka.

In Fig. 2 the Byrd $\delta^{18}O$ record is shown together with the ^{10}Be data. Peak 1 is clearly present. A detailed comparison of the peaks (Fig. 4) shows that even the shapes look quite similar. However, there is no indication of peak 2. Since the length of the odd tubes which were not analyzed is 1.5 m corresponding to 2700 years we cannot exclude that peak 2 falls in one of these gaps and is therefore not shown by our measurements.

The only continuous ^{10}Be record of the northern hemisphere is the one of Camp Century (Beer et al. 1984). Since the main purpose of analyzing the Camp Century core was to get a rough overview over the whole core, the time resolution is only of the order of 500 years at 35 ka B.P. and 4000 years at 60 ka B.P..

The agreement between the $\delta^{18}O$ curves of Vostok and Camp Century (Fig. 1) is good enough to identify the most likely intervals of the ^{10}Be peaks in the Camp Century core. In fact the ^{10}Be at the expected position of peak 1 (35 ka B.P.) turned out to be the second highest concentration found in the whole core (the highest concentration is a single point at 95 ka B.P. not shown here) (Fig. 3). Unfortunately the ^{10}Be peak is adjacent to a gap in the data. In Fig. 4 the Camp Century peak 1 is compared with the ones from Vostok and Byrd. All peaks are simultaneous within the accuracy of dating. A new unpublished chronology for the Byrd core, based on annual layer counting, dates the peak 1 at 39 ka B.P. (Clausen, private communication). The agreement between Vostok (Fig. 4a) and Byrd (Fig. 4b) is very good regarding duration, amplitude and shape of the two peaks. In the case of Camp Century (Fig. 4c) the agreement is worse due to a low ^{10}Be value at 34 ka B.P. which is not observed in Vostok and Byrd. It corresponds to high $\delta^{18}O$ values which are not seen in the $\delta^{18}O$ record of Dye 3. Since in this time interval the agreement between the $\delta^{18}O$ records of Camp Century and Dye 3 is otherwise excellent we cannot exclude a mix-up of samples. The comparison is further hampered by the missing data between 36 and 47 ka B.P..

Peak 2 at 60 ka B.P. is not observed. The time resolution is so poor (4000 years per sample) that, based on the peak size at Vostok, one might not expect to see a pronounced peak in this record at Camp Century. But in this case the record is continuous and at least an indication of a peak should be visible. There is only one single high concentration value at 95 ka B.P. which, without further confirmation by additional measurements, should not be considered. However, using a different

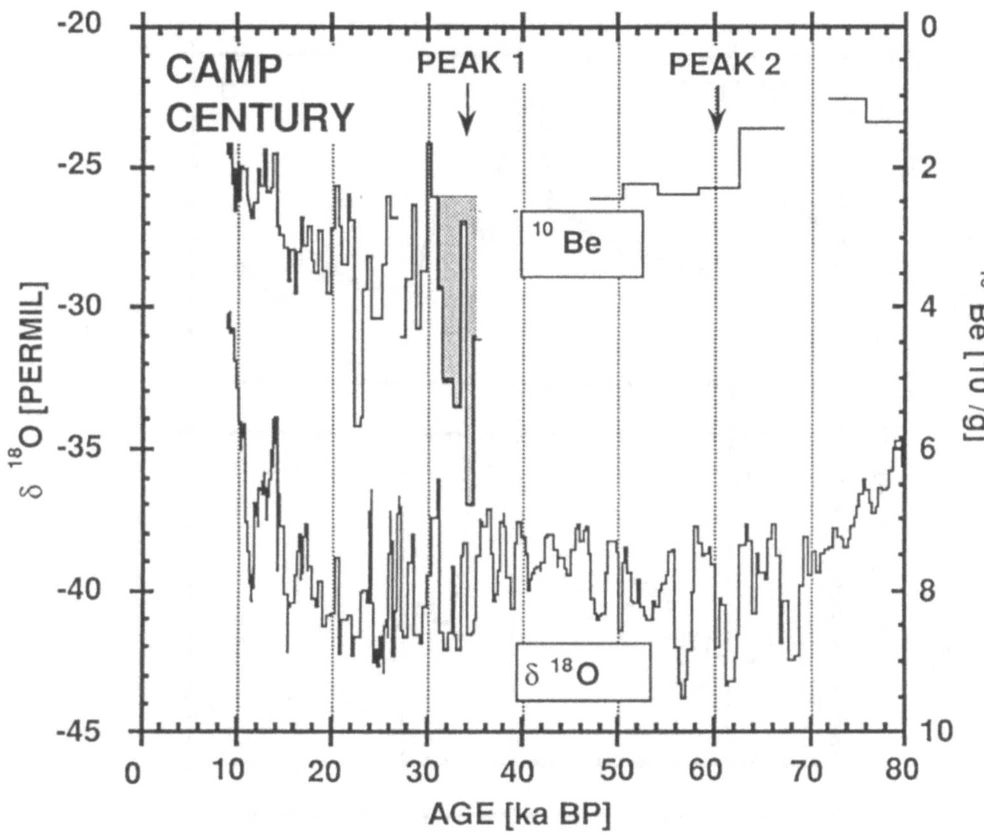

Fig. 3 Comparison of the $\delta^{18}O$ and the ^{10}Be records from Camp Century. There is clear indication for ^{10}Be peak 1 at 35 ka B.P. but not for peak 2 at 60 ka B.P.. Typical errors (1 σ) of the ^{10}Be concentration are 4-7%.

time scale Reeh et al. (1991) conclude that this high value represents the 60 ka B.P. peak.

An additional rather narrow ^{10}Be peak is observed at 23 ka B.P. in the Byrd and the Camp Century core.

Discussion

The data shown in Figs. 3-4 strongly suggest, although they do not decisively prove, because of limitations in sampling interval and temporal resolution, that the

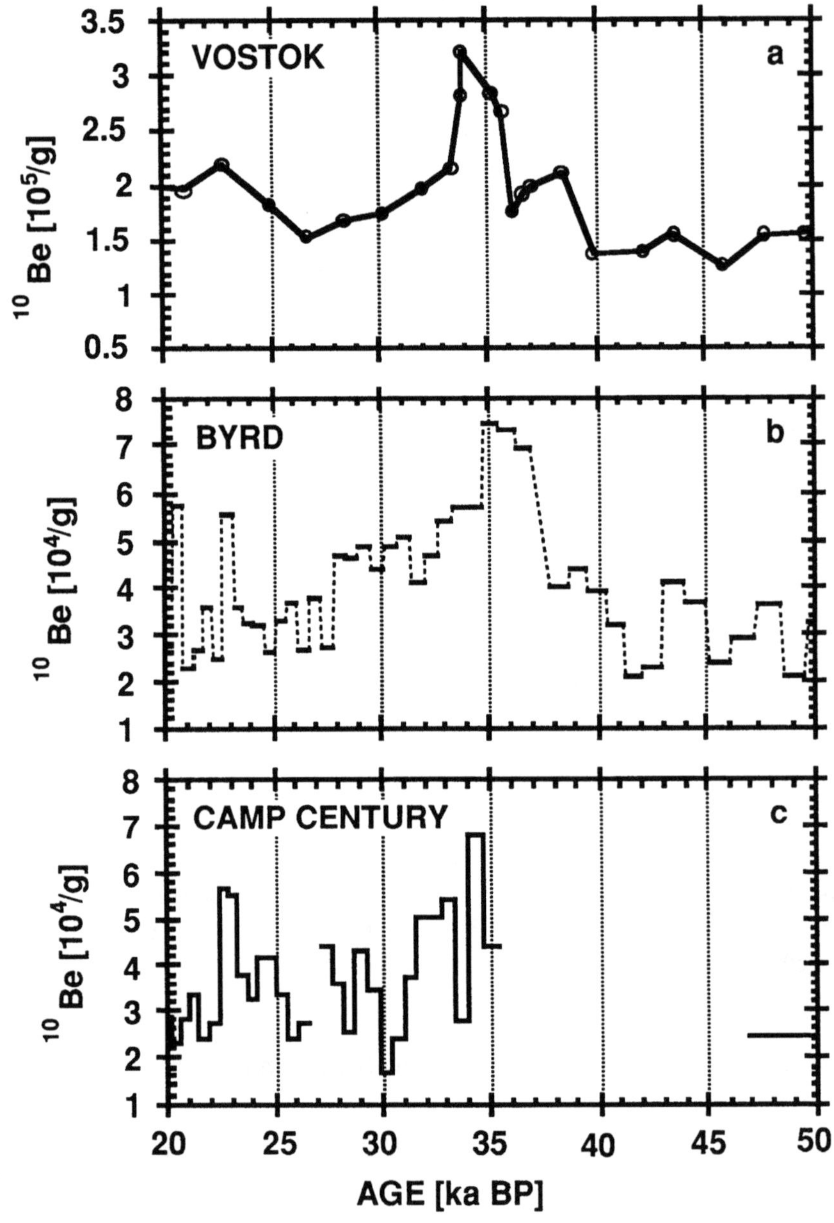

Fig. 4 Detailed comparison of the ^{10}Be peak 1 for Vostok (a), Byrd (b) and Camp Century (c). The small time shifts between the peaks is due to uncertainties (± 5 ka) of the time scales. The relatively low ^{10}Be value at 34 ka B.P. in the Camp Century record is discussed in the text.

35 ka ^{10}Be peak is present in the northern hemisphere. The question still remains of whether the peaks are caused by an increased production rate or by concentration effects due to special climatic conditions. The $\delta^{18}O$ and δD data of the Vostok core do not show the correspondingly large fluctuations which would be expected if the observed peaks result from precipitation rate change (Raisbeck et al. 1987). However, $\delta^{18}O$ and δD reflect basically temperature variations which control to some extent the water cycle and therefore the accumulation rate. Changes in atmospheric transport processes and in the exchange between stratosphere and troposphere could also cause strong variations of the fallout of cosmogenic radioisotopes without significantly changing the $\delta^{18}O$ and δD signal. A possible avenue to investigate such effects is to look for another non-cosmogenic atmospheric tracer. SO_4 is one possible candidate which is transported by aerosols. Its main sources are marine and biotic activity and volcanic emissions. We measured sulfate concentrations in the Camp Century core using ion chromatography at the Ice Core Laboratory of the University of Buffalo. In Fig. 5 the relationships between ^{10}Be, SO_4 and $\delta^{18}O$ from Camp Century are shown using scatter plots. All three parameters have been determined on the same samples. The best correlation is obtained between SO_4 and $\delta^{18}O$ ($r = 0.81$). The highest ^{10}Be concentrations (filled circles in Fig. 5) deviate clearly from the regression line indicating that they can only partly be explained by SO_4 or $\delta^{18}O$. In Fig. 6 the same ^{10}Be, SO_4 and $\delta^{18}O$ data are plotted as a function of depth corresponding to the time interval between the transition (10 ka B.P.) and peak 1 (35 ka B.P.). Both SO_4 and $\delta^{18}O$ show similar shapes to ^{10}Be except for the two peaks at 35 ka B.P. and 23 ka B.P.. This is confirmed by Fig. 6a and 6b where the measured ^{10}Be concentration is compared with the calculated one based on the linear regressions shown in Fig. 5. Both figures show that most of the ^{10}Be variations can be explained by $\delta^{18}O$ and SO_4, except for those occurring at two peaks around 35 ka B.P. and 23 ka B.P.. In Fig. 6c however, the SO_4 variations during the period of peak 1 is well explained by $\delta^{18}O$ variations. Based on SO_4 and $\delta^{18}O$ there is no indication that changes in atmospheric transport processes caused the observed peaks, and therefore the most likely explanation for the ^{10}Be peak at 35 ka B.P. is an increase of the ^{10}Be production rate. Whether the higher production rate was caused by a higher cosmic ray flux, a lower geomagnetic dipole moment or a period of low solar activity cannot be decided from the present data. The only direct information available on the subject so far are some reports which indicate a geomagnetic excursion during this period (Levi et al. 1990). However, it is still a matter of debate if this event was global (Tauxe and Valet 1989). The younger peak at 23 ka B.P. in the Camp Century core is less pronounced, but it also shows up in

the Byrd core. However, there is no indication of the peak in the Vostok core. (Raisbeck, personal communication). Higher levels in the atmospheric ^{14}C concentration have been found in stalagmites around 35 ka B.P. (Vogel 1983) and in corals between 10 and 30 ka B.P. (Bard et al. 1990).

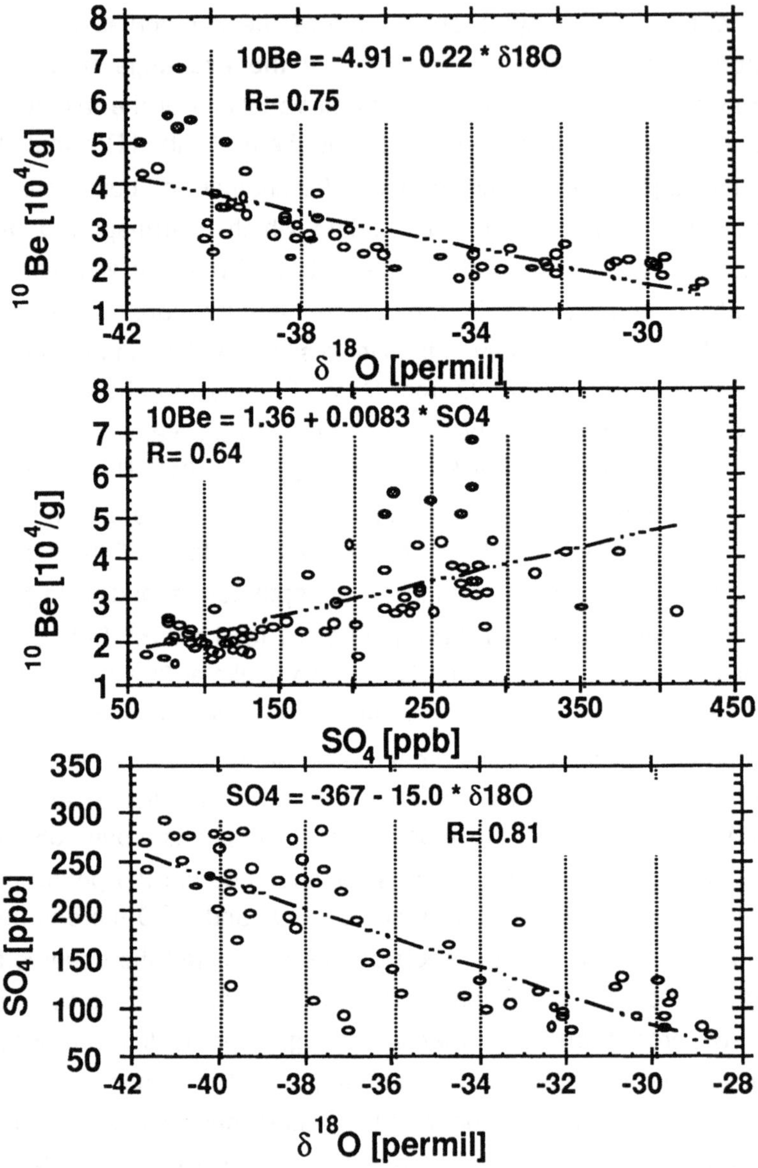

Fig. 5 Correlation between ^{10}Be, $\delta^{18}O$ and SO_4 for the depth interval 1100 to 1220 m corresponding to about 10 to 40 ka B.P. of the Camp Century record. Filled circles represent the ^{10}Be concentrations with the largest deviations from the linear regression lines.

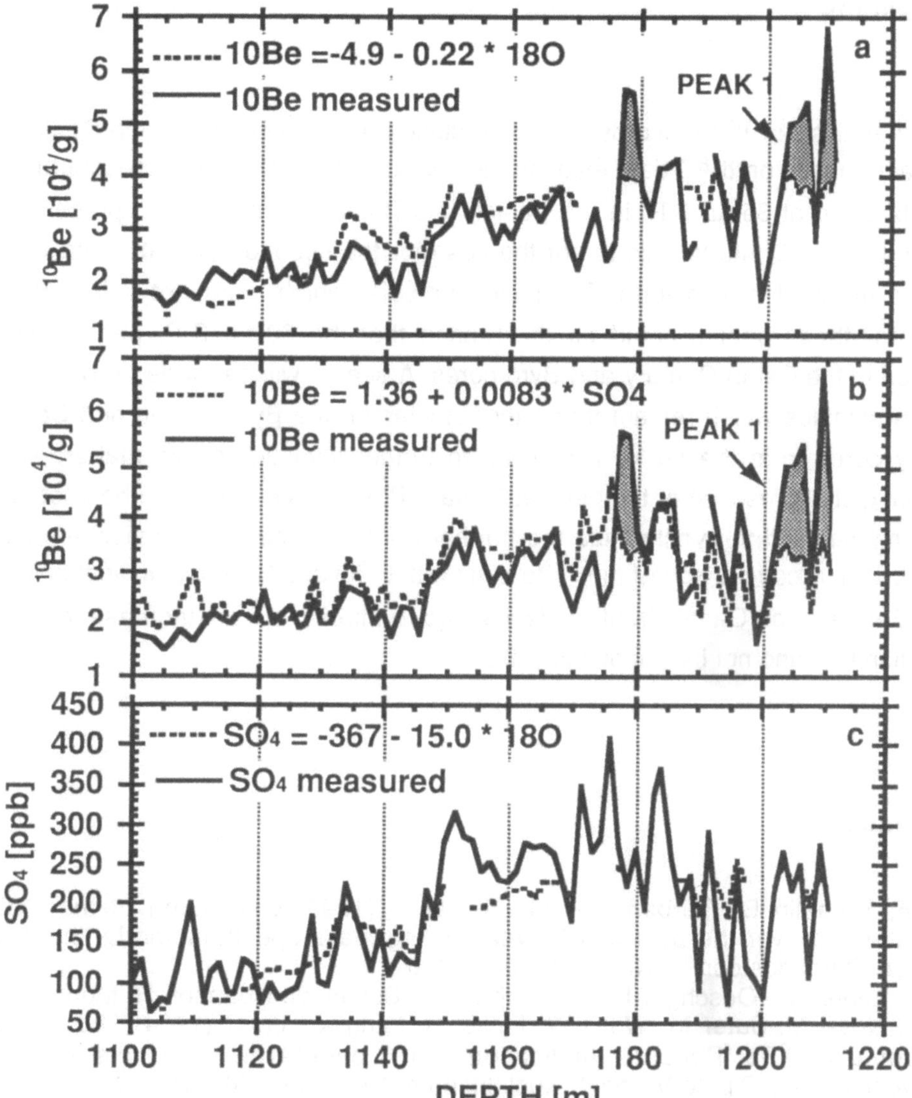

Fig. 6 ^{10}Be and SO_4 records from Camp Century for the depth interval 1100 to 1220 m corresponding to about 10 to 40 ka B.P.

a) shows the measured ^{10}Be concentration (solid line) together with the calculated one (dotted line) based on $\delta^{18}O$ measured on the same samples. There are several gaps due to missing data.

b) shows the measured ^{10}Be concentration (solid line) together with the calculated one (dotted line) based on SO_4 measured on the same samples.

c) shows the measured SO_4 concentration (solid line) together with the calculated one (dotted line) based on $\delta^{18}O$ measured on the same samples.

In contrary to most other variations the ^{10}Be peaks at 1210 m (35 ka B.P.) and 1175 m (23 ka B.P.) cannot be explained either by $\delta^{18}O$ or by SO_4.

Conclusions

Detailed analysis of ^{10}Be records from Byrd station, Antarctica, and Camp Century, Greenland, based on $\delta^{18}O$ data leads to the conclusion that the ^{10}Be concentration peak observed at 35 ka B.P. in the Vostok and Dome C cores is also present at Byrd and Camp Century. This is not the case for the second peak at 60 ka B.P. which is of smaller amplitude. Sampling intervals, which were not selected to search for these peaks, could be the reason that the 60 ka B.P. peak is not observed in the Camp Century and Byrd cores. A search will be made for possible ice core samples which remain from other studies on the Byrd station and Camp Century cores from the 60 ka B.P. level to further investigate this matter. The apparent global presence of the peak at 35 ka B.P. suggests that it can be used as a time marker and to synchronize the northern and the southern hemisphere ice core records. A comparison of the ^{10}Be peak with $\delta^{18}O$ and SO_4 data indicates that the ^{10}Be peak at Camp Century is caused primarily by an increase of the production rate and not by climatic effects.

References

Bard E, Hamelin B, Fairbanks R G, Zindler A (1990) Calibration of the 14C timescale over the past 30,000 years using mass spectrometric U-Th ages from Barbados corals, Nature 345: 405-410

Beer J, Andrée M, Oeschger H, Siegenthaler U, Bonani G, Hofmann H, Morenzoni E, Nessi M, Suter M, Wölfli W, Finkel R, Langway Jr. C (1984) The Camp Century ^{10}Be Record: Implications for Long-Term Variations of the Geomagnetic Dipole Moment, Nucl. Instrum. Meth. B5: 380-384

Beer J, Bonani G, Hofmann HJ, Suter M, Synal A, Wölfli W, Oeschger H, Siegenthaler U, Finkel RC (1987) ^{10}Be Measurements on Polar Ice: Comparison of Arctic and Antarctic Records, B29: 203-206

Clausen HB, Gundestrup NS, Johnsen SJ, Bindschadler R and Zwally J (1988) Glaciological investigations in the Crete area, Central Greenland: A search for a new deep-drilling site, Annals of Glaciology 10: 10-15.

Dansgaard W, Johnsen SJ (1969) flow model and a time scale for the ice core from Camp Century, Greenland, Journal of Glaciology 8: 215-223

Dansgaard W, Oeschger H (1989) Past environmental long term records from the arctic. In: (ed.) Oeschger H, Langway CC The Environmental Record in Glaciers and Ice Sheets. New York: John Wiley & Sons, 287-317

Hammer CU (1984) Traces of Icelandic Eruptions in the Greenland Ice Sheet. Jökull 34 : 51-65

Jouzel J, Raisbeck G, Benoist JP, Yiou F, Raynaud D, Petit JR, Barkov NI, Korotkevitch S, Kotlyakov VM (1989) A Comparison of Deep Antarctic Ice Cores and Their Implications for Climate between 65,000 and 15,000 Years Ago, Quaternary Research 31: 135-150

Kocharov GE (1990) Investigation of astrophysical and geophysical problems by AMS: successes achieved and prospects, Nucl. Instr. Meth. B52: 583-587.

Levi S, Audunsson H, Duncan RA, Kristjansson L, Gillot PY, Jakobsson P (1990) Late Pleistocene geomagnetic excursion in Icelandic lavas: confirmation of the Laschamp excursion, Earth and Planetary Science Letters 96: 443-457

Lorius C, Raisbeck G, Jouzel J, Raynaud D (1989) Long-term Environmental Records from Antarctic Ice Cores. In: (ed.) Oeschger H, Langway CC The Environmental Record in Glaciers and Ice Sheets. New York: John Wiley & Sons 343-361

Ram M, Gayley, RI (1991) Long-range transport of volcanic ash to the Greenland ice sheet, Nature 349: 401-404

Raisbeck GM, Yiou F, Bourles D, Lorius C, Jouzel J, Barkov NI (1987) Evidence for Two Intervals of Enhanced ^{10}Be Deposition in Antarctic Ice During the Last Glacial Period, Nature 326: 273-277

Reeh N, Oerter H, Letréguilly A, Miller H, Hubberten, HW (1991) A new, detailed ice-age oxygen-18 record from the ice-sheet margin in central west Greenland, Global and Planetary Change, in press

Sonett CP, Morfill GE, Jokipii JR (1987) Interstellar Shock Waves and ^{10}Be from Ice Cores, Nature 330: 458-460

Tauxe L, Valet JP (1989) Relative paleointensity of the Earth's magnetic field from marine sedimentary records: a global perspective, Physics of the Earth and Planetary Interiors 56: 59-68

Vogel JC (2 1983) ^{14}C Variations During the Upper Pleistocene, Radiocarbon 25: 213-218

Yiou F, Raisbeck GM, Bourles D, Lorius C, Barkov NI (1985) ^{10}Be in ice at Vostok Antarctica during the last climatic cycle, Nature 316: 616-617

VARIATION OF GEOMAGNETIC FIELD INTENSITY FROM 8-60 Ky BP, MASSIF CENTRAL, FRANCE

J.SALIS, N.BONHOMMET
Département de Géotectonique
Université de Paris VI (T.25,1°E)
4 place Jussieu
F-75252 PARIS CEDEX 05

ABSTRACT. Paleointensity data have been obtained from a set of lava flows on the Chaîne Des Puys, France. The volcanoes have been dated giving ages from 7 ky to 60 ky. The Thellier method was used to recover the absolute field intensity.The results show a low field intensity in the interval 10ky to 40ky. More over, at 40 ky the intensity collapses to the fifth of the normal value.

(1) Introduction

It is well known that the direction of the earth magnetic field (EMF) is subjected to variations in time. As an example, data for the last two thousand years at Paris (france) are illustrated in figure 1. In this example, the two parameters declination (D) and inclination (I), represent the change in

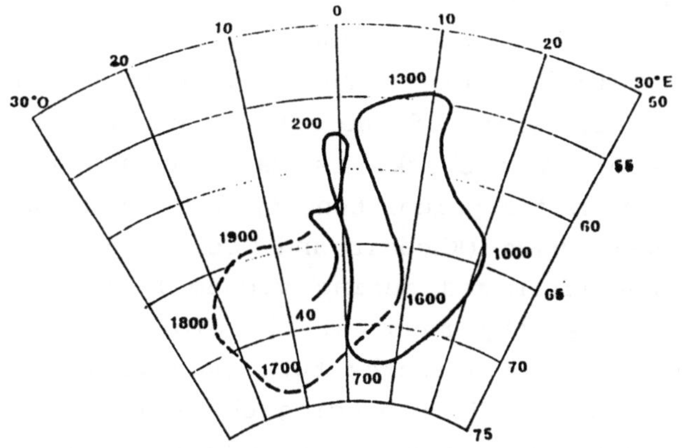

Figure 1: Variation of declination and inclination at Paris for the last two thousand years (Thellier, 1981).

NATO ASI Series, Vol. I 2
The Last Deglaciation: Absolute and Radiocarbon Chronologies
Edited by E. Bard and W. S. Broecker

direction of the field with time, also called "secular variation"; The field direction (D,I) can be obtained from observatory data for the last two hundred years; for older periods this direction can be obtained by using baked earth (pottery) or dated volcanic rocks, which have acquired, during cooling, a thermal remanent magnetization (TRM) parallel to the EMF.

On the contrary, the intensity of the past geomagnetic field is not well known, and methods to recover the true intensity rely on complex laboratory techniques. The best technique, up to now, for recovering the magnetic field intensity from lava flows is the Thellier method. The choice in selecting specimens must follow two criteria: (1) The natural remanent magnetization (NRM) must be a TRM and (2) the thermal stability of the ferromagnetic minerals.

(2) Method

We have obtained paleointensity data using the Thellier method (Thellier and Thellier, 1959) modified by Coe (1967). The procedure followed here is well known; The double heating temperature steps procedure of Thellier is increased progressively from room temperature to the maximum blocking temperature. If the partial TRMs (PTRMs) are independant and additive, then, each temperature step provides a unique paleointensity determination. However along the experiments chemical alteration may became apparent beyound a certain temperature. This change in the magnetic properties can be detected by non linearity of the PNRM/PTRM plot. It follows that extensive use of PTRM checks is made to assess the sample's continued integrity for paleointensity determinations.

It has been shown (Khodair and Coe, 1975) that temperature alterations are minimized during vacuum heating yielding more accurate paleointensity results. Therefor, in our experiments samples were heated in a quartz tube evacuated to pressure less than 0,01 Torr. The field producing the PTRM was applied during the entire temperature cycle. At each temperature step the first measurement was the partial NRM (PNRM) left after heating and cooling in the laboraory field at the same temperature. Vector substraction gave values of PNRM lost and PTRM acquired.

(3) Application to the lava flows of the Chaîne Des Puys.

The original Thellier (1937) method was restricted to potteries and brick clays which have stable magnetizations, and high Curie temperatures. Using lava flows, we cannot expect to get results as good as for potteries and brick clays, but we are able to sample a much longer time range. For instance, on the Chaîne des Puys (Massif Central, France) it is possible to recover,by spots, the field intensity from 8000 y BP to 60000 yBP

Lava flows do not behave like potteries or baked earth; however, we can select the samples which approach this behavior. Mechanisms which cause a non ideal behavior, have been studied by R.Coe (1967) and Khodair,(1975). Khodair, for instance,shows the importance of working in vacuum to determine the intensity. Levi (1975) compares different methods and check the additivity law (1979). Coe (1967) analyses the mechanism which causes non ideal behavior in the thellier experiments.

(4) Results.

The samples selected for paleointensity measurements (SALIS.J et al.1989), come from lava flows in the "Chaîne Des Puys", a volcanic chain in the Massif Central in France (figure 2). Most of the lavas were dated at the Centre Des Faibles Radioactivités of Gif/Yvette, France, using thermoluminescence, U-Th, K-Ar, Ar-Ar and radiocarbon techniques .In some cases these techniques were systematically used on different specimens of the same sample to crosscheck the reliability of the dating methods.(see table 1).

It can be seen from Table 1 that a great deal of work has been focussed on the two lava flows of Laschamp and Olby, because of their negative magnetic inclination which is opposite to that of the Brunhes period of normal polarity. However, the paleofield intensity obtained from these two flows (4 μT) is one tenth of the present field value (ROPERCH P.et al., 1988). The interpretation of these data is in favor of a collapse of the geomagnetic field. This drop in intensity 40,000 y ago, as recorded by the well dated lavas of Lashamp and Olby

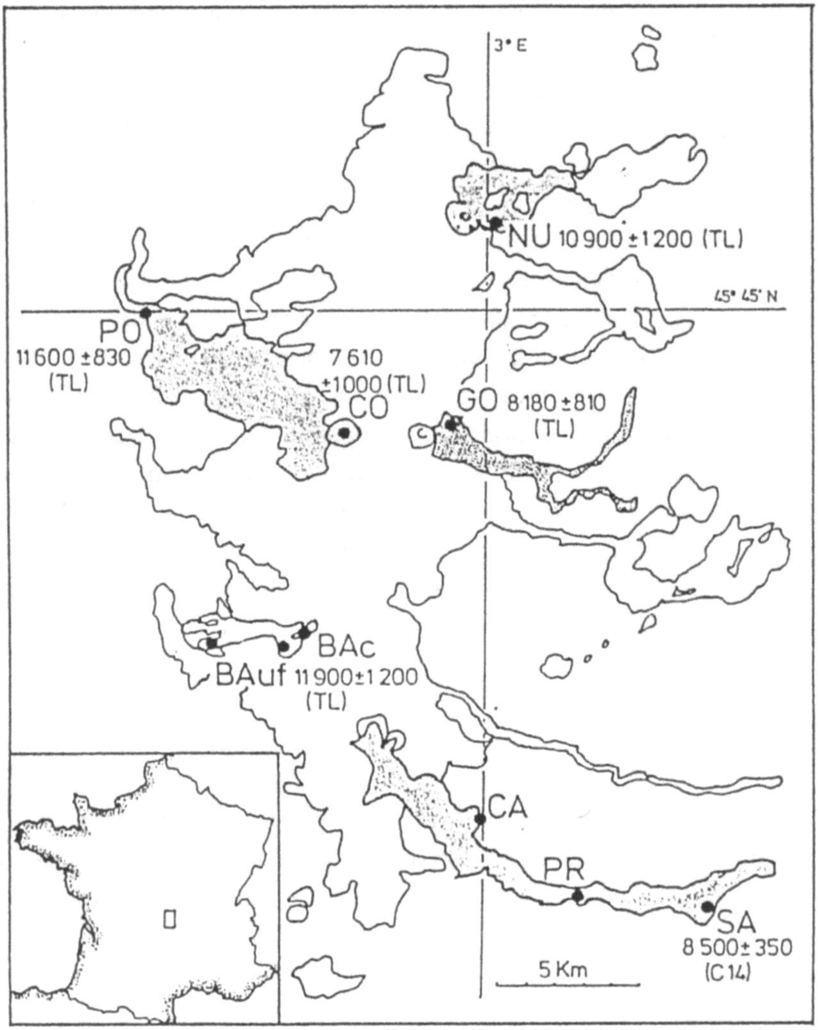

Figure 2. Location map of the Chaîne Des Puys, France. Shaded regions indicate major volcanic units with age estimates.

characterise the "Laschamp event" (BONHOMMET and BABKINE J.,1966,1967. GILLOT et al(1979), HALL and YORKD (1978). Confirmation of this event has been found in south west iceland (Marchall et al., 1988) where late Wisconsin basalts in the active rift zone of Reykjanes peninsula give a mean intensity of 4.3 +/- 0.6 μT wheras the mean paleointensity (PI) of the five normal samples is 30 μT. Recent paleointensity from sea cores confirms on a large scale such a low intensity as a global phenomena over the period 10,000-50,000 y BP.

The main result of this work is that the magnetic field intensity was lower than the present day field from 40 ky to 10 ky (figure 3 and table 2).In that case, changes in the non-dipole field sources will have much more significant effects on the magnetic field observed at the earth surface than if it were its normal value.(MERRILL and McELHINNY, 1983)

On the other hand a low field intensity will modify the C14 production rate over the last 40,000 y BP which in turn modifies the radiocarbon timescale (BARBETTI and FLUDE ,1979. MAZAUD et al. this volume.)

Dating units using different methods

site	âge(BP)	method	reference	année
LA	8000+-2000	K-Ar	Bonhommet and Zahringer	1969
	45400+-2500	K-Ar	Hall and York	1978
	47400+-1900	Ar-Ar	Hall and York	1978
	35400+-4000	TLq	Gillot et al.	1979
	31900+-2850	TLpl	Guérin and Valladas	1980
	32500+-3100	TLpl	Guérin and Valladas	1980
	30200+-2900	TLpl	Guérin	1983
	29800+-3300	TLpl	Guérin	1983
LB	39000+-6000	U-Th	Condomines	1978
	42000+-5000	Tlq	Gillot et al.	1979
	37900+-3500	TLpl	Guérin and Valadas	1980
	39200+-3000	TLpl	Guérin	1983
SA	8500+- 350	C14	Pelletier et al.	1982
	8100+-1100	TLq	Huxtable et al.	1978
	8820+- 870	TLpl	Guérin	1983
	9150+- 550	TLpl	Guérin	1983
MA	13300+-3500	U-Th	Condomines et al.	1982
	17300+-2000	TLpl	Guérin	1983

Table 1

SITE	age (ky)	paleointensity (μT)	VADM
CO	7.6 +- 1.0	29. +-4	4.9
GO	8.2 +- 0.8	45. +-3	7.5
SA	8.5 +- 0.4	64. +-5	10.4
NU	10.9 +- 1.2	37. +-4	5.8
PO	11.6 +- 0.8	33. +-2	5.3
BAc	11.9 +- 1.2	30. +-4	5.0
FG	13.3 +- 3.5	42.8 +-5	7.0
GA	16.4 +- 1.5	20.3 +-1	3.5
MA	17.3 +- 2.0	43.5 +-11	7.1
QR	21.6 +- 3.6	28.4 +-4	4.6
GN	28.7 +- 4.0	34.	5.5
LY	29.6 +- 2.7	27.0 +-2	4.4
LP	30.4 +- 2.1	37.2	6.1
MM	30.5 +- 3.5	35.5 +-4	5.8
MA	37.8 +- 3.8	20.0 +-4	3.3
LA	37.2 +- 5.0	7.7 +-1.6	-
LB	42.8 +- 7.0	7.7 +-1.6	-

Table 2

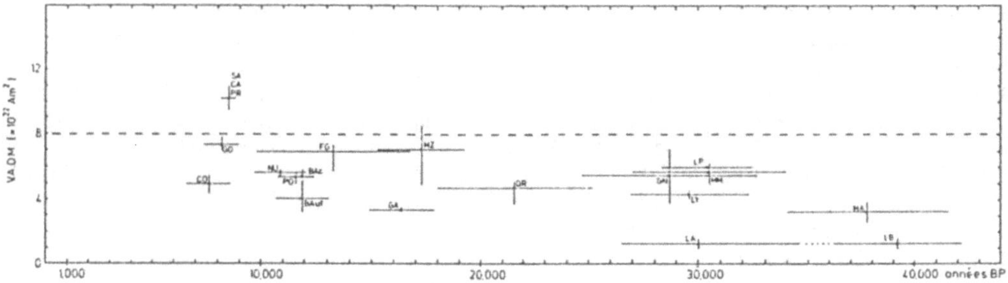

figure 3

Barbetti M.F., Flude K. (1979). Geomagnetic variation during the late pleistocene and changes in the radiocarbon time scale. Nature, vol. 279, p. 202-205.
Bonhommet N., Babkine J., (1966).Sur une direction anormale du champ magnétique terrestre au cours du Quaternaire récent. C.R. Acad. Sc. Paris, t. 262, p. 919-921.
Bonhommet N.,Babkine J.(1967). Sur la présence d'aimantations inversées dans la Chaîne des Puys. C.R.Acad.Sc.,Paris,t.264-B, p.92-994.
Coe R.S.,(1967 a).The determination of paleointensities of the earth magnetic field with emphasis on mechanisms which could cause non-ideal behavior in the Thellier methode. J.Geomagn., Geoelectr.,vol.19,p.157-159.
Gillot P.Y.,Labeyrie J.,Valladas.G.,Guerin G.,Poupeau.G., Delibrias G.(1979). Age of the Laschamp geomagnetic excursion revisited. Earth Pl.Sc.Lett. vol.42, p. 444-450.
Hall C.M.,York D.,(1978). K-Ar and 40Ar/39Ar age of the Laschamp geomagnetic polarity reversal. Nature, vol. 274, p. 462-464.
Khodair A. and R.S. Coe (1975). Determination of geomagnetic intensities in vacuum. Geophys.J.Roy.Astr.Soc.v42, p.107-115.
Levi S.,(1975) Comparison of two methods of performing the Thellier experiment (Or how the Thellier experiment should not be done).J. Geomag. Geoelec. vol. 27, p.245-255.
Levi S.(1979).The additivity of partial thermal remanent magnetization in magnetite.Geophys.J.Roy.Astr.Soc.v.59, p.205-218.
Marchall ét al.(1988). Preliminary paleointensity measurements and detailed magnetic analysis of basalts from the Skalamaelifell excursion, south west Iceland.
Mazaud A.,Laj C.,Bard E.,Arnold M.,E.Tric.(1991). A geomagnetic calibration of the radiocarbon time scale.(This volume).
Merrill and McElhinny (1983).The earth's magnetic field.Its History, Origin and Planetary perspectives.Int. Geophys. series.32. Academic press (1983)
Roperch P.,Bonhommet N.,Levi S.(1988). Palaeointensity of the earth magnetic field during the Laschamp excursion and its geomagnetic implications.Earth Planet.Sc. Letters,88,p.209-219.
Salis J.S.,Bonhommet N.,Levi S.(1989).Palaeointensity of the earth magnetic field from dated lavas of the Chaîne Des Puys, 7-12 Thousand years BP.J. Geophys. Res. v.94..
Thellier E.,(1937).Aimantation des terres cuites; application à la recherche de l'intensité du champ magnétique terrestre dans le passé. C. R. Acad. Sc. Paris, vol.204, p.184-186
Thellier et Thellier (1959). Sur l'intensité du champ magnétique terrestre dans le passé historique et géologique. Ann. Géophys.15,3,p.285-376
Thellier E.,(1981). Sur la direction du champ magnétique terrestre, en France, durant les deux derniers millénaires. Phys. Earth Planet. Int. 24: 89-132.

A GEOMAGNETIC CALIBRATION OF THE RADIOCARBON TIME-SCALE

A. Mazaud, C. Laj, E. Bard, M. Arnold, E. Tric
Centre des Faibles Radioactivités
Laboratoire Mixte CNRS-CEA
Avenue de la Terrasse
91198 Gif-sur-Yvette Cedex
France.

A high resolution record of the intensity of the geomagnetic field during the past 80 kyr has been recently obtained from marine cores. Since the upper atmospheric production of cosmogenic isotopes is modulated by dipole field intensity changes, this new intensity record offers an opportunity to assess variations in the abundance of ^{14}C in the atmosphere related to field changes, and consequently for calibrating the radiocarbon time scale, in particular for age older than 20 kyr, when no precise calibration exists from other geochronometers.

Introduction.

Radiocarbon dating is based on the knowledge of the fraction of ^{14}C remaining in a closed system, after natural disintegrations. It is necessary to know the initial concentration of ^{14}C, when the considered system was exchanging atoms of carbon with external environment. The ratio of atoms remaining is readily accessible by using counting or mass spectrometry techniques, but the initial ratio cannot be directly measured. Therefore, in order to obtain dates, it is supposed as a basic hypothesis, that the atmospheric $^{14}C/^{12}C$ ratio has remained constant through time, equal to the pre-industrial value [Libby, 1952]. This assumption, also expressed as $\Delta^{14}C=0$, defines the standard radiocarbon age, which differs slightly from the true age when initial $\Delta^{14}C$ was not zero.

It is now well established that atmospheric $^{14}C/^{12}C$ ratio has undergone significant variations in the past, independent of the recent artificial perturbations from the use of fossil fuels and atomic bombs. For the last 10 kyr, calibration of ^{14}C ages and $\Delta^{14}C$ values have been deduced from comparison between ^{14}C age determinations and direct dating of tree rings and varves. Atmospheric $\Delta^{14}C$ has decreased from about 100‰ at 10 kyr BP to -20‰ at 1.8 kyr BP [Stuiver et al., 1986]. For older ages, when direct dating by tree ring or varve counting is no longer possible, few spot calibrations have been obtained from thermal ionization mass spectrometry (TIMS) U-Th dating [Edwards, 1988]. It shows that the atmospheric $\Delta^{14}C$ was about 300-400‰ at the end of the last glacial period [Bard et al., 1990a and b].

To obtain a more complete calibration of ^{14}C ages, some authors have attempted to use geomagnetic dipole intensity changes [Barbetti and Flude, 1979; Barbetti, 1980; Bard et al., 1990a]. Since galactic and solar rays are mainly composed of charged particles, a large fraction are deflected by the Earth's magnetic field.

NATO ASI Series, Vol. I 2
The Last Deglaciation: Absolute and Radiocarbon Chronologies
Edited by E. Bard and W. S. Broecker

Changes in field strength will thus efficiently control the fraction of cosmic rays which impinge on the Earth's atmosphere and consequently, the rate of production of cosmogenic nucleides [Lal, 1988]. Because this deflection occurs at the edges of the magnetosphere, several Earth radii away from the Earth's surface, only changes of the dipole field strength are responsible for variation in production of cosmogenic nucleides. A precise knowledge of the history of the geomagnetic field intensity variations is needed to produce a record of cosmogenic ^{14}C production. Other difficulties in computing a synthetic $\Delta^{14}C$ signal arise because phenomena other than geomagnetic screening have affected the isotopic balance of carbon in the atmosphere. In particular, changes in the carbon cycle have to be considered. The deep ocean absorbs a large fraction of the ^{14}C produced in the atmosphere and conversely re-introduces previously created ^{14}C atoms to the atmosphere. There is evidence that during glaciation deep waters of the world ocean were renewed at a slower rate than at present, enhancing the atmospheric $\Delta^{14}C$ during glacial time [Shackleton et al., 1988]. Also the CO_2 contents of the atmosphere has increased since the last glacial episode. It is generally accepted that $\Delta^{14}C$ changes due to climatic changes do not exceed 100-150 ‰ [Siegenthaler et al.,1980; Shakleton et al., 1988; Bard al., 1990a].

The geomagnetic dipole intensity and the ^{14}C atom production over the past 50 ky.

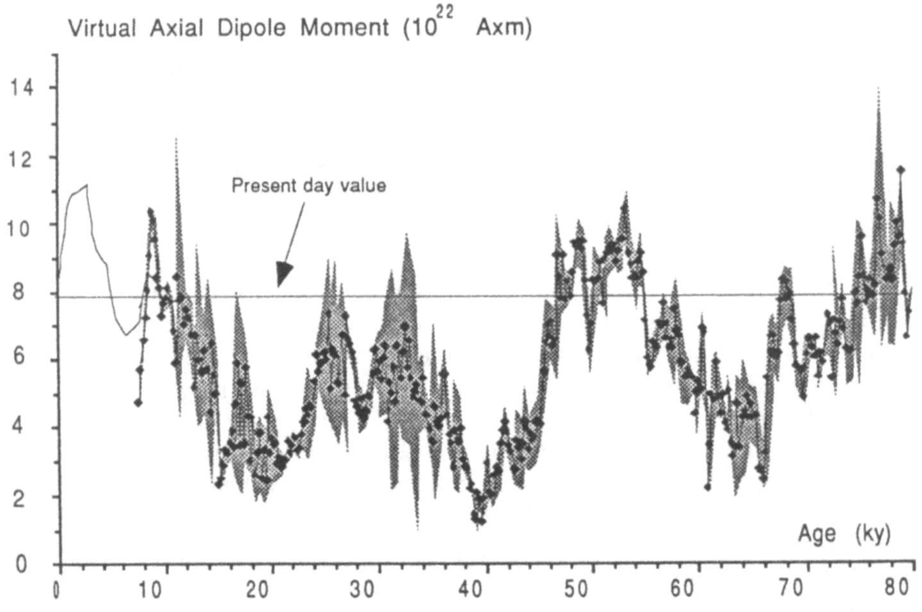

Fig. 1. Variations of geomagnetic intensity during the past 80 kyr (fromTric et al., in press). Shaded area corresponds to uncertainties in virtual axial dipole moment determinations. The full line (0-10 kyr) shows McElhinny and Senanayake data [1982].

We have used a new record of the geomagnetic dipole strength over the last 50 ky [Tric et al. in press] (Fig. 1) to produce a synthetic atmospheric $\Delta^{14}C$ record covering the past 80 kyr. A detailed description of this record, which was established using five marine cores, from the Tyrrhenian Sea and the Eastern Mediterranean, is given elsewhere [Tric et al., in press]. There is a remarkable agreement between this record and different published Virtual Axial Dipole Moments (VADMs) determined from volcanic formations from widely separated geographical regions [Tric et al., in press]. This record also agrees with record obtained from piston core in the Pacific [Constable and Tauxe, 1987; Tauxe and Wu, in press]. All these results suggest that the record of Tric et al [in press] is one of global dipole field variation. This record extends and adds detail to absolute paleointensity determinations [McElhinny and Senanayake, 1982] and shows that the intensity of the geomagnetic field has been on average lower than its present day value, particularly between 15 and 45 kyr, The 39 ky minimum is probably related to the Laschamp event [Bonhommet, 1972; Chauvin et al., 1989].

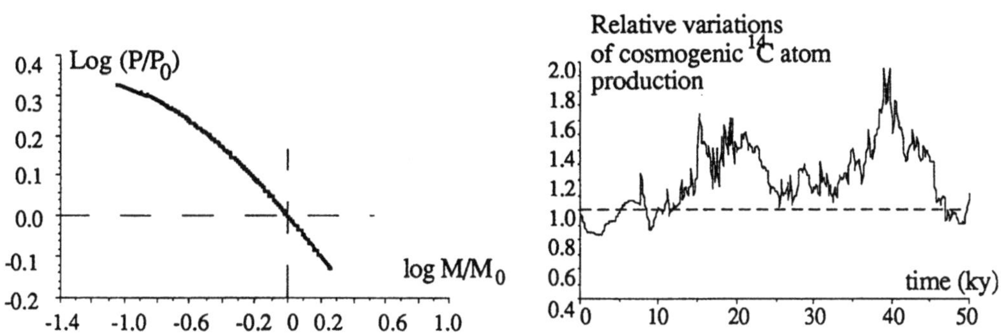

Fig. 2 a) Variations in the production P of ^{14}C atoms versus changes in the geomagnetic dipole moment M. P_0 and M_0 are present day values of the production and of the geomagnetic dipole moment [Lal, 1988]. b) Relative variations in cosmogenic ^{14}C production deduced from the intensity record of Tric et al. [in press].

We have calculated an instantaneous ^{14}C production rate from this record using Lal's calculations [Lal, 1988]. A curve of production versus intensity of the geomagnetic dipole is shown in Fig. 2a. At high values of the field, it can be approximated by $(Mo/M)^{1/2}$, M being the dipole moment at the time considered and Mo its present day value. At lower dipole intensities, the apparent saturation indicates removal of the geomagnetic screening [Merril and McElhinny, 1983]. A curve of variation of cosmogenic ^{14}C production is presented in Fig. 2b. It has been obtained from the geomagnetic record of Fig. 1 by considering only geomagnetic changes. High production between 18 and 45 kyr is a direct consequence of a weaker geomagnetic dipole than observed today.

The atmospheric ^{14}C in the past 50 ky and the calibration of the radiocarbon time scale.

The production signal (Fig. 2b) does not represent the variations of the atmospheric ^{14}C because of natural disintegrations, and of the participation of ^{14}C atoms to the CO_2 exchanges between the different natural reservoirs of carbon. We have therefore converted the instantaneous production record of Fig. 2b into an atmospheric $\Delta^{14}C$ signal, using a two box model developed for the carbon cycle [Houterman et al., 1973], already used by Bard et al. [1990a]. In this model the first box comprises the atmosphere and a fraction of the oceanic mixed layer, the second reservoir being the deeper parts of the world ocean. The response to a pulse input is composed of two exponentials. The first, characterized by a rapid relaxation time corresponds to the absorption of the produced ^{14}C atoms into the deeper ocean, whilst the second, associated with a relaxation time of 8267 years, corresponds to the radioactive decay. This model acts as a low pass filter which smooths out short term production varations, such as that related to the 11 yr solar cycle [Houterman et al., 1973].

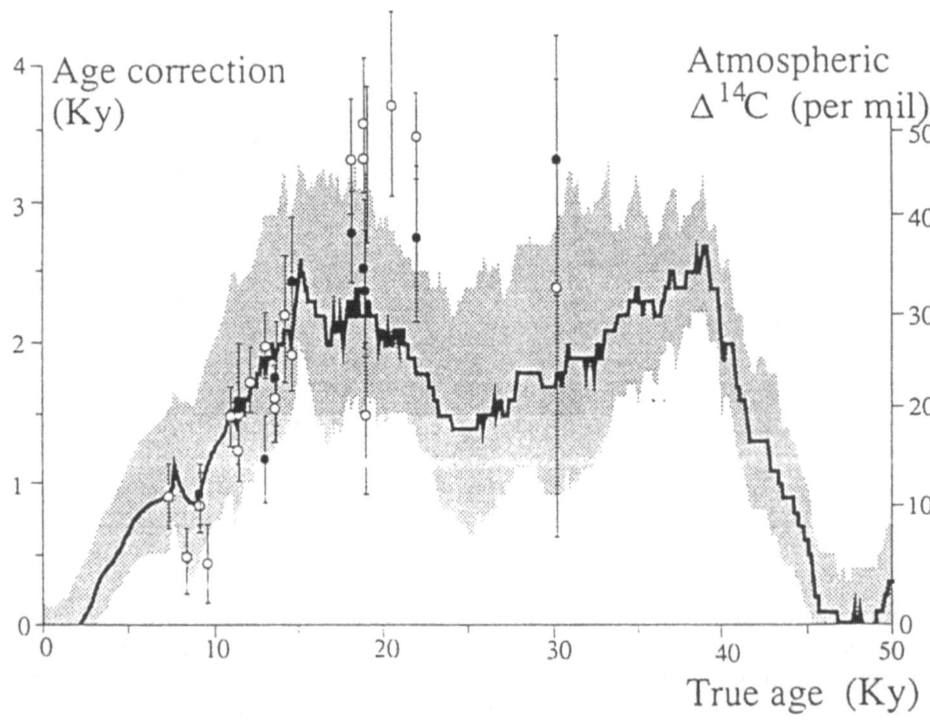

Fig. 3. Magnetic calibration of the radiocarbon time-scale, together with spot U-Th calibrations of ^{14}C ages. Open symbols show results of Bard et al. [1990a], Dark symbols show determinations of Bard et al. [1990b]. Right vertical axis gives atmospheric $\Delta^{14}C$. Shaded area corresponds to the uncertainties of the magnetic intensities given by Tric et al. [in press].

The $\Delta^{14}C$ record, also expressed in terms of shift of ^{14}C ages toward older ages is presented, together with spot comparisons of ^{14}C and U-Th ages of Barbados corals [Bard et al., 1990a and b], in Fig. 3. In these latter measurements, U-Th ages were obtained using the U-Th TIMS technique. Because this method is more precise than the classical counting technique, and is not affected by the usual ^{14}C bias caused by changes in the production of cosmogenic isotopes and/or changes in oceanic circulation, it is assumed that the U-Th ages represent true ages [Edwards, 1988; Bard et al., 1990b]. There is remarkable agreement between the magnetic calibration of ^{14}C ages deduced from the data of Tric et al. [in press] and U-Th dating in the 7-15 kyr interval. For older ages, the ^{14}C ages associated with U-Th dating depend on how the ^{14}C ages were determined [Bard et al., 1990b]. The U-Th calibrations give corrections of slightly higher amplitude than the magnetic calibration. This difference may result from climatic contribution, which is estimated to be lesser than 150‰ by several authors [Siegenthaler et al.,1980; Shakleton et al., 1988; Bard al., 1990a]. $\Delta^{14}C$ as large as 300 - 400 ‰ cannot be explained solely by climatic changes, or/and by changes in the repartition of the carbon in the different reservoirs.

Conclusion.

Although other causes, such as solar modulation suggested by peaks in ^{10}Be profiles from antarctic ice [Raisbeck et al., 1987], may have induced short term $\Delta^{14}C$ variations, the observed agreement between the geomagnetic and the U-Th calibration of ^{14}C ages strongly reinforces the hypothesis that geomagnetic variations are the major source of millenar variations in the abundance of ^{14}C in the atmosphere and the upper part of the ocean. It agrees with the conclusion reached by Stuiver et al for the last 10 ky [1991]. This magnetic calibration of the radiocarbon time-scale predicts that low intensities of geomagnetic dipole were responsible for a shift of ^{14}C ages toward young age of about 2 to 3 kyr in the interval from 20 to 40 kyr. It also predicts a good agrement between the ^{14}C and true ages in the interval 45-50 kyr.

Aknowledgements:

This work was supported by the CEA, the CNRS and the INSU program DBT-Messages sédimentaires. CFR Contribution 1225.

References:

Adams, J.M., Faure, H., Faure-Denard, L., McGlade, J.M., and Woodward, F.I., Increases in terrestrial carbon storage from the Last Glacial Maximum to the present, *Nature* 348, 711-714, (1990).

Barbetti, M., and Flude, K. Geomagnetic variation during the late Peistocene period and changes in the radiocarbon time scale, *Nature* 279, 202-205, (1979).

Barbetti, M., Geomagnetic strength over the last 50,000 years and changes in atmospheric ^{14}C concentration: emerging trends, *Radiocarbon*, 22, 192-199, (1980).

Bard, E., Hamelin, B., Fairbanks, R. G., and, Zinder, A., Calibration of the ^{14}C timescale over the past 30,000 years using mass spectrometric U-Th ages from Barbados Corals, *Nature* 345, 405-410, (1990a).

Bard, E., Hamelin, B., Fairbanks, R. G., Zinder, A., Mathieu, G., and Arnold, M., U-Th and ^{14}C ages of corals from Barbados and their use for calibrating the ^{14}C timescale beyond 9000 years B.P., *Nuclear Instruments and Methods in Physics Research* B52, 461-468, (1990b).

Bonhommet, N., Sur la direction d'aimantation des laves de la Chaîne des Puys, et le comportement du champ terrestre en France au cours de l'événement du Laschamp, *Thèse, Université Louis Pasteur de Strasbourg*, 1972.

Chauvin, A., Duncan, R. A., Bonhommet, N., and Levi, S., Paleointensity of the Earth's magnetic field and K-Ar dating of the Louchardière volcanic flow (central France): new evidence for the Laschamp excursion,*Geophys. Res. Lett.* 16, 1189-1192, (1989).

Constable, C. G., and Tauxe, L., Paleointensity in the pelagic realm: marine sediment data compared with archeomagnetic and lake sediment records, *Geophys. J. R. Astr. Soc.*, 90,43-59, 1987.

Edwards, R. L., High precision Thorium-230 ages of corals and the timing of sea level fluctuations in the late Quaternary.*Thesis, Cal. Inst. Technol.*, (1988).

Houterman, J. C., Suess, H. E. and Oeschger, H. Reservoir models and production rate variations of natural radiocarbon, *J. Geophys. Res.* 78, 1897-1907, (1973).

Prentice, K. C., and Fung, I. Y., The sensitivity of terrestrial carbon storage to climate change, *Nature* 346, 48-51, (1990).

Keir, R. S., Reduction of thermohaline circulation during the deglaciation: the effect on atmospheric radiocarbon and CO_2,*Earth Planet. Sci. Lett.* 64, 445-456, (1983).

Lal, D., Theoretically expected variations in the terrestrial cosmic-ray production rates of isotopes, *Soc. Italiana di Fisica-Bologna-Italy*, XCV corso, (1988).

Libby, W. F., Radiocarbon dating, *Univ. Chicago Press* 124p, (1952).

McElhinny, M. W., and Senanayake, W. E., Variations in the geomagnetic dipole I: the past 50,000 years, *J. Geomag. Geolectr.* 34, 39 (1982).

Merril, R. T., and McElhinny, M.W., The Earth's magnetic field *International Geophysics Series, Academic Press*, 400 pages, (1983).

Raisbeck, G. M., Yiou, F., Bourles, D., Lorius, C., Jouzel, J. & Barkov, N. I., Evidence for two intervals of enhanced ^{10}Be deposition in Antarctic ice during the last glacial period, *Nature* 326, 273-277, (1987).

Shakleton, N. J., Duplessy, J. C, Arnold, M., Maurice, P., Hall, M. A., and Cartlidge J., Radiocarbon age of last glacial deep water, *Nature* 335, 708-711, (1988).

Siegenthaler, U., Heimann, M. and Oeschger, H., ^{14}C variations caused by changes in the global carbon cycle, *Radiocarbon* 22, 177-191, (1980).

Stuiver, M., Kromer, B., Becker, B., and Ferguson, C. W., Radiocarbon age calibration back to 13 300 years BP and the ^{14}C age matching of the German oak and U.S. bristlecone pine chronologies, *Radiocarbon* 28, 969-979, (1986).

Stuiver, M., Braziunas, T. F., Becker, B., and Kromer, B., Climatic, solar, oceanic, and geomagnetic influences on Late-Glacial and Holocene atmospheric $^{14}C/^{12}C$ change, *Quaternary Research* 35, 1-24, (1991).

Tauxe, L., and Wu, G. Normalized remanence in marine sediment data from the western equatorial Pacific: relative paleointensity of the geomagnetic field? *J. Geophys. Res.*, in press.

Tric, E., Valet, J. P., Tucholka, P., Paterne, M., Labeyrie, L., Guichard, F., Tauxe, L., Fontugne, M., Paleointensity of the geomagnetic field during the last eighty thousand years, *J. Geophys. Res.*, accepted.

PART III:

CLIMATIC CHANGES DURING THE LAST DEGLACIATION

THE STRENGTH OF THE NORDIC HEAT PUMP

W.S. Broecker
Lamont-Doherty Geological Observatory of
Columbia University
Palisades, NY USA 10964

ABSTRACT

The Allerod-Younger Dryas cycle and also earlier cycles in the $\delta^{18}O$ record for Greenland ice cores have an asymmetric shape; the warmings are abrupt and the coolings gradual. A possible explanation for this shape is that the warmings are caused by abrupt turn-ons of the Atlantic's conveyor circulation. Upon turn-on, the conveyor runs at maximum strength (and hence releases a maximum amount of heat to the atmosphere over the northern Atlantic). Export of salt by the conveyor and dilution of salt by meltwater gradually drive down the salinity of the Atlantic, thereby reducing the buoyancy gradient between deep waters formed in the northern Atlantic and deep waters in the Antarctic. This reduction results in a gradual slow down of the conveyor lessening the amount of heat released to the atmosphere over the northern Atlantic. Eventually the buoyancy drive diminishes to the point where the conveyor shuts down.

NATO ASI Series, Vol. I 2
The Last Deglaciation: Absolute and Radiocarbon Chronologies
Edited by E. Bard and W. S. Broecker

INTRODUCTION

It has been my contention that heat released by the Atlantic Ocean's conveyor circulation maintains the relatively mild climate of northern Europe. The reason is that while the temperature of the water carried northward to the region around Iceland by the upper limb of the conveyor averages about 10°C, that of the water carried southward by the lower limb of the conveyor averages about 3°C. Hence during deepwater formation, about seven calories of heat per cubic centimeter of water transported in this great current system is released to the atmosphere over the northern Atlantic. As the flux of deep water production is about 20 Sverdrups, this release totals 5×10^{21} calories per year and equals about 25% of the heat received from the Sun by the troposphere over the northern Atlantic. Wolf Berger has aptly named this process the Nordic Heat Pump.

During the Holocene, air temperatures over Greenland remained amazingly constant (see figure 1) suggesting that the conveyor has run steadily for the last nine thousand years. By contrast, during glacial time the Greenland ice core record is punctuated by a number of large (5-7°C) temperature changes (see figure 2). I attribute these changes to the turning on and off of the conveyor and its associated heat pump. Any lingering suspicion that the temperature changes recorded in the ice cores might be the product of ice tectonics rather than climate has now been removed by the oxygen isotope record for the Renland ice core presented at this meeting by S. Johnsen. As the same sequence of events appear in the record for this new core as in those for Camp Century and Dye 3 cores, it is difficult to envision a scenario which would build the same false message into the ice in northern Greenland (Camp Century), in southern Greenland (Dye 3) and in a separate cap on an island east of central Greenland (Renland).

That temperature changes of this magnitude and rapidity are created by turning on and off the Nordic Heat Pump presents no problem to ocean and atmosphere dynamists. Ocean modeling experiments by Manabe and Stauffer (1988) show that under the same meteorologic forcing the Atlantic's circulation can assume two quite different quasi-stable modes of operation, one characterized by a strong thermohaline circulation akin to the conveyor and another devoid of thermohaline circulation. As expected these two modes lead to quite different temperature for the surface waters of the northern Atlantic (see figure 3). Maier-Reimer and Mikolajewicz (1988) have shown that the conveyor circulation in the Hamburg ocean model can be abruptly terminated by the addition of excess fresh water to the high latitudes of the Atlantic. Finally, Rind et. al., 1986 show that the geographic pattern of the winter air temperature anomaly produced by changing the temperature of the surface waters in the northern Atlantic in the GISS GCM is confined to the western fringe of the Americas (north of about 40°N) and to Europe north of the Pyrenees and Alps. This pattern conforms nicely to that for the actual Younger Dryas cooling. So also does the temperature change suggested by the models.

To account for the repetitive character of these climatic events Broecker et. al. (1990) propose that when surrounded by large ice sheets, the Atlantic's conveyor circulation is prone to oscillate. Under this circumstance, if the conveyor comes on, two things happen which tend to drive down the salinity of the Atlantic Ocean. First, the lower limb of the conveyor sweeps excess salt out of the Atlantic into the Antarctic and second, the heat released to the atmosphere over the northern Atlantic causes a meltback of the surrounding ice sheets. It is assumed that together these two tendencies toward salinity reduction more than compensate for the tendency toward salinity buildup resulting from vapor export from the Atlantic Ocean and its drainage basin. The consequent reduction in salt content causes the density of Atlantic waters to drop. Eventually a point is reached where the conveyor can no longer function; the density of deep waters produced in the northern Atlantic is too small to push back waters resident in the deep Antarctic.

The cessation of the conveyor is self stabilizing for it causes the salinity of surface waters in the northern Atlantic to drop by two or so grams per liter thereby preventing the conveyor from turning back on. The reason is that the ~0.3 Sverdrups of fresh water entering the Atlantic north of 40°N is no longer efficiently swept away by the fast flowing conveyor. Rather it tends to pool at the surface as currently happens in the northern Pacific. The ocean model of Manabe and Stauffer clearly shows this pooling effect (see figure 3).

Once the conveyor stops, salt begins once again to accumulate in the Atlantic. The reason is that both the export of excess salt by the conveyor and the meltback of ice by the Nordic Heat Pump cease. In this situation we envision that the buildup of salt resulting from the net export of water vapor from the Atlantic Ocean and its drainage basin dominates the Atlantic's salt budget. Eventually the density of Atlantic waters builds back to a level which permits the conveyor to reinitiate. This turn-on is self stabilizing; for the conveyor sweeps away the fresh water pooled in the northern Atlantic. As long as the external conditions remain more of less the same this cycle can repeat itself over and over again (Birchfield and Broecker, 1990).

All this has been said before. The new thoughts have to do with my thoughts regarding the shape of these climate cycles and the role of Fairbank's (1989) rapid sea level rises (see figure 4).

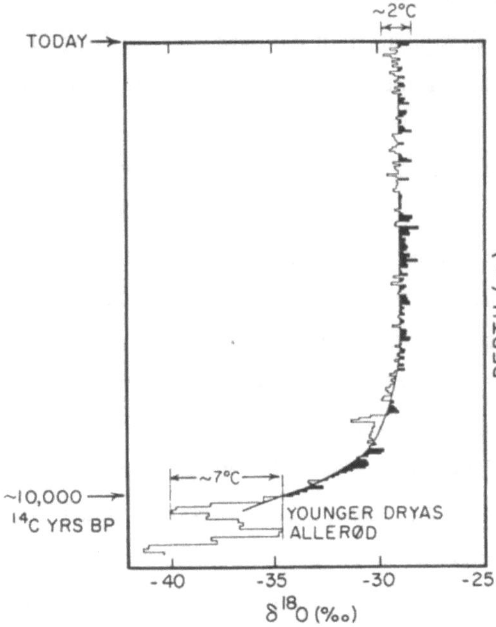

Figure 1. Oxygen isotope record for the Camp Century Greenland ice core (Dansgaard et al., 1971) covering the period from about 13,000 radiocarbon years B.P. to the present.

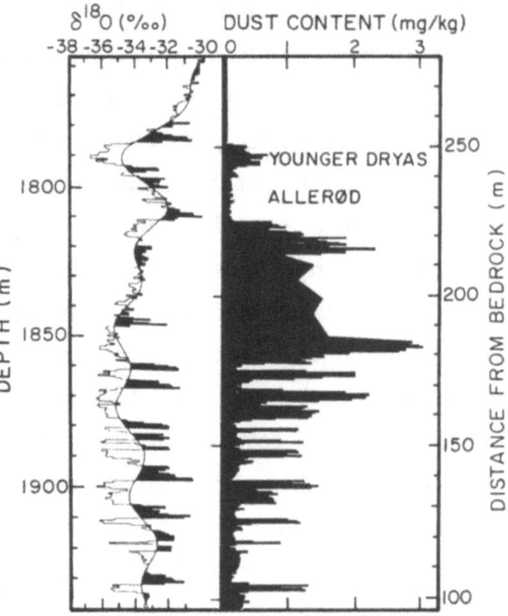

Figure 2. Oxygen isotope and dust records for the Dye 3 Greenland ice core covering the time period ~45,000 to ~8000 years B.P. (Hammer et al., 1985). Note that many of the oygen isotope events are characterized by rapid warmings followed by more gradual coolings.

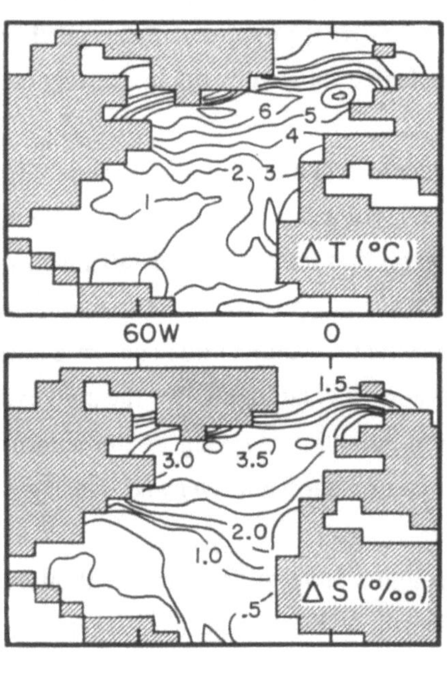

Figure 3. Maps showing the difference in North Atlantic surface water temperature (upper) and surface water salinity (lower) between the Manabe and Stauffer (1988) model circulation scheme with and scheme without thermohaline circulation. As can be seen, a turnoff of thermohaline circulation leads to a pronounced cooling and freshening of the waters in the Northern Atlantic.

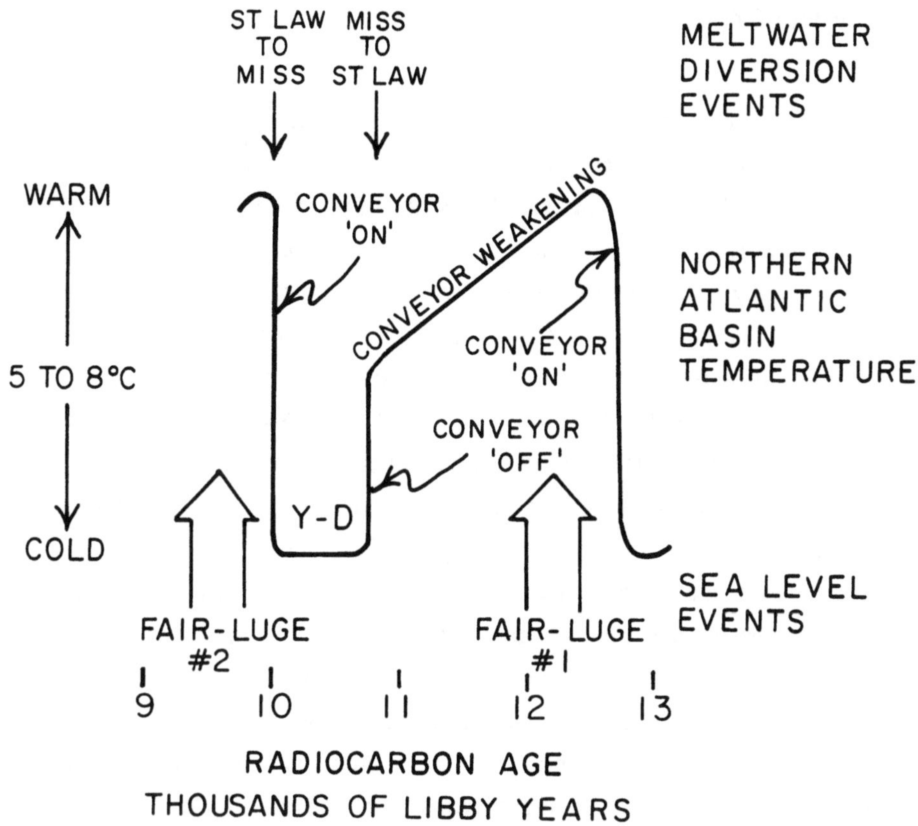

Figure 4. First order features of the climate record for northern Atlantic and its surrounding lands: About 12,700 Libby years ago, air and water temperatures suddenly increased presumably as the result of a turn on of the conveyor belt and hence, also the Nordic Heat Pump. This warming caused the rate of ice sheet melt back to increase leading to the first of Fairbank's dramatic rises in sea level. During the next roughly 2000 Libby years, a combination of export of salt (via the conveyor's lower limb) and dilution of salt (via ice melting) caused the salinity of the Atlantic's waters to steadily drop. This drop lessened the density contrast between the deep waters formed in the northern Atlantic and waters in the deep sea outside the Atlantic and thereby reduced the strength of the conveyor and hence also that of the Nordic Heat Pump. The climate in the northern Atlantic basin steadily cooled. By 10,800 Libby years ago the conveyor system had become sufficiently vulnerable to shut down that the diversion of meltwater from the Mississippi to the St. Lawrence River caused it to terminate bringing on the Younger Dryas cold event. Reduced dilution by ice melting and reduced export of salt allowed the salt content of Atlantic waters to build until about 10,000 years ago the density contrast between the Atlantic and the remaining ocean was high enough to permit the conveyor to turn on. Perhaps the diversion of meltwater from the St. Lawrence back to the Mississippi triggered this rejuvenation by reducing the influx of meltwater into the critical region where deep water forms. The heat released by the conveyor brought the Younger Dryas to a close and initiated the second episode of rapid sea level rise.

SHAPE OF THE CYCLE

The geological evidence in hand points to a characteristic shape for the air temperature changes associated with these cycles. The cycles show an assymetry akin to that for the 100,000 year cycle, namely, an extended period of progressive cooling is brought to a close by an abrupt warming. This shape is best documented for the last of these cycles. About 12,600 Libby years ago a very rapid warming occurred in Europe. Palynologists refer to this event as the onset of the Bölling. Lötter (1991) shows from the records for two Swiss lakes, Rötensee and Söppensee, that the transition in vegetation from tundra to forest and the 3‰ rise in the $\delta^{18}O$ in the authigenic calcite occurred in quite rapidly (i.e. in a few decades). I attribute this warming to a turn on of the conveyor.

This warming was followed by a period of generally declining temperature which culminated in a sharp drop in temperature about 10,800 Libby years ago. I see this decline as a natural consequence of the cycle outlined above. The logic is as follows. In order for the conveyor to turn on, the Atlantic's ambient salinity must be high enough to overcome the inhibition created by the pooling of fresh water at the surface of the northern Atlantic. Thus, when the conveyor does turn on, the buoyancy contrast between the deep water generated in the northern Atlantic and that in the Antarctic is unusually large. Because of this, when first turned on, the conveyor runs with unusual vigor. This leads to an enhanced strength for the Nordic Heat Pump and hence enhanced warming of northern Europe. The strength of the conveyor and hence also of the Nordic Heat Pump gradually drops as the Atlantic's excess salt is exported by the conveyor and diluted by glacial melting. This accounts for the progressive cooling following the onset of the Bölling. Then when the excess salt content of Atlantic waters becomes sufficiently small the conveyor turns off, accounting for the sharp cooling which initiated the Younger Dryas about 10,800 Libby years ago.

During the course of the Younger Dryas the salt content of the Atlantic builds back up to the point where the conveyor reinitiates. This reinitiation brings the Younger Dryas to an abrupt close. A warming occurs which is similar in magnitude and rapidity to that which occurred at the onset of the Bölling. That this second warming was also abrupt is shown by the oxygen isotope record in the Dye 3 core. Based on this Dansgaard et al., 1989 estimate that the warming which marks the close of the Younger Dryas occurred in 50 years. As reported by M. Pazdur and K. Rozanski at this meeting, the oxygen isotope record for $CaCO_3$ from the sediments in Gosciaz Lake in Poland show exactly the same pattern for the end of the Younger Dryas as that for Dye 3. They estimate the duration of the warming to be less than a century.

THE ROLE OF MELTWATER EVENTS

Our original explanation for the Younger Dryas was that it was triggered by a diversion of meltwater from the Mississippi River to the St. Lawrence drainage (Broecker et al., 1989). When Fairbanks (1989) produced a sea level record for the period of deglaciation showing two episodes of rapid melting, we were forced to rethink our mechanism. The reason is that the magnitude of meltwater flux associated with the Fair-luges exceeds that associated with the diversions. As a result we developed the salt oscillator hypothesis. In this new concept we downgrade the Mississippi to St. Lawrence meltwater diversion from a first order causal mechanism to that of a trigger (Broecker et al., 1989). The idea is that at the time of the Mississippi to St. Lawrence diversion the Atlantic's salinity had already dropped to the point where the conveyor was already poised for turn off. The diversion merely pushed it over the brink.

We also initially attributed the turn on of the conveyor which brought the Younger Dryas to an end to a rediversion of the southern Laurentide meltwater from the St. Lawrence back to the Mississippi. Again we back away from this primary forcing status and relegate the rediversion to trigger status. During the course of the Younger Dryas the Atlantic's salinity built to the point where turn-on of the conveyor was imminent. The reduction, by the rediversion of meltwater discharge directly into the region where deep water is produced, allowed the conveyor to restart.

The relegation of the diversion from primary to secondary status eliminates a problem which has long bothered me. Namely under the original hypothesis, if the Dansgaard-Oeschger events in the glacial portion of the Greenland ice core record were assumed to be akin to the Younger Dryas event then each would require a river diversion. Such a requirement seems unrealistic.

But what caused the two episodes of rapid sea level rise? Quite likely both are related to the extra strength attained by the Nordic Heat Pump following the reinitiations of the conveyor. Perhaps these blasts of heat led to catastrophic collapses of marine-based portions of the ice caps. What role did these Fair-luges play in the salt oscillator scenario? Clearly the fresh water released to the Atlantic as a result of these melting episodes contributed greatly to the diminution of the Atlantic's salt content. While the first of the Fair-luges contributed greatly to the reduction of salinity it did not shutdown the conveyor circulation. Another 1000 years passed before this happened. Why neither the second Fair-luge and the continuing millwater release which followed failed to cause a second Younger Dryas event remains an unresolved question.

By presenting only positive arguments here, I do not mean to leave the impression that no flaws are to be found in the salt oscillator argument. As pointed out by Eystein Jansen who reviewed this paper, the ocean sediment record is not entirely consistent with the ice core and European Lake records. For example, the oceanic warming associated with the onset of the Bölling warm appears to have taken place in two steps. The first of these at about 14,500 Libby years ago appears to correspond to the warming which caused the deglaciation of the major alpine valleys (Schlüchter 1988). The second step appears to correspond to warming which allowed trees to replace shrubs in these valleys. But, as Jansen asks, which step corresponds to the turn-on of the conveyor? My guess is that it was the second. But if so, what caused the first? I don't know. Jansen also points out that the carbon isotope and cadmium content records for benthic foraminifera obtained by Keigwin and Boyle shows Younger Dryas-like peaks within the Böling-Allerod warm. What event does this peak record? Again, I don't know. Clearly the marine sediment record must be examined more closely before any firm conclusions regarding the history of the Great Ocean Conveyor's past operation can be drawn. What I present here is a target at which my detractors can shoot.

ACKNOWLEDGEMENTS

This research was made possible by grants from the Exxon Corporation and the Lawrence Livermore National Laboratory.

REFERENCES

Birchfield, G. E., and W.S. Broecker, (1990) A salt oscillator in the glacial Atlantic? 2. A "Scale Analysis" Model, Paleoceanography, 5, 835-843.

Broecker, W.S., D. Peteet, D. Rind, (1985) Does the ocean-atmosphere system have more than one stable mode of operation? Nature, 315. 21-25.

Broecker, W.S., and G. Denton, (1989) The role of ocean-atmosphere reorganizations in glacial cycles, Geochim. Cosmochim. Acta, 53, 2465-2501.

Broecker, W.S., J. Kennett, S. Trumbore, G. Bonani, and W. Wolfli, (1989) The routing of Laurentide ice-sheet meltwater during the Younger Dryas cold event, Nature, 314, 319-321.

Broecker, W.S., (1990) Salinity history of the northern Atlantic during the last deglaciation, Paleoceanography, 5, 459-467.

Broecker, W.S., G. Bond, M. Klas, G. Bonani, and W. Wolfli, (1990) A salt oscillator in the glacial Atlantic? 1. The Concept, Paleoceanography, 5, 469-477.

Dansgaard, W., S.J. Johnsen, H.B. Clausen, and C.C. Langway, Jr., (1971) Climatic record revealed by the Camp Century ic core, in The Late Cenozoic Glacial Ages, edited by K. Turekian, New Haven and London, Yale University Press, pp. 38-56.

Dansgaard, W., J.W.C. White, and S.J. Johnsen, (1989) The abrupt termination of the Younger Dryas climate event, Nature, 339, 532-533.

Fairbanks, R. G., (1989) Glacio-eustatic sea level record 0-17,000 years before present: Influence of glacial melting rates on Younger Dryas "event" and deep ocean circulation, Nature, 3422, 637-642.

Hammer, C., H.B. Clausen, W. Dansgaard, A. Neftel, A. Kristindottir, and E. Johnson, (1985) Continuous impurity analysis along the Dye 3 deep core. In: Greenland Ice Core Geophysics, Geochemistry and Environment (eds C.C. Langway, H. Oeschger and W. Dansgaard): Amer. Geophys. Un. Mon., 33, pp. 90-94.

Lötter, A.F., (1991) Absolute dating of the late-glacial period in Switzerland using annually laminated sediments, Quaternary Research, 35, 321-330.

Maier-Reimer, E., and U. Mikolajewicz, (1989) Experiments with an OGCM on the cause of the Younger Dryas. In: A. Ayala-Castanares, W. Wooster, A. Yanez-Arancibia (eds) Oceanography 1988, UNAM Press, Mexico, pp. 87-100.

Manabe, S., and R.J. Stauffer, (1988) Two stable equilibria of a coupled ocean-coupled ocean-atmosphere model. J. Climate 1, 841-866.

Rind, D., D. Peteet, W. Broecker, A. McIntyree, and W. Ruddiman, (1986) The impact of cold North Atlantic sea surface temperatures on climate: Implications to the Younger Dryas cooling (11-10K) Climate Dynamics 1, 3-33.

Schlüchter, C., (1988) The deglaciation of the Swiss-Alps: A paleoclimatic event with chronological problems. Bull. de l'Assocation francaise pour l'etude du Quaternaire, 141-145.

$\delta^{18}O$ TIME-SLICE RECONSTRUCTION OF MELTWATER ANOMALIES AT TERMINATION I IN THE NORTH ATLANTIC BETWEEN 50 AND 80 °N

M. Sarnthein[1)], E. Jansen[2)], M. Arnold[3)], J.C. Duplessy[3)], H. Erlenkeuser[4)], A. Flatoy[2)], T. Veum[2)], E. Vogelsang[1)] and M.S. Weinelt[1)]

[1)]Geologisch-Paläontologisches Institut, Universität Kiel, Olshausenstr. 40, D-2300 Kiel, Germany
[2)]Geologisk Institutt avd.B, Universitetet i Bergen, Allégaten 41, N-5007 Bergen, Norway
[3)]Centre de Faibles Radioactivités, Laboratoire mixte CNRS-CEA F-91190 Gif sur Yvette, France
[4)]C14-Labor, Universität Kiel, Olshausenstr. 40, D-2300 Kiel, Germany

ABSTRACT

Based on about 55 time-calibrated planktonic $\delta^{18}O$ records, we reconstructed five time slices, deciphering massive changes in surface-water paleoceanography of the Norwegian-Greenland Sea and northeastern North Atlantic across glacial Termination I. The Last Glacial Maximum (LGM; about 15,000-18,000 YBP) showed a uniformly cold Norwegian-Greenland Sea, whereas the eastern North Atlantic margin was marked by either lowered salinity, implying meltwater, or slightly increased sea-surface temperatures, suggesting a weak incursion of warm water from the south. The first deglacial meltwater event culminated about 13,600 YBP. Clockwise and counter-clockwise flow patterns are discussed for explaining the distribution of meltwater and icebergs, emanating from the Barents ice shelf. Here a second meltwater event culminated near 12,300 YBP, then linked to a clearly counter-clockwise circulation and various eddies. This current pattern continued during the Younger Dryas, which thus markedly differed from the LGM, despite the general extreme cooling. The Holocene circulation and temperature regime started near 9,000 YBP, locally still linked to weak meltwater influence. Maximum warmth was reached only about 7,000 YBP. The two meltwater

NATO ASI Series, Vol. I 2
The Last Deglaciation: Absolute and Radiocarbon Chronologies
Edited by E. Bard and W. S. Broecker

incursions about 13,600 YBP and after the Younger Dryas were associated with warming and paralleled a marked short-term reduction in deep-water ventilation of the Atlantic, which, however, is not observed for the time prior to the Younger Dryas.

INTRODUCTION

The modern oceanography of the Norwegian-Greenland Sea has two major climatic implications. Firstly, it forms the nordic heat pump, providing an exceptionally warm climate for Europe. Moreover, it is the main location for deep-water renewal of the Atlantic and global ocean, which to a large extent controls the global atmospheric CO_2 budget.

In this paper we try to reconstruct the surface-water paleoceanography of the northern North Atlantic and Norwegian-Greenland Sea by means of the oxygen-isotope ($\delta^{18}O$) record obtained from the planktonic foraminifer Neogloboquadrina pachyderma (sinistral) (125-250 μm fraction). $\delta^{18}O$ oscillations record complex variations in both sea-surface temperatures (SST) and salinities (SSS), the latter being especially variable in high latitudes where the advection of freshwater from rivers and melting ice sheets are common features. Mapping of narrowly defined $\delta^{18}O$ time slices of past oceanic regimes may help to distinguish between the two factors SST and SSS, since several features of low or high $\delta^{18}O$ values can be traced to local and regional source points and hence identified as either temperature and/or salinity driven. Moreover, the time-slice presentation of data from narrowly defined intervals enables us to exclude the problems with separating the ice volume effect in the $\delta^{18}O$ data, which has changed by about 1.1-1.3 $^{o}/_{oo}$ across glacial Termination I (Labeyrie et al., 1987; Chappell & Shackleton 1987). A more detailed history of the ice-effect variation was deduced from the coral reef record by Fairbanks (1989).

Our reconstructions are based on isotopic data from more than 55 deep-sea cores. Sedimentological studies, carried out on most of our cores, show that the sedimentary sequences are largely continuous (e.g., Bischof et al., 1990; Henrich et al., 1989; Kassens, 1990). Isotopic data were measured by standard techniques (Duplessy, 1978; Jansen et al., 1989; Vogelsang, 1990) in the isotope laboratories of Bergen, Gif-sur-Yvette, and Kiel (Table 1) and will be documented in detail in a separate paper (Sarnthein et al., in prep.). Out of the 55 $\delta^{18}O$ records, about 30-55 records each were used for reconstructing the various time slices. If any time interval (Fig. 1) was not identified with certainty in a particular core, we omitted it from the time-slice compilation. Most records are based on closely sampled core sections, more than 25 cores have high sedimentation rates exceeding 5-8 cm/1000 years.

AGE CONTROL

Stratigraphic correlations were based on well defined events in $\delta^{18}O$ records that were AMS-^{14}C dated from *N. pachyderma* (s) tests in seven core sections (Table 1; Fig. 1; locations in Fig. 2, LGM). When interpolating ages of Termination I events between the datings (by the assumption of linear sedimentation rates), we first applied the U/Th ages of Bard et al. (1990) for converting the radiocarbon years of glacial Termination I into calendar years. This conversion was necessary because the datings of Bard et al. (op. cit.), Becker et al. (1991) and Lotter (1991) showed a telescopic and, in part, *non-linear* relationship between the two time scales, including "^{14}C plateaus"; thus the use of radiocarbon years results in biased estimates of sediment ages and sedimentation rates. The linear interpolations, using calendar years, were then reconverted into ^{14}C years, so that they are comparable to other studies, with ages given in ^{14}C years BP.

The five time slices presented in this paper were defined by the following features in the $\delta^{18}O$ records:

The time slice "Last Glacial Maximum (LGM)" does not rely on single $\delta^{18}O$ values but averages the $\delta^{18}O$ maximum range found between 14,800/15,200 and 17,000/18,500 YBP (= ^{13}C corrected radiocarbon years before present minus the 400 years ocean-reservoir effect; age intervals marked as bar on the AMS-^{14}C dated $\delta^{18}O$ records in Fig. 1). The LGM time slice ends at the main inflection point in the $\delta^{18}O$ curves marking the first strong $\delta^{18}O$ depletion defined as the onset of Termination IA.

The time slice "13,600 YBP" shows the subsequent first great meltwater event in the Norwegian Greenland Sea (interpreted sensu Jones and Keigwin, 1988) and northern North Atlantic (approximately matching the center of "Phase II" of Ruddiman and McIntyre, 1981). In most cases, the event is displayed by a marked isolated $\delta^{18}O$ minimum spike, particularly pronounced along the northeastern continental margin and less marked, but recognizable in the Greenland Sea. A smaller precursor spike occurs, where sedimentation rates are high. In some cases, the event is reduced to a shoulder in the $\delta^{18}O$ decrease of Termination IA, for example, at cores V 23-81 west of Ireland and V 28-14 west of Iceland (Fig. 1). The maximum meltwater signal was directly dated to 13,600$\pm$300, 14,010$\pm$290, and 13,660$\pm$210 YBP in the high-resolution cores M 23071, HM 52-43, and V 23-81, where sedimentation rates exceed 7-10 cm/1000 years. The age was interpolated to 13,560 and 13,660 YBP in the high-sedimentation rate cores M 23259 and CH 73-139c. The average age of about 13,600 YBP contrasts with a higher age of 14,480 YBP in core PS 21295 (Jones and Keigwin, 1989), an estimate, however, that is based on low sedimentation rates of less than 2 cm/1000 years and low $\delta^{18}O$ sampling resolution (Fig. 1). We thus prefer the younger age dated in high sedimentation and sampling rate cores.

The time slice "12,300 YBP" depicts a second major meltwater event, just prior to the first incursion of warm subpolar faunas into the Norwegian Greenland Sea. This second $\delta^{18}O$ minimum spike is distinguished from the preceding $\delta^{18}O$ minimum by a pronounced $\delta^{18}O$ maximum near 13,100-12,500 YBP in most cases (Late Bölling; Andersen, 1979; see cores 23071 and 23259 in Fig. 1). The spike has an interpolated age of 12,500 YBP in core HM 52-43, 12,300 YBP in cores M 23259, V 28-14, and V 23-81 (well dated $\delta^{18}O$ minimum of *N. pachyderma* dextral), and of 12,010-12,400 YBP in core M 23071, all cores having sedimentation rates of 5-16 cm/1000 years across this interval (Fig. 1). The two meltwater events at about 13,600 and 12,300 YBP are consistent with the two-spike signal reported by Keigwin et al. (Erice NATO conference; this vol.). A subsequent, generally strong $\delta^{18}O$ enrichment was dated in our cores (Fig. 1) to about 11,900 YBP.

The time slice "10,600 YBP" represents the regime of the Younger Dryas, i.e., the last $\delta^{18}O$ increase prior to the $\delta^{18}O$ decrease of Termination IB (Duplessy et al., 1981). This should not be mixed up with the earlier, more pronounced maximum at 11,900 YBP (Fig. 1). In many cores from the southern Norwegian Greenland Sea and the northern North Atlantic this level was identified by means of the Vedde Ash layer (Mangerud et al. 1984). At the Vöring Plateau and off the Barents Shelf, it was defined only by means of the oxygen isotope record.

The time slice "9000 YBP" is well defined by the first $\delta^{18}O$- minimum spike or shoulder at or slightly below the beginning of the Holocene $\delta^{18}O$ plateau (Fig. 1). The event represents both the last, mostly minor meltwater influence felt in the Norwegian Greenland Sea and a major warming of SST. Based on a slight subsequent decrease in the $\delta^{18}O$ values, the temperatures, however, reached their maximum at many sites only as late as 7000 YBP. The age of this event was directly dated and/or closely interpolated to 8500-9300 YBP in all cores plotted in Fig. 1, except for core CH 73-139c, where the $\delta^{18}O$ signal of *N. pachyderma* (sinistral) is shifted slightly upcore due to bioturbation.

RESULTS AND DISCUSSION (Fig. 2)

At the LGM, the sea region under discussion shows a relatively zonal $\delta^{18}O$ pattern in the south, as compared with the present meridional regime: The North Atlantic south of the Iceland Faroer Ridge displays values of 4.0-4.1 $^o/_{oo}$, whereas the Norwegian Greenland Sea shows values of 4.45-4.96 $^o/_{oo}$, the maxima of >4.7 $^o/_{oo}$ being centered both in the southern central basin and in a belt extending N-S off the eastern margin. The strong north-south gradient across the Island Faroer Ridge differs from the more meridional oceanographic pattern found in this region today. It may correspond to a north-south SST increase by 2-3°C or a salinity decrease by 0.7-1.4 $^o/_{oo}$.

In the Norwegian Greenland Sea lower $\delta^{18}O$ values (4.3-4.5 $^o/_{oo}$) are linked to a narrow band along the eastern margin up to its southeastern-most corner, and to the area west of Jan Mayen. Based on the complete absence of subpolar planktonic foraminifers tracing warmer SST (Kellogg, 1980; Jansen & Erlenkeuser, 1985), these local $\delta^{18}O$ values probably indicate lowered salinity due to modest meltwater advection, spreading along the ice fronts off the Barents shelf, Norway, and west of Jan Mayen.

In the northern North Atlantic, the $\delta^{18}O$ distribution pattern of the LGM (Fig. 2) largely matches the pattern presented by Keigwin & Boyle (1989). The composite $\delta^{18}O$ pattern of Fig. 2 and the data produced by Keigwin & Boyle reveals a large-scale east-west increase in $\delta^{18}O$ values from 4.0 to 4.5 $^o/_{oo}$ up to 50°W. According to Ruddiman & McIntyre (1981) and Bard et al. (1987) subpolar foraminifers then were extremely rare (less than 10-20%) in the sediments west of Ireland, indicating low August temperatures near 2-6°C, which increase rather than decrease from east to west (McIntyre et al., 1976). In contrast, Pujol (1980) depicted a narrow belt of increased August SST immediately west of Ireland. Hence it cannot be clearly decided now whether the $\delta^{18}O$ decrease towards Europe may reflect a temperature increase (implying a weak south-north current) or a salinity reduction in the glacial surface water, such as further north along the Norwegian Sea margin, possibly implying a north-south current.

At 13.600 YBP, the major feature of the $\delta^{18}O$ time slice is a marked minimum (1.6 $^o/_{oo}$ $\delta^{18}O$) west of the Barents ice sheet, at 72°N and 9-11.5°E (arrow in Fig. 2). Based on an almost complete absence of subpolar foraminifers as tracers of warmer SST in the Norwegian-Greenland Sea, the extreme local $\delta^{18}O$ deviation by more than 2 $^o/_{oo}$ is interpreted as a meltwater-induced salinity reduction by more than 2.0 ‰ (Duplessy, 1982; glacial regression coefficient of 1.04 by Vogelsang, 1990). The salinity minimum is confined by a steep gradient to the west and north and a weak gradient toward the south, where the low-salinity surface water is found, at least, as far south as 65°N.

The low-salinity water probably originated from the melting of icebergs. This may also account - at least in part - for the offshore $\delta^{18}O$ gradient seen off the Barents shelf and in the eastern Norwegian Sea, indicating that iceberg melting took place after the icebergs had drifted away from the shelf ice margin.

One may interpret the surface ocean circulation pattern during this time, when normal thermohaline circulation patterns were reduced, in two different ways. On one hand, one can interpret the $\delta^{18}O$ distribution pattern as a counterclockwise circulation analogous with the modern counterclockwise flow patterns. The main meltwater flux is noted in the north, but a significant meltwater influence is read from the data along the whole eastern boundary of the area. The very strong local anomalies may indicate, if they are not produced by artifacts of data resolution, melting of icebergs confined to eddy-like features. The meltwater is rapidly reduced and diluted in north- and westward directions. The heavy and least meltwater-influenced values are observed in the Iceland and central Norwegian Sea, possibly indicating a situation with a subtle counterclockwise gyre advecting some icebergs and meltwater westward toward Jan Mayen and a meltwater-devoid current passing eastward north of Iceland.

On the other hand, the $\delta^{18}O$ distribution pattern can be interpreted as representing a (wind-driven) clockwise/anti-cyclonic circulation system, opposed to the modern system. Such a model is in harmony with the lack of subpolar foraminifers even in the southeasternmost part of the Norwegian Sea (Jansen & Erlenkeuser, 1985). These faunas could be expected because a cyclonic gyre would bring warmer water from the Atlantic into this region. In the case of the anti-cyclonic model, the main flow of icebergs and meltwater originated at the Barents ice shelf such as in the other model. Most icebergs, however, went longshore south with an ongoing lateral addition of icebergs from Northern Norway. Distribution patterns of ice-rafted debris, the evidence of which may be still controversial, also suggest a clockwise circulation of icebergs west of Norway during that time (Bischof et al., 1990). Taking account for the general longevity of icebergs (about 30 years; Harvey, 1989), their transport and that of meltwater can be regarded as passing far south from its source on several tracks, possibly across the Island Faroer Ridge south to west of Ireland and Portugal. Indeed, also in this area Bard et al. (1987b) found a $\delta^{18}O$ minimum 13,580 YBP, paired with an extreme low in foraminifer-based paleotemperatures.

The heavy $\delta^{18}O$ values N and NE of Iceland, slightly increasing from east to west, are least influenced by meltwater and interpreted by the second model as a region that is most remote from the tracks of meltwater dilution by north-south drifting icebergs. Possibly this area received a weak surface-water advection from west of Iceland, since also the patches of weak meltwater dilution near Jan Mayen and east Greenland do not advance into this region. Finally, the small-scale but striking offshore $\delta^{18}O$ gradient off the Barents-Norwegian margin is interpreted as a signal of coastal upwelling of isotopically heavier subsurface water, in addition to the outlined effects of offshore directed iceberg drift. This upwelling would be characteristic of the clockwise north-south circulation, inferred by this model.

Both models explain the somewhat lower $\delta^{18}O$ values in the Fram Strait as signs of meltwater probably flowing out of the Arctic. Significant meltwater must have been routed this way from the ice sheets in the northern Barents, Kara and East Siberian Seas and the northern sector of the Laurentide ice sheet.

At 12,300 YBP, again meltwater advection was pronounced in the Norwegian Greenland Sea. Reduced $\delta^{18}O$ values representing meltwater concentrated to the cold polar water west of the Barents Sea and east of Greenland. Within these, northeast to southwest running, wedge shaped areas with $\delta^{18}O$ differences of 1.5 $^{o}/_{oo}$, corresponded to more than 1.5-2.0 $^{o}/_{oo}$ local salinity reduction. By contrast, the tongue of lowered $\delta^{18}O$ values in the southeastern Norwegian Sea match the first advection of subpolar foraminifers, hence indicating the postglacial start of warm-water incursion from the eastern North Atlantic, as also deduced from sediments in the northern North Atlantic by Ruddiman & McIntyre (1981; Phase IIIa). Heavy values, i.e., the least warming and meltwater influence occured north and northeast of Iceland.

In contrast to 13,600 YBP, the surface circulation in the Norwegian Greenland Sea now followed more clearly a counterclockwise gyre as today, with a noticeable Greenland Current. In contrast to today, the $\delta^{18}O$ meltwater signal (and the underlying iceberg tracks) west off the Barents Shelf depicted small-scale eddy-like features that may have blocked a further northward penetration of the Norwegian Current. A rather large meltwater lense area was located southeast of Jan Mayen. Subsequent to the 12,300 YBP event, at 11,900 YBP (not depicted in Fig. 2), the meltwater influence disappeared (Fig. 1), but signs of the meridional warm water advection continued in the southeastern-most corner of the Norwegian Sea (Jansen and Erlenkeuser, 1985).

At 10,600 YBP, we note a zonal $\delta^{18}O$ gradient (0.55-0.85 $^{o}/_{oo}$), separating the central Norwegian Sea from the northeastern North Atlantic, equal to a temperature gradient of 2-3°C near the Norwegian margin and increasing to 1.3 $^{o}/_{oo}$ toward Iceland. Two areas of maximum values lay northeast of Iceland and in the central Norwegian Greenland Sea, each possibly depicting a cyclonic gyre. Except for a weak meltwater anomaly off the Barents Sea, only little meltwater signals occured. Hence, the time slice of the Younger Dryas, 10,600 years ago, strongly differed from those at 13,600 and 12,300 YBP, where meltwater was a dominant feature.

Ruddiman & McIntyre (1981; Phase IIIb) found that the Younger Dryas was linked to a general marked cooling episode all over the northeastern Atlantic, resembling almost the regime of the LGM. The $\delta^{18}O$ distribution pattern of the Younger Dryas, however, differed from the LGM in the eastern portion of the Norwegian Greenland Sea (anomaly map in Fig. 2). Main characteristics of the Younger Dryas were two branches of slightly, up to 1.5 $^{0}/_{00}$ ^{18}O depleted surface water near the Norwegian margin reaching from the south up to the Barents Sea and up to south of Jan Mayen. A further light $\delta^{18}O$ anomaly lay west of Iceland. It is unclear whether the anomaly also resulted from enhanced warm water advection, such as the present Irminger Current, or from a local meltwater source in South Greenland. Ruddiman & McIntyre (1973) postulated a counterclockwise ice-bearing subarctic gyre in the North Atlantic, starting approximately from this position west of Iceland during that time.

At 9000 YBP, the oceanographic regime was fairly similar to today. Low $\delta^{18}O$ values representing a strong advection of warm water marked the southeast and east of the Norwegian Sea up to 75°N, where, however, slight meltwater anomalies probably still were present. A branch of lighter $\delta^{18}O$ values, i.e. of warmer and/or less saline water was directed toward Jan Mayen. On the other hand, heavy $\delta^{18}O$ values of the cold Iceland Current pointed toward the Faroer Islands similar to today.

The surface waters of the Norwegian and Iceland Sea were colder at 9000 YBP than at 7000 YBP (compare Fig. 1; not depicted in Fig. 2), when the peak Holocene warming penetrated further north than today and the last $\delta^{18}O$ meltwater anomalies west of the Barents Sea fully disappeared.

CONCLUSIONS and IMPLICATIONS

In this paper we tried to interpret various $\delta^{18}O$ time slices from Termination I as past ocean circulation patterns. The glacial temperature distribution in the northern North Atlantic and Norwegian Greenland Sea was characterized by a marked zonal pattern and south-north temperature gradient across the Island-Faroer Ridge. The onset of melting and the first meltwater fluxes in the Norwegian Greenland Sea occured after 14,800 YBP. The influence of meltwater, particularly that originating from the Barents ice shelf, culminated at 13,600 YBP. A second main phase of melting that influenced both the eastern and western margins with meltwater, culminated at 12,300 YBP. Alternative clockwise and counterclockwise circulation patterns were proposed to explain the meltwater distribution pattern during the first event. A possible clockwise flow of icebergs and meltwater would imply an ongoing strong climatic cooling of Northwest Europe, opposing the rising summer insolation during Termination IA, and an enhanced Irminger Current west of Iceland. Later, after about 13,000-12,500 YBP, cyclonic meridional currents can be clearly identified.

The Younger Dryas showed only minor meltwater anomalies, but clear signs of increasingly meridional current patterns, i.e, ongoing warm-water incursion in the southeast. The same holds true for 9000 YBP, where the distribution of warm surface water was similar to today.

The meltwater events culminating at 13,600 YBP and after the Younger Dryas, at about 9600 YBP, are directly correlated with benthic $\delta^{13}C$ minima in the North Atlantic Deep Water, lasting for a few 100 to 1000 years (cores EN 120GGC1; Boyle & Keigwin, 1987; Fairbanks, 1989; V23-81; Jansen and Veum, 1990; ODP Site 658; Sarnthein & Tiedemann, 1990; M12347; Winn, unpubl. data). No ventilation minimum, however, parallels the meltwater spike near 12,300 YBP. Hence, the two breakdowns in the North Atlantic "salinity conveyor belt" (Birchfield & Broecker, 1990; Broecker, Erice conference) parallel the actual phases of maximum ice decay between 14,800 and 13,600 YBP and about 10,500 and 9600 YBP, respectively. These are linked to warming phases and not to cooling, which may instead occur subsequent to the meltwater anomalies, such as after the meltwater spike 13,600 YBP. This finding does not fit the salt oscillator theory, since peak warming and peak meltwater forcing coincide with a turned-off conveyor belt. The Younger Dryas cooling does not follow, but preceed a meltwater event and the reduction in deep-water ventilation during Termination IB. Accordingly, it turns out that different meltwater events produced from different ice margins had a different impact on oceanic deep-water circulation, probably related to the proximity of the meltwater flow to areas crucial for deep-water formation.

ACKNOWLEDGMENTS

We thank Ed Bard for stimulations and patience with this paper. Our studies were supported by the BMFT (07KF021) and DFG in Bonn (Sonderforschungsbereich SFB 313 contribution no. 143), by the Norwegian Research Council for Science and the Humanities (NAVF) (POC contribution no 45), and by CNRS, INSU, and CEA, Paris.

Table 1. Core locations and data sources used for time slice reconstructions in Fig. 2.

Core	Latitude	Longitude	Water Depth	Isotope Laboratory/Reference
CH 73-110	59°30'N	08°56'W	1366m	Labeyrie & Duplessy (1985)
CH 73-139C	54°38'N	16°21'W	2209m	Bard (1987)
AMS - ^{14}C ages				CNRS, Gif sur Yvette
CH 77-07	66°36'N	10°31'W	1487m	Duplessy et al. (1988)
FRAM I/4	84°30'N	08°59'W	3820m	Zahn et al. (1985)
FRAM I/7	83°53'N	06°57'W	2990m	Zahn et al. (1985)
HM 25-09	63°03'N	04°47'E	600m	Jansen & Erlenkeuser (1985)
HM 28-18	62°56'N	02°44'E	770m	Bergen University
HM 31-33	63°38'N	01°46'E	1580m	Jansen & Erlenkeuser (1985)
HM 31-36	64°15'N	00°31'E	2620m	Jansen & Erlenkeuser (1985)
HM 52-43	64°31'N	00°44'E	2781m	Bergen University
AMS - ^{14}C AGES				CNRS, Gif sur Yvette
HM 71-12	68°26'N	13°52'W	1547m	Bergen University
HM 71-14	69°50'N	18°05'W	1624m	Bergen University
HM 71-19	69°29'N	09°31'W	2210m	Bergen University
K 11	71°47'N	01°36'W	2900m	Duplessy et al. (1988)
K 708-1	50°00'N	23°45'W	4053m	Ruddiman & McIntyre (1981)
KN 714-15	58°46'N	25°57'N	2598m	Keigwin & Jones (1989)
M 17045	52°26'N	16°39'W	3663m	Winn (unpubl.)
M 17048	54°18'N	18°10'W	1859m	Winn (unpubl.)
M 17049	55°16'N	26°44'W	3331m	Jung (unpubl.)
M 17724	76°00'N	08°20'E	2354m	Weinelt et al. (1991)
M 17728	76°00'N	03°57'E	2485m	Kiel University
M 17730	72°03'N	07°19'E	2769m	Kiel University
M 17732	71°37'N	04°13'E	3103m	Weinelt et al. (1991)
M 23041	68°41'N	00°14'E	2258m	Vogelsang (1990)
M 23043	70°16'N	03°21'W	2133m	Vogelsang (1990)
M 23055	68°25'N	04°01'E	2311m	Vogelsang (1990)
M 23059	70°18'N	03°07'W	2283m	Vogelsang (1990)
M 23062	68°44'N	00°10'E	2244m	Vogelsang (1990)
M 23063	68°45'N	00°00'E	2299m	Vogelsang (1990)
M 23064	68°40'N	00°19'E	2571m	Vogelsang (1990)
M 23065	68°30'N	00°49'E	2804m	Vogelsang (1990)
M 23068	67°50'N	01°30'E	2230m	Vogelsang (1990)
M 23071	67°05'N	02°55'E	1308m	Vogelsang (1990)
AMS - ^{14}C ages				CNRS, Gif sur Yvette
M 23074	66°40'N	04°55'E	1157m	Vogelsang (1990)
M 23256	73°10'N	10°57'E	2061m	Kiel University
M 23258	75°00'N	13°58'E	1781m	Weinelt et al. (1991)
M 23259	72°02'N	09°16'E	2518m	Weinelt et al. (1991)
AMS - ^{14}C ages				CNRS, Gif sur Yvette
M 23260	72°09'N	11°27'E	2084m	Weinelt et al. (1991)
M 23261	72°11'N	13°58'E	1625m	Weinelt et al. (1991)
M 23262	72°11'N	14°26'E	1131m	Weinelt et al. (1991)
M 23323	67°46'N	05°56'E	1286m	Blaume (unpubl.)
M 23331	66°52'N	07°50'E	834m	Blaume (unpubl.)
M 23353	70°34'N	12°44'W	1394m	Piotrowski (1990)
MG 123	79°16'N	00°48'E	3050m	Morris (1988)

continued next page

Core	Latitude	Longitude	Water Depth	Isotope Laboratory/Reference
PS 21295	78°00'N	02°25'W	3112m	Jones & Keigwin (1988)
AMS - ^{14}C ages				Jones & Keigwin (1988)
PS 21906	76°30'N	02°05'W	2939m	Hamich (1991)
PS 23199	68°23'N	05°14'E	1968m	Vogelsang (1990)
PS 23205	67°37'N	05°46'E	1411m	Vogelsang (1990)
PS 23243	69°23'N	06°33'W	2715m	Vogelsang (1990)
PO 16343	66°56'N	07°28'E	1043m	Blaume (1990)
PO 16396	61°52'N	11°15'W	1145m	Dorn, Werner, Winn (unpubl.)
V 23-42	62°11'N	27°56'W	1514m	Keigwin & Boyle (1989)
V 23-81	54°02'N	16°08'W	2393m	Jansen & Veum (1989)
AMS - ^{14}C ages				CNRS, Gif sur Yvette
V 28-14	64°47'N	29°34'W	1855m	Lehman et al. (1991)
AMS - ^{14}C ages				Lehman et al. (1991)
V 28-38	69°23'N	04°24'W	3411m	Duplessy et al. (1988)

FIGURE CAPTIONS

Fig. 1. AMS-^{14}C dated oxygen isotope records from *Neogloboquadrina pachyderma sin. in the Norwegian Greenland Sea and northern North Atlantic. Levels of* ^{14}C datings are indicated by age numbers. Bar and arrows mark isotopic time slices at about 18,500-14,800, 13,600, 12,300, 10,600, and 9000 YBP, used for time slice reconstructions in Fig. 2. Core locations and data sources are listed in Table 1 and depicted in Fig. 2 (LGM).

Fig. 2. Distribution patterns of $\delta^{18}O$ in *N. pachyderma* sin. (small numbers) in the Norwegian Greenland Sea and the northern North Atlantic, compiled for five time slices at about 14,800/15,200 - 17,000-18,500 YBP (LGM), 13,600, 12,300, 10,600 (Younger Dryas), and $\delta^{18}O$ anomaly map for LGM minus 10,600 YBP. Stars in LGM map mark locations of sediment cores with AMS-^{14}C datings. Arrow in map for 13,600 YBP marks center of presumed meltwater source.

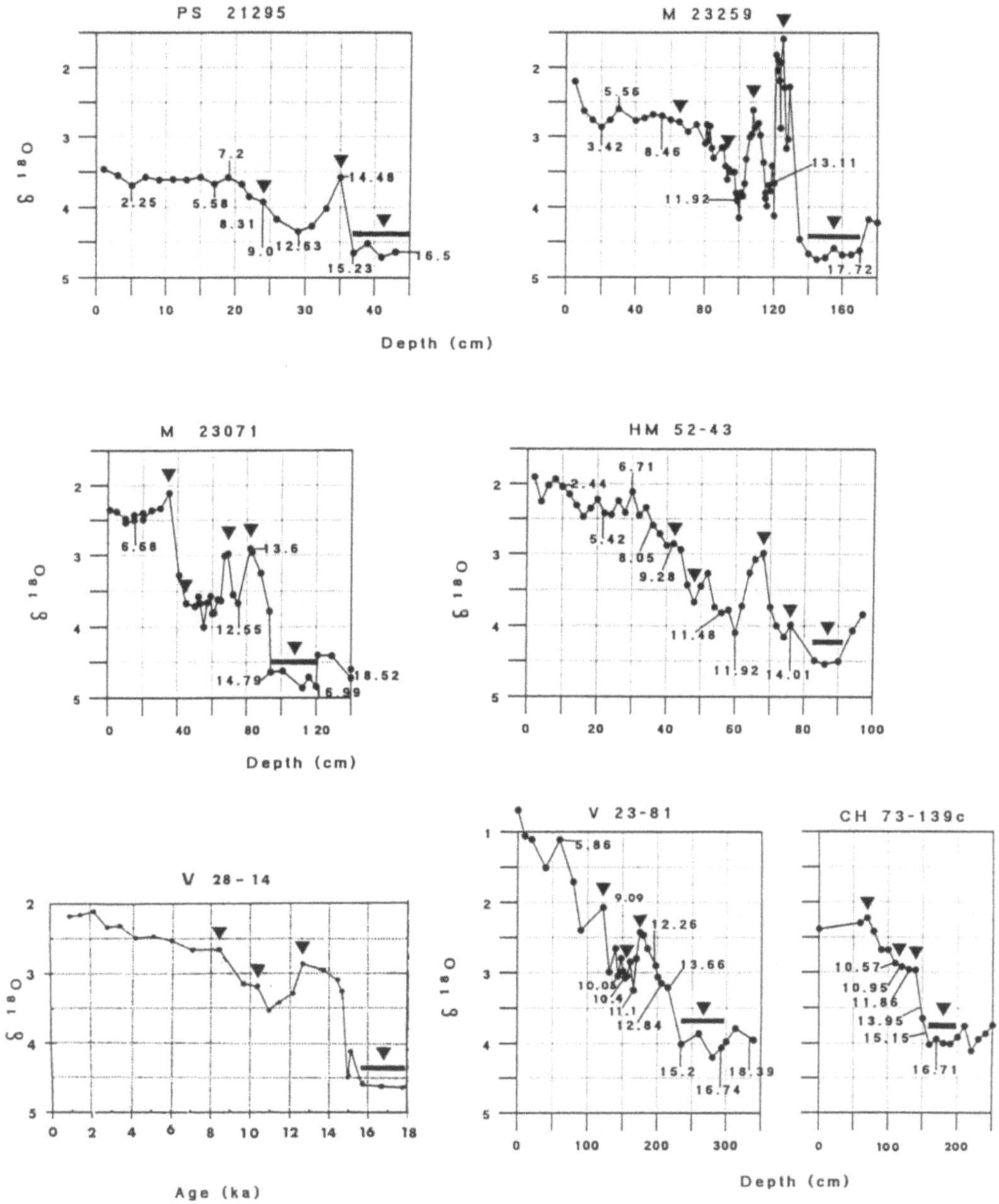
PS 21295
M 23259
δ 18O
2.25
5.58
7.2
8.31
9.0
12.63
14.48
15.23
16.5
3.42
5.56
8.46
11.92
13.11
17.72
Depth (cm)
M 23071
HM 52-43
6.68
12.55
13.6
14.79
16.99
18.52
2.44
5.42
6.71
8.05
9.28
11.48
11.92
14.01
Depth (cm)
V 28-14
Age (ka)
V 23-81
CH 73-139c
5.86
9.09
10.05
10.4
11.1
12.26
12.84
13.66
15.2
16.74
18.39
10.57
10.95
11.86
13.95
15.15
16.71
Depth (cm)

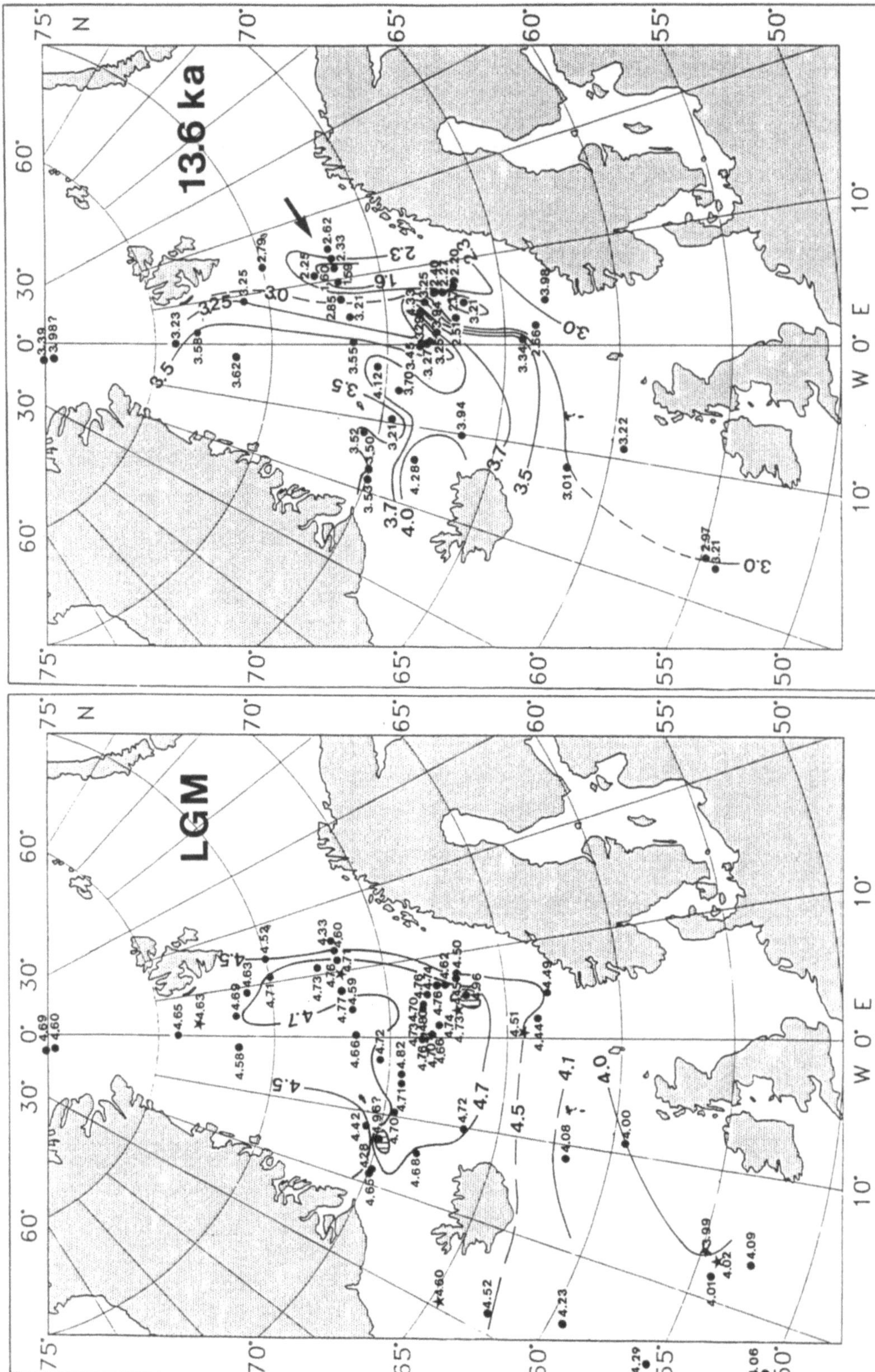
13.6 ka
LGM

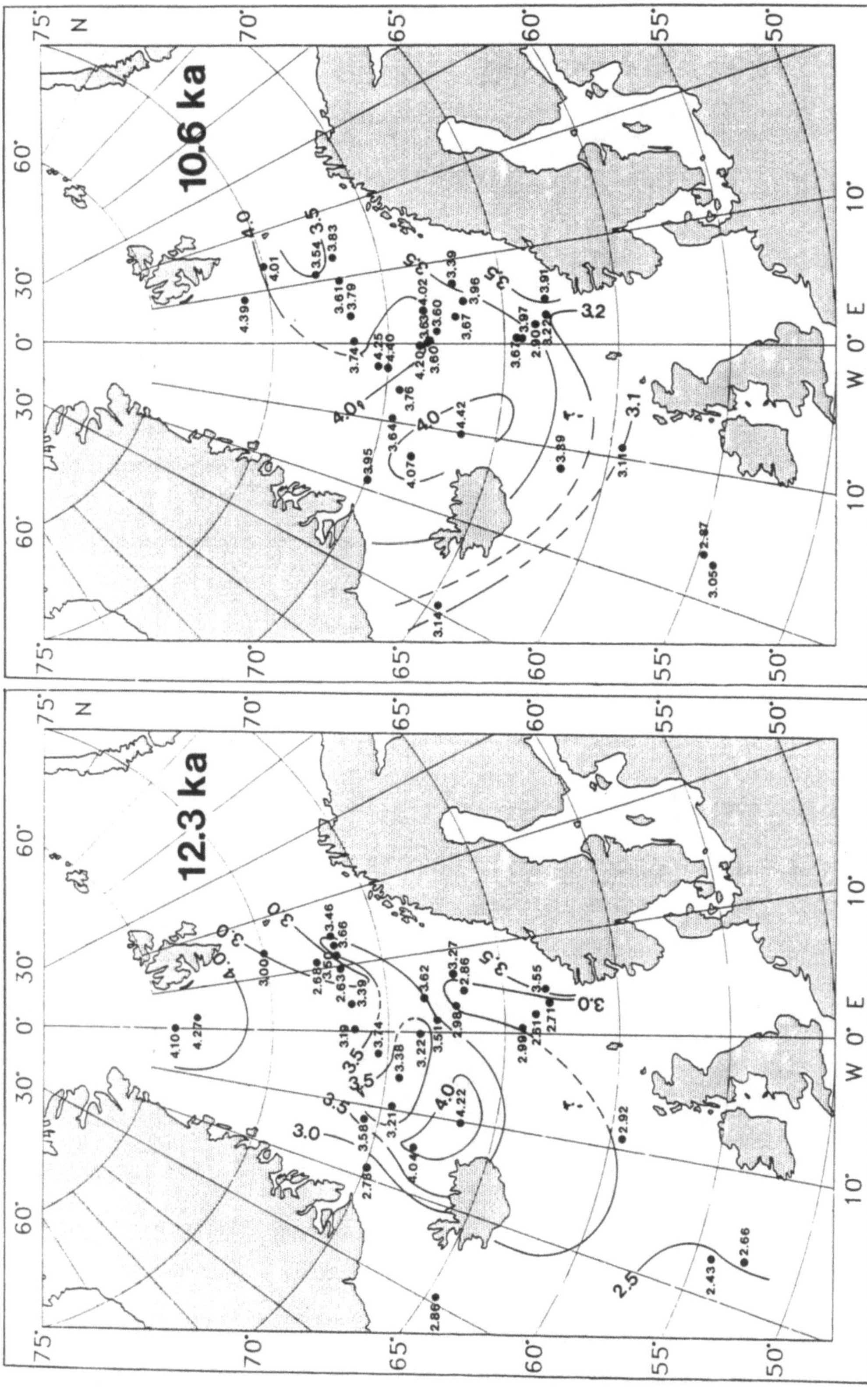
10.6 ka
12.3 ka

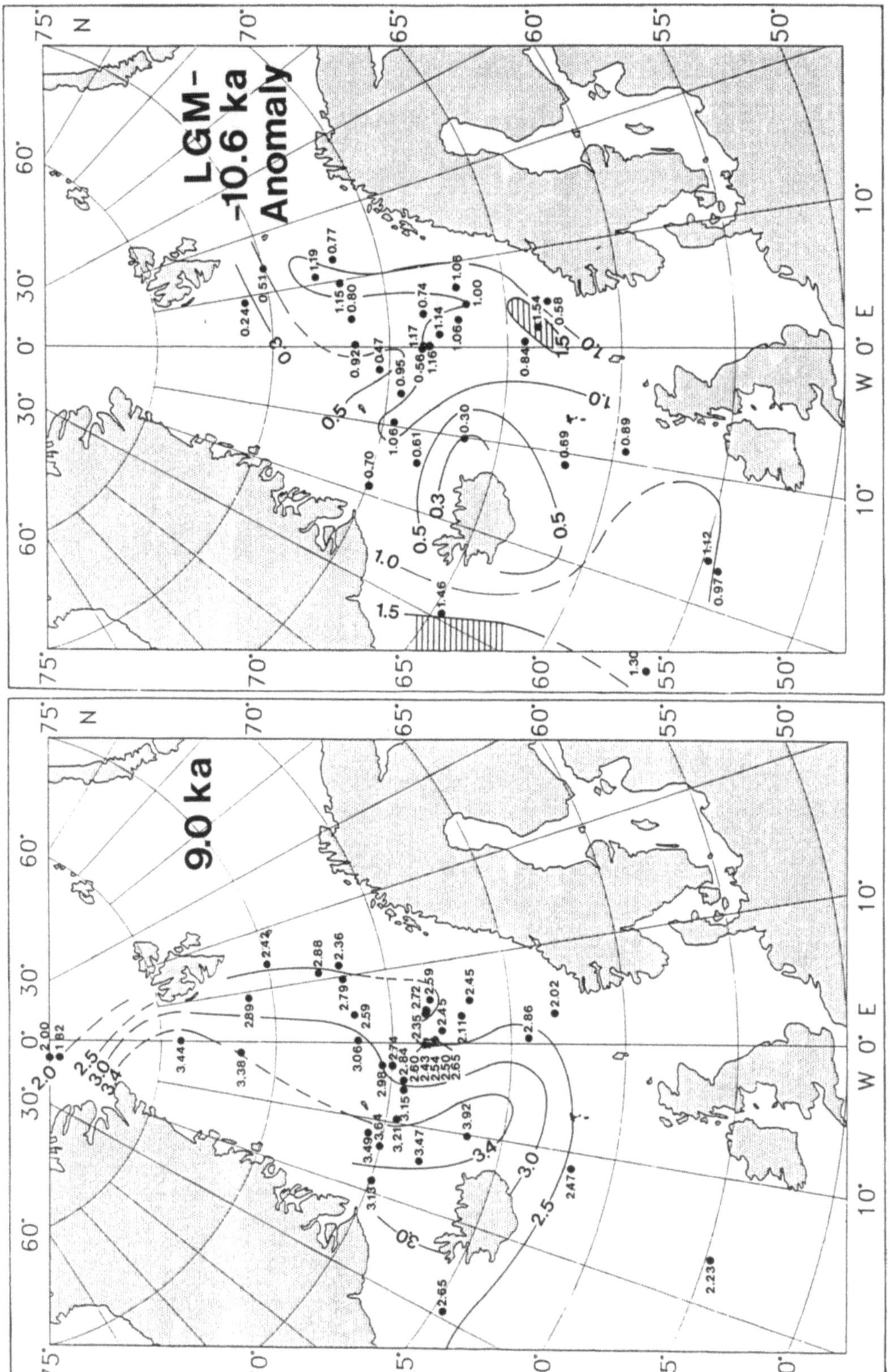
LGM - 10.6 ka Anomaly
9.0 ka

REFERENCES

Andersen, B.G., 1979: The deglaciation of Norway 15,000-10,000 B.P., *Boreas* 8, 79-87.

Bard, E., Arnold, M., Duprat, J., Moyes, J., and Duplessy, J.-C., 1987a: Reconstruction of the last deglaciation: deconvolved records of $\delta^{18}O$ profiles, micropaleontological variations and accelerator mass spectrometric ^{14}C dating, *Climate Dynamics* 1, 101-112.

Bard, E., Arnold, M., Maurice, P., Duprat, J., Moyes, J. & J.-C. Duplessy, 1987b: Retreat velocity of the North Atlantic polar front during the last deglaciation determined by ^{14}C accelerator mass spectrometry. *Nature* 328, 791-794.

Bard, E., Hamelin, B., Fairbanks, R.G., and Zindler, A., 1990: Calibration of the ^{14}C timescale over the past 30,000 years using mass spectrometric U-Th ages from Barbados corals, *Nature* 345, 405-410.

Becker, B., Kromer, B. & Trimborn, P., 1991: A stable isotop tree-ring timescale of the late Glacial/ Holocene boundary, *Nature* (in press).

Birchfield, G.E. and Broecker, W.S., 1990: A salt oscillator in the glacial Atlantic?, 2, A "scale analysis" model, *Paleoceanogr.* 5 (6), 835-845.

Bischof, J., 1990: Dropstones in the Norwegian Greenland Sea - Indications of a late Quaternary Circulation Pattern?, *in:* Bleil, U. & Thiede, J. (eds.). Geologic History of the Polar Oceans: Arctic versus Antarctic, NATO ASI Series C, Kluwer, Doordrecht, 499-518.

Blaume, F., 1990: Postglaziale Hochakkumulationszentren am Kontinentalhang der Norwegischen See: Vergleich Barentsseehang und Gebiet 67° Nord, *SFB-313, Berichtsband* 88-90, 242-248.

Boyle, E.A., and Keigwin, L., 1987: North Atlantic thermohaline circulation during the last 20,000 years: Link to the high latitude surface temperature, *Nature* 330, 35-40.

Chapell, J., and Shackleton, N.J., 1986: Oxygen isotopes and sealevel, *Nature* 324, 137-140.

Dansgaard, W. and Oeschger, H., 1989: Past environmental long-term records from the Arctic. *In: "The Environmental Record in Glaciers and Ice Sheets", (eds. H. Oeschger & C.C. Langway, Jr.) 287-318 (Dahlem Workshop Reports) Wiley-Interscience. Duplessy, J.-C., 1982: Glacial to interglacial contrasts in the northern Indian Ocean, Nature,* 295, 494-498.

Duplessy, J.-C., 1978: s. Isotope Studies. *In: Climatic change,* (J. Gribbin, ed.), Cambridge, Univ. Press, 46-67.

Duplessy, J.-C., Delibrias, G., Turon, J.L., Pujol, C. & Duprat, J., 1981: Deglacial warming of the northeastern Atlantic Ocean. Correlation with the paleoclimatic evolution of the European continent. *Paleo. Paleo. Paleo.* 35, 121-144.

Duplessy, J.-C., Labeyrie, L., and Blanc, P.L., 1988: Norwegian Sea deep water variations over the last climatic cycle: paleo-oceanographical implications, *In:* H. Wanner and U. Siegenthaler et al. (edits), *Long and Short Term Variability of Climate,* Lecture Notes in Earth Sciences, Vol. 16, 83-116, (Springer Verlag).

Fairbanks, R.G., 1989: A 17,000-year glacio-eustatic sea level record: influence of glacial melting rates on the Younger Dryas event and deep-ocean circulation, *Nature,* 342, 637-642.

Harvey, L.D.D., 1989: Modelling the Younger Dryas, *Quater. Sci. Rev.* 8, 137-149.

Henrich, R., Kassens, H., Vogelsang, E., Thiede, J., 1989: Sedimentary facies of glacial/interglacial cycles in the Norwegian Sea during the last 350 ky. *Mar. Geol.* 86, 283-319.

Jansen, E., and Erlenkeuser, H., 1985: Ocean circulation in the Norwegian Sea during the last deglaciation: isotopic evidence, *Paleo. Paleo. Paleo.* 49 (3/4), 189-206.

Jansen, E., Slettemark, B., Bleil, U., Henrich, R., Kringstad, L., and Roysen, S., 1989: Oxygen and carbon isotope stratigraphy and magnetostratigraphy of the last 2.8 Ma: Paleoclimatic comparisons between the Norwegian Sea and the North Atlantic, *Proc. ODP Scientific Results* 104, 255-269.

Jansen, E., and Veum, T., 1990: Two-step deglaciation: timing and impact on North Atlantic deep water circulation, *Nature* 343, 612-616.

Jones, G.A,. and Keigwin, L.D., 1988: Evidence from Fram Strait (78°N) for early deglaciation, *Nature* 336 (6194), 56-59.

Kassens, H., 1990: Verfestigte Sedimentlagen und seismische Reflektoren: Frühdiagenese und Paläo-Ozeanographie in der Norwegischen See, *Unpubl. Ph.D. Thesis*, Univ. of Kiel, 120 pp.

Keigwin, L.D., and Boyle, E., 1989: Late Quaternary paleo-chemistry of high-latitude surface waters, *Paleo. Paleo. Paleo.* 73, 85-106.

Kellogg, T.B., 1980: Paleoclimatology and paleo-oceanography of the Norwegian and Greenlandseas: glacial-interglacial contrasts, *Boreas* 9, 115-137.

Labeyrie, L.D., and Duplessy, J.C., 1985: Changes in the oceanic C^{13}/C^{12} ratio during the last 140,000 years: high latitude surface water records, *Paleo. Paleo. Paleo.* 50, 217-240.

Labeyrie, L.D., Duplessy, J.C., and Blanc, P.L., 1987: Variations in mode of formation and temperature of oceanic deep waters over the past 125,000 years, *Nature* 327, 477-483.

Lehmann, S.J., Jones, G.A., Keigwin, L.D., Andersen, E.S., Butenko, G., and Östmo, S.-R., 1991: I nitiation of Fennoscandian ice-sheet retreat during the last deglaciation, *Nature* 349, 513-516.

Lotter, A.F., 1991: Absolute dating of the Late-Glacial period in Switzerland using annually laminated sediments, *Quater. Res.* 35, 321-330. Mangerud, J., Lie, S.E., Furnes, H. Kristiansen, I.L., and Lömo, L., 1984: A Younger Dryas ash bed in western Norway and its possible correlations with tephra in cores from the Norwegian Sea and the North Atlantic, *Quater. Res.* 21, 85-104.

McIntyre, A., Kipp, N.G., Bé, A.W.H., Crowley, T., Kellog, T., Gardner, J.V., Prell, W., Ruddiman, W.F., 1976: Glacial North Atlantic 18,000 years ago: A CLIMAP reconstruction, *GSA*, Mem. 145, 43-76.

Morris, T.H., 1988: Stable isotope stratigrphy of the Arctic Ocean: Fram Strait to central Arctic, *Paleo. Paleo. Paleo.* 64, 201-219.

Piotrowski, A., 1990: Beitrag zur Rekonstruktion der paläoozeanographischen Entwicklung des Islandsee während der letzten 30,000 Jahre, *Unpubl. M.sc. thesis, Kiel University.*

Pujol, C., 1980: Les foraminiferes planctoniques de l'Atlantique nord au Quaternaire. Ecologie-Stratigraphie-Environnement, *Mém. de l'Institut de Geol. du Bassin d'Aquitaine,* No 10, 1-254.

Ruddiman, W.F., and McIntyre, A., 1973: Time-transgressive deglacial retreat of polar waters from the North Atlantic, *Quater. Res.,* 3, 117-130.

Ruddiman, W.F., and McIntyre, A., 1981: The North Atlantic Ocean during the last deglaciation, *Paleo. Paleo. Paleo.* 35, 145-214.

Sarnthein, M., and Tiedemann, R., 1990: Younger Dryas-style cooling events at glacial Terminations I-VI: associated benthic $\delta^{13}C$ anomalies at ODP Site 658 constrain meltwater hypothesis, *Paleoceanography,* 6, 1041-1055.

Sarnthein, M., Duplessy, J.-C., Hahn, M., Jansen, E., Arnold, M., Erlenkeuser, H., Flatoy, A., Veum, T., and Vogelsang, E. (in prep.): Stable isotope record of the evolution of surface and deep-water circulation in the Norwegian Greenland Sea since the Last Glacial Maximum, *Paleoceanography.*

Vogelsang, E., 1990: Paläo-Ozeanographie des Europäischen Nordmeeres an Hand stabiler Kohlenstoff- und Sauerstoffisotope, *(Unpubl.) Ph.D. Thesis University Kiel,* p. 1-137.

Weinelt, M.S.., Sarnthein, M., Vogelsang, E., Erlenkeuser, H., 1991: Early decay of Barents shelf ice sheet-spread of stable isotope signals across the Eastern Norwegian Sea, *Norsk Geol. Tidskr.* (in press), 6 ms-p.

A NEW METHOD TO RECONSTRUCT SEA SURFACE SALINITY: APPLICATION TO THE NORTH ATLANTIC OCEAN DURING THE YOUNGER DRYAS

J-C DUPLESSY, L. LABEYRIE, A. JUILLET-LECLERC, and J. DUPRAT (*)
Centre des Faibles Radioactivités
Laboratoire mixte CNRS-CEA,
91198 GIF SUR YVETTE Cedex
France.

ABSTRACT. A new method to reconstruct surface water salinity is based on a comparison between transfer function estimates of sea surface temperature and the oxygen isotope ratio of two common planktonic foraminiferal species, *N. pachyderma* (left coiling) and *G. bulloides*. This method is used to show that during the Younger Dryas cold event, the surface salinity of the North Atlantic Ocean, corrected for the mean salinity increase due to the presence of continental ice sheets over Canada and northern Europe, was significantly smaller than at present. During the last glacial to interglacial transition, superimposed over the salinity decrease associated with the melting of continental ice sheets, episodes of warming in the North Atlantic were associated with saline waters whereas cold episodes were associated with lower salinity/$\delta^{18}O$ waters.

INTRODUCTION

During the last deglaciation, the climate of the northern hemisphere has been marked by abrupt changes in air and sea surface temperature (Dansgaard et al., 1982 ; Mangerud et al., 1974 ; Ruddiman and McIntyre, 1973; 1981 ; Duplessy et al., 1981 ; Bard et al., 1987a,b) and the global ocean circulation experienced major changes (Berger and Vincent, 1986 ; Boyle and Keigwin, 1987 ; Duplessy et al., 1991a ; Broecker et al., 1985 ; Jansen and Veum, 1990). However, contradictory

(*) Département de Géologie et Océanographie, Université de Bordeaux-1, Avenue des Facultés, 33405 TALENCE, France.

NATO ASI Series, Vol. I 2
The Last Deglaciation: Absolute and Radiocarbon Chronologies
Edited by E. Bard and W. S. Broecker

evidences have been derived from the sediment record and occurrence of sporadic shutdowns of North Atlantic Deep Water (NADW) production during the glacial to Holocene transition is still a bone of contention. Oceanic circulation is driven by the wind field at the sea surface and the density of the water. In order to obtain a complete reconstruction of the ocean in the geological past, it is necessary therefore, to estimate both the temperature and the salinity of the various water masses. Sea-surface temperature can be estimated by planktonic fauna analysis (Imbrie and Kipp, 1971), but a similar method cannot be used to reconstruct surface salinities in the high latitude ocean, because temperature and salinity are too closely correlated in these areas.

Another possible approach to estimate past salinity relies on oxygen isotope analysis. Polar water has lower $^{18}O/^{16}O$ ratios than does tropical and temperate waters. These differences are closely tied to the ocean surface salinity distribution (Craig and Gordon, 1965) ; the higher the salinity, the higher the $^{18}O/^{16}O$ ratio. Excess evaporation concentrates the less volatile $H_2{}^{18}O$ along with salt, while precipitation, more depleted in $H_2{}^{18}O$ than ocean water, decreases both salinity and $^{18}O/^{16}O$ ratio of the sea surface water. The $^{18}O/^{16}O$ ratio of planktonic foraminifera records these variations since it depends on both the $^{18}O/^{16}O$ ratio and the temperature of the sea water in which foraminifera have grown. The $^{18}O/^{16}O$ ratio of $CaCO_3$ directly reflects variations of the $^{18}O/^{16}O$ ratio of the ambient water and the oxygen isotope fractionation between $CaCO_3$ and water increases by about 0.25 ‰ for each degree the water is cooled (Epstein et al., 1953; Shackleton, 1974). This is expressed by the paleotemperature equation :

$$T = 16.9 - 4.38 * (\delta^{18}O_{carbonate} - \delta^{18}O_{water}) + 0.10 * (\delta^{18}O_{carbonate} - \delta^{18}O_{water})^2$$

If the temperature of deposition of foraminiferal calcite is known (for instance through a transfer function estimate), it becomes feasible to calculate the $^{18}O/^{16}O$ of the water in which the foraminifera have grown. Using a large set of cores from the North Atlantic and the Southern oceans, Duplessy et al. (1991b) have statistically established the relationship linking the $^{18}O/^{16}O$ ratios of *Neogloboquadrina pachyderma* and *Globigerina bulloides,* two species common in the North Atlantic, to the summer sea-surface water temperature (SST) and the ambient $^{18}O/^{16}O$ ratio. This relationship is briefly summarized below. We then use this information to compare summer SST estimates with $^{18}O/^{16}O$ measurements made on planktonic foraminifera from North Atlantic cores to derive estimates of the surface water $^{18}O/^{16}O$ ratio and salinity during the Younger Dryas cold event. All the isotope measurements are reported using the conventional $\delta^{18}O$ notation, the standards being PDB for the carbonates and SMOW for the water.

ISOTOPIC CALIBRATION OF *G. BULLOIDES* AND *N. PACHYDERMA*

$\delta^{18}O$ measurements were performed on *G. bulloides* (>150 µm) from 83 Holocene core tops distributed throughout the North Atlantic Ocean and the Southern Ocean. The $\delta^{18}O$ values for surface sea water (δw) were calculated from summer salinity data gathered in the Levitus Atlas (1982) and by applying the δw/salinity relationships derived from the GEOSECS (1987) data for the North Atlantic Ocean and the Southern Ocean. The comparison of the Levitus summer SST with that derived from the isotopic measurements (Figure 1A) shows that the $\delta^{18}O$ values of *G. bulloides* is a linear function of the summer SST only in the range 7 to 22°C. In the following, we shall call this temperature range the "optimal temperature range" for this species. Fig. 1A also shows that the isotopic temperature of *G. bulloides* is generally lower than the summer SST. A statistical analysis performed on core tops raised from locations where the summer SST is in the range 7 to 22°C shows that the isotopic temperature given by *G. bulloides* is highly correlated with the summer SST ($r^2 = 0.89$) and that the slope of the regression line (0.91 with $\sigma = 0.04$) is not significantly different from 1 at the 2σ level. This indicates that *G. bulloides* probably records the whole of the summer SST variations and we assume that the best correlation between foraminiferal isotopic temperature and observed temperature is a linear relationship with a slope of 1. Over the whole "optimal temperature range" of *G. bulloides,* the simple relationship : $T_{isotope}$ = summer SST minus 1°C (Fig. 1B) differs by less than 0.7°C from the statistically derived regression line. Whereas the estimate of the isotopic temperature provided by the regression line has a standard deviation of 1.17°C, that obtained with the simple linear relationship has a standard deviation of 1.21°C. We shall therefore use the relationship $T_{isotope}$ = summer SST minus 1°C as a calibration for *G. bulloides*.

We calibrated *N. pachyderma* (left coiling) by analyzing 69 Holocene samples from cores distributed throughout the North Atlantic Ocean and the Southern Ocean. Comparison of the Levitus (1982) summer SST with that derived from the $\delta^{18}O$ value of *N. pachyderma* (Fig. 2) shows trends similar to those observed for *G. bulloides*. In the temperature range 3 to 10°C, the isotopic temperature given by *N. pachyderma* is a linear function of the summer SST, whereas in warmer waters, isotopic temperature values exhibit a noticeable scatter (Fig. 2A). Within the temperature range -1.5°C to 3°C, the isotopic temperatures given by *N. pachyderma* are in the range 0-3°C, and are often warmer than the atlas mean summer values. This is true even when a permanent ice cover is generally present at the surface. This indicates that, in very cold water, *N. pachyderma* develops only when it can find a water mass of which the temperature is close to 0-1°C.

A statistical analysis performed on core tops raised from locations where the summer SST is in the range +3 to + 10°C shows that the isotopic temperature given by *N. pachyderma* is linearly correlated to the summer SST ($r^2 = 0.70$) and that the best fit line has a slope (0.947 with $\sigma = 0.04$) which is not significantly different from 1 at the 2σ level. Over the whole "optimal temperature range" of *N. pachyderma* (left coiling), this regression line differs by less than 0.4°C from the value calculated by substracting 2.5°C from the summer SST (Fig. 2B). Whereas the estimate of the isotopic temperature provided by the regression line has a standard deviation of 1.50°C, that obtained by substracting 2.5°C from the summer SST has a standard deviation of 1.51°C. In the following, we consider the optimal temperature range for *N. pachyderma* as +3°C to +10°C and the simple linear relationship $T_{isotope}$ = summer SST minus 2.5°C will be used as a calibration for *N. pachyderma* (left coiling).

We therefore define a parameter, which we call T*, and which represents a statistical estimate of the mean calcification temperature for each species : T* = Summer SST -1°C for *G. bulloides* and T* = Summer SST -2.5°C for *N. pachyderma*. Taking all the *N. pachyderma* $\delta^{18}O$ values from sites where the summer SST is higher than 3°C and lower than 10°C and all the *G. bulloides* $\delta^{18}O$ values from sites where the summer SST is higher than 7°C and lower than 22°C, we find that the isotopic temperatures calculated from the $\delta^{18}O$ values of these species are not significantly different from the temperature T* (Fig. 3). They are highly correlated ($r^2 = 0.918$) and the slope of the regression line (0.959 with σ = 0.04) is not significantly different from 1 at the $2\ \sigma$ level. We shall assume in the following section that this statistical relationship is also valid under different climatic conditions.

SURFACE WATER $\delta^{18}O$/SALINITY DURING THE YOUNGER DRYAS

We consider (Table 1) only cores located in areas where the Younger Dryas (YD) summer SST estimate was lower than 22°C and in which *G. bulloides* or *N. pachyderma* have lived in the respective optimum temperature ranges defined above. We selected cores in which the YD cold event was well-documented in order to minimize the effects of bioturbation, which are particularly severe when abrupt events of short duration are to be reconstructed (Bard et al., 1987a). The summer surface-water $\delta^{18}O$ value were calculated assuming that the foraminifera grew their shells in isotopic equilibrium at T*.

In order to compare with the modern pattern, we calculated the $\delta^{18}O$ anomaly as the difference between the YD $\delta^{18}O$ values minus the modern $\delta^{18}O$ value and corrected this difference from the mean oceanic $\delta^{18}O$ change (Δ) due to the growth of continental ice sheets :

$$\delta^{18}O\ (\text{anomaly}) = \delta^{18}O\ (\text{YD}) - \delta^{18}O\ (\text{modern}) - \Delta\ (\text{ice volume})$$

Recent estimates for the $\delta^{18}O$ change associated with the glacial-interglacial transition range from 1.1‰ to 1.3‰ (Labeyrie et al., 1987 ; Shackleton, 1987 ; Fairbanks, 1989) and we adopted here the mean value of 1.2‰. As the YD is about midway of the complete glacial to interglacial transition (Fairbanks, 1989), we adopted a Δ value of 0.6 ‰ to compute the $\delta^{18}O$ anomaly (Table 1).

In order to estimate the salinity of the surface waters in the past, we have to take into account two effects which add together : first, the salinity increased because a significant amount of freshwater was used to form ice sheets over the continents. Second, the salinity also varied because the evaporation/precipitation ratio (E/P) was locally different from the modern one in response to the climatic change.

We display in Figure 4 two examples showing the simultaneous changes affecting both the sea water $\delta^{18}O$ and salinity for an hypothetical sample of YD water. We calculate first the ice volume effect. The present mean ocean salinity is 34.7. As the present mean ocean water depth is 3900 m and as the sea level was lower than today by about 60 m during YD conditions (Fairbanks, 1989), the mean glacial salinity was higher than the modern one by 34.7x60/3840 = 0.5. Therefore the ice volume effect is characterized by a salinity increase of 0.5 and a $\delta^{18}O$ increase of 0.6‰ as calculated previously. We then take into account the effect of changes in the E/P ratio, which produced an additional salinity anomaly. Since the vapor pressure is higher for H_2O^{16} than for H_2O^{18}, an increase in the E/P ratio will result in both a salinity and $\delta^{18}O$ increase. At present we have no way to estimate precisely the slope of the salinity/$\delta^{18}O$ relationship during the YD and we adopt here the mean value measured for the modern ocean by Craig and Gordon (1965) : we thus assume that the salinity increased by 2‰ when the sea water $\delta^{18}O$ value increased by 1‰ (Fig. 4A). Following the same rationale, if the E/P ratio was lower during YD than at present, salinity and $\delta^{18}O$ decreased and varies linearly (Fig. 4B). The salinity anomaly is thus calculated by applying the salinity/$\delta^{18}O$ relationship to the $\delta^{18}O$ anomaly reported in Table 1 and we calculated the summer surface water salinity for the YD North Atlantic Ocean as

$$S_{YD} = S_{modern} + 0.5 + (2 \times \delta^{18}O_{anomaly})$$

DISCUSSION

The accuracy of the surface water $\delta^{18}O$ reconstruction depends on the accuracy of both the oxygen isotope measurements performed on foraminiferal samples and summer sea surface temperature estimates derived from transfer functions. We also assume that the relationship between the summer sea surface temperature and the oxygen isotopic composition of *N. pachyderma* and *G. bulloides*, determined for modern conditions, was also valid during cold conditions (see a detailed discussion in Duplessy et al., 1991b). Finally, this method assumes that bioturbation does not prevent a precise reconstruction of sea surface temperatures and foraminiferal $\delta^{18}O$ values during an event as short as the Younger Dryas. In fact, bioturbation constitutes a severe limitation. In Table 1, we used cores either with high sedimentation rates or in which the YD event was well-marked in both the SST and $\delta^{18}O$ records. For the 5 cores raised from the Bay of Biscay, it is clear that the scatter in the reconstructed value is primarily related to that in the sea surface temperature estimates, whereas the $\delta^{18}O$ values are much more homogenous. This is easily explained because *N. pachyderma* was the dominant species during the YD and exhibited a peak of abundance in the sedimentary record. Regardless of the bioturbation amplitude, the analyzed shells have effectively lived in the cold waters. By contrast, the estimates of the sea surface temperature depend critically on the amplitude of the bioturbation, which introduces warm foraminiferal species shells within the cold fauna. This phenomenon is clearly evident in core CH 67-19, in which the temperature estimates for the YD exhibit only a narrow peak (Duplessy et al., 1981), which prevents accurate reconstruction of the temperature and also in core SU 81-47, in which the $\delta^{18}O$ values of *G. bulloides* are much lighter than those of *N. pachyderma* in the same sediment level (Table 1). As bioturbation is important in core CH 73-139C, the raw faunal and $\delta^{18}O$ data have been rerstored by deconvolving the signal using an equivalent bioturbation depth of 10-12.5 cm (Bard et al., 1987a). The resulting estimate is in good agreement with that derived from nearby core V 23-81, which has a larger sedimentation rate.

When salinity has to be estimated, an additional uncertainty is added, since the slope of the salinity/$\delta^{18}O$ relationship for climatic conditions different from modern ones is not known. We assume that it was similar to the mean value for the modern climatic pattern. Surface sea water $\delta^{18}O$ and salinity estimates are therefore linearly correlated and below we only discuss the $\delta^{18}O$ pattern.

Table 1 shows that the $\delta^{18}O$ anomaly is negative in most cores, indicating that the North Atlantic surface water was comparatively more diluted by freshwater than under modern conditions. Such a pattern would tend to reduce winter convection in the high latitudes of the North Atlantic Ocean and in the adjacent seas and to slow down the production of NADW. This data is however not sufficient to determine whether NADW formation ceased completely during the Younger Dryas (Broecker et al., 1985 ; Berger and Vincent, 1986 ; Boyle and Keigwin, 1987 ; Jansen and Veum, 1990).

In order to place these low $\delta^{18}O$ values within the perspective of changes associated with the deglaciation, we calculated the variation of surface sea water $\delta^{18}O$ with time in the following four cores for which we had simultaneously sea surface temperature estimates and $\delta^{18}O$ analyses for either *G. bulloides* or *N. pachyderma* in their optimal temperature range. Norwegian Sea core HM 52-43 has been analyzed by Veum (1990) for both faunal and isotope measurements. Isotope analyses for North Atlantic core V 28-14 have been made by Shackleton at Cambridge (Kellogg et al., 1978) and faunal counts have been performed by Kellogg (1973). Sea surface temperature estimates for this core like for cores SU 81-18 and CH 73-139C were calculated by J. Duprat following the Imbrie and Kipp (1971) technique. Calibration of discrete foraminiferal assemblages from surface sediment samples against modern sea surface temperature was obtained at Bordeaux from faunal counts in the >149µm fraction of 261 core-top samples (Pujol, 1980). Isotope records of cores SU 81-18 and CH 73-139C were obtained at Gif (Duplessy et al., 1981; Bard et al., 1987b).

The variations of sea surface water $\delta^{18}O$ values and temperature versus time are reported in figure 5A - 5D. In core SU 81-18, except at the 10 and 80 cm levels which were discarded from this analysis, *G. bulloides* has lived permanently in the temperature range 8 to 22°C. In North Atlantic core V 28-14 and in Norwegian Sea core HM 52-43, summer sea surface temperature estimates are lower than 10°C only before the Holocene. We used the *N. pachyderma* record only when this species has lived in its optimal temperature range, and we reconstructed sea water $\delta^{18}O$ variations before 9,000 years B.P.. In North Atlantic core CH 73-139C, temperature variations have been so large that neither *N. pachyderma* nor *G. bulloides* have lived continuously in their optimal temperature range (Duplessy et al., 1981). We therefore generated a continuous isotopic record for the last 15,000 years by selecting *G. bulloides* during the warm episodes and *N. pachyderma* during the cold episodes. In order to reduce the influence of bioturbation, the raw faunal and $\delta^{18}O$ data have been restored by deconvolving the signal using an equivalent bioturbation depth of 10-12.5 cm (Bard et al., 1987a).

The four records show that, superimposed over the general decreasing trend associated with the melting of the continental ice sheets, the sea water $\delta^{18}O$ (and the salinity) tend to increase during warm episodes and to decrease during cold episodes. This suggests that the sea surface temperature variations in the North Atlantic were primarily associated with oscillations of the polar front separating warm saline waters from cooler and fresher polar waters. As high salinity favours convection and deep water formation, this reconstruction supports also theories, which tend to link the warmth of the North Atlantic to the poleward flux of warm surface water and to the strength of the flux of NADW (Broecker and Denton, 1989).

ACKNOWLEDGEMENTS

Thanks are due to J. Antignac, B. Le Coat and H. Leclaire for analytical help. This research is supported by EEC (EPOCH), PNEDC, CNRS and CEA. This is CFR contribution N° 1216.

Core	Latitude	Longitude	Foram	YD summer SST	YD foram δO-18	YD SSW δO-18	modern SSW salinity	modern SSW δO-18	YD SSW δO-18 anomaly	YD SSW salinity	data source
V28-14	64.47	-29.34	pachy	6.9	3.47	0.67	34.60	0.05	0.02	35.14	1,2
HM 52-43	64.31	0.44	pachy	6.2	3.67	0.68	35.10	0.33	-0.25	35.11	3
CH 73-139C	54.38	-16.21	pachy	7.4	2.95	0.28	35.40	0.49	-0.81	34.29	4
V23-81	54.15	-16.50	pachy	7.5	3.03	0.39	35.39	0.49	-0.70	34.49	5
V27-116	52.50	-30.20	bulloides	8.0	2.35	0.23	34.92	0.23	-0.60	34.22	6
K 708-1	50.00	-23.45	bulloides	13.0	1.3	0.42	35.46	0.53	-0.71	34.54	6
CH 72-101	47.28	-8.34	pachy	7.0	2.75	-0.02	35.56	0.58	-1.20	33.66	7
CH 72-104	46.54	-8.05	pachy	6.8	2.77	-0.06	35.52	0.56	-1.22	33.59	7
CH 67-19	45.45	-3.57	bulloides	10.9	2.3	0.91	35.13	0.34	-0.03	35.57	7
SU 81-47	44.53	-3.18	pachy	8.2	2.92	0.46	34.93	0.23	-0.37	34.70	8
SU 81-47	44.53	-3.18	bulloides	8.2	2.23	0.16	34.93	0.23	-0.67	34.09	8
V 30-101K	44.06	-32.30	bulloides	10.2	1.6	0.04	34.85	0.19	-0.75	33.84	6
SU81-18	37.46	-10.11	bulloides	14.0	1.62	0.98	36.10	0.88	-0.50	35.60	8,9

Table 1

Summary of the data used to reconstruct the North Atlantic sea surface salinity during the Younger Dryas.

Latitude and Longitude are expressed as degrees and minutes (+ indicates N or E ; - indicates S or W).
The Younger Dryas sea surface delta O-18 value was calculated assuming isotopic equilibrium with T* defined in the text.
Modern sea surface delta O-18 was calculated from the Levitus (1982) salinity data and the regression line derived from GEOSECS data.
The Younger Dryas surface delta O-18 anomaly was calculated as difference between Younger Dryas and modern values minus the global ice volume effect.
Data Sources : 1 = Kellogg thesis, 1973 ; 2 = Kellogg et al., 1978 ; 3 = Veum thesis, 1990 ;
4 = Bard et al., 1987a ; 5 = Jansen and Veum, 1990 ; 6 = Ruddiman and McIntyre, 1981 ;
7 = Duplessy et al., 1981 ; 8 = This work ; 9 = Bard et al., 1987b.

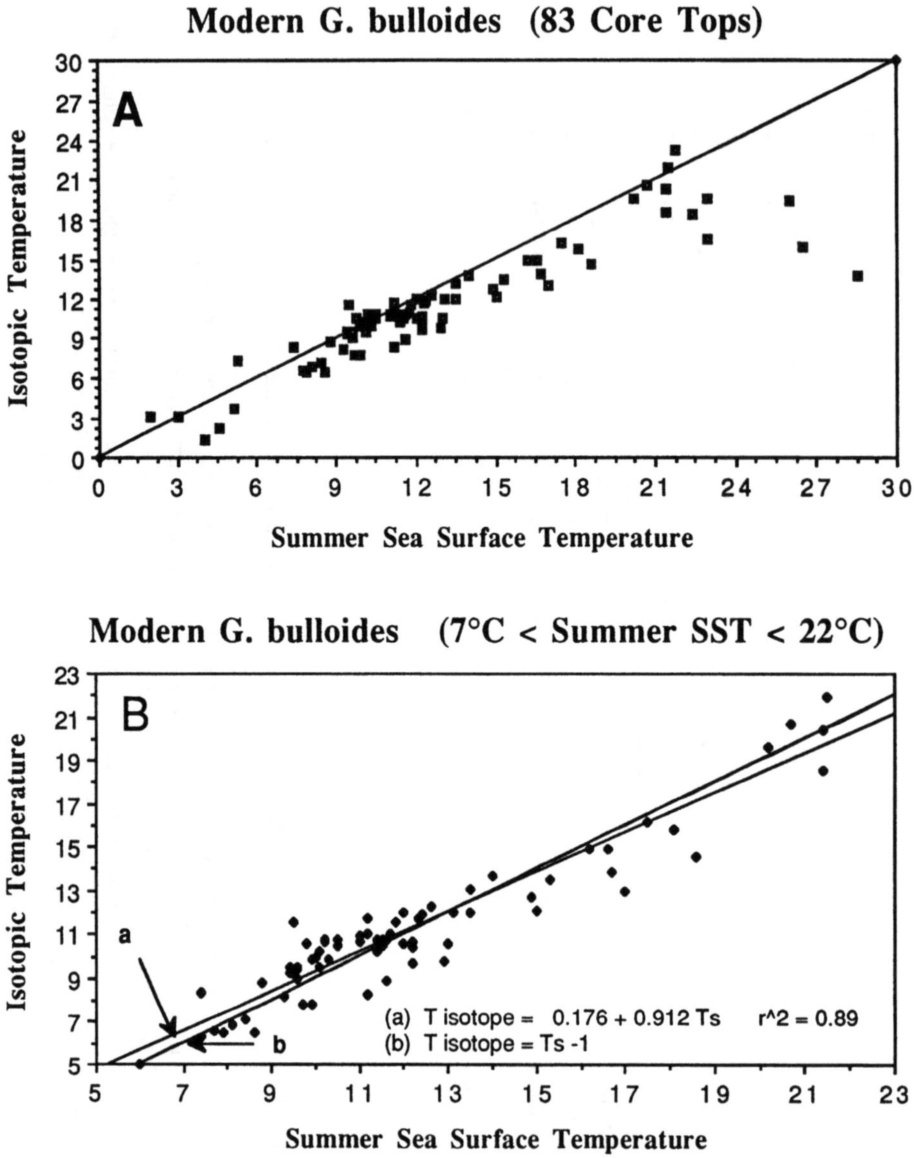

Figure 1

A : Plot of the isotopic temperature calculated from *G. bulloides* $\delta^{18}O$ values against summer sea surface temperature for 83 core tops from the North Atlantic and the Southern oceans.

B : Comparison of the regression line (a) calculated for samples for which the summer SST is in the range 7-22°C, with the linear relationship (b) : $T_{isotope}$ = summer SST - 1°C.

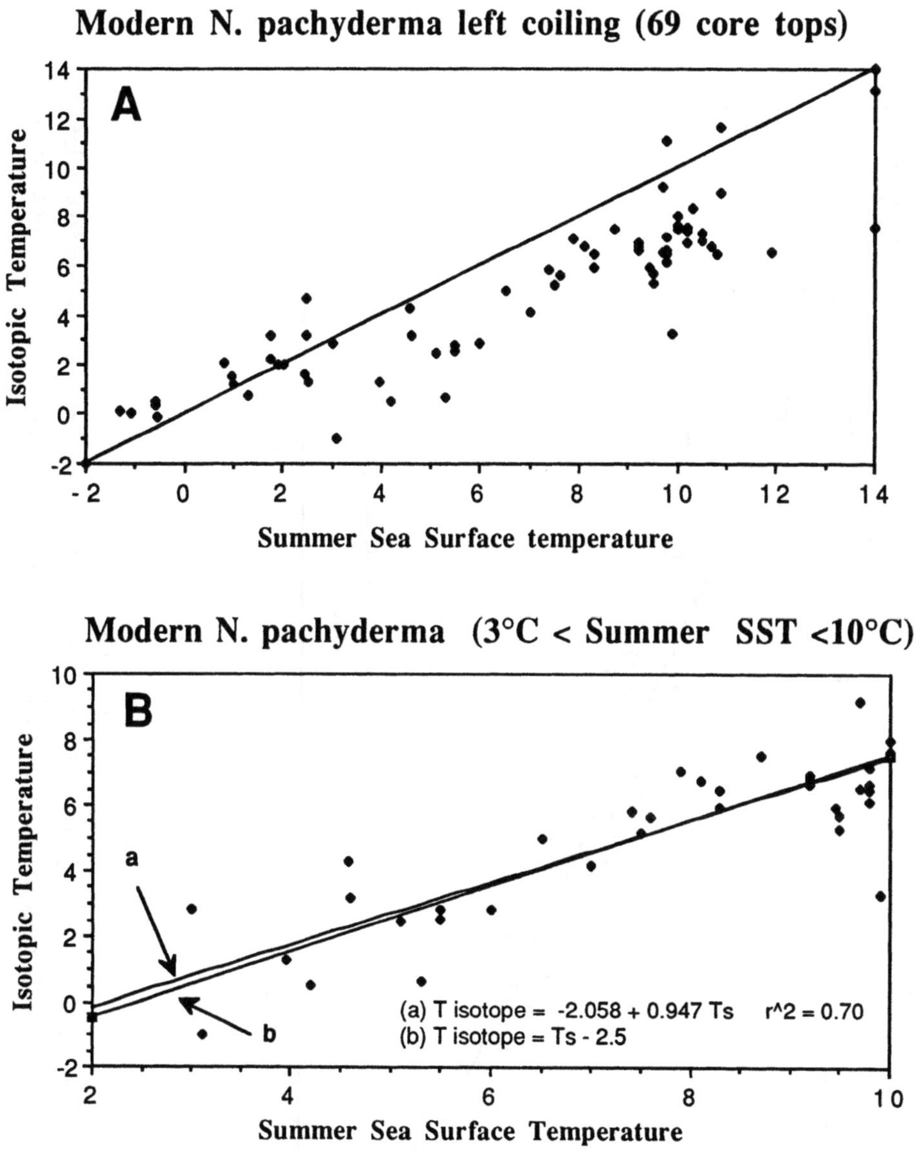

Figure 2

A : Plot of the isotopic temperature calculated from *N. pachyderma* (left coiling)) $\delta^{18}O$ value against summer sea surface temperature for 69 core tops from the North Atlantic and Southern oceans.

B : Comparison of the regression line (a) calculated for samples for which the summer SST is in the range 3-10°C, with the linear relationship (b) : $T_{isotope}$ = summer SST - 2.5°C.

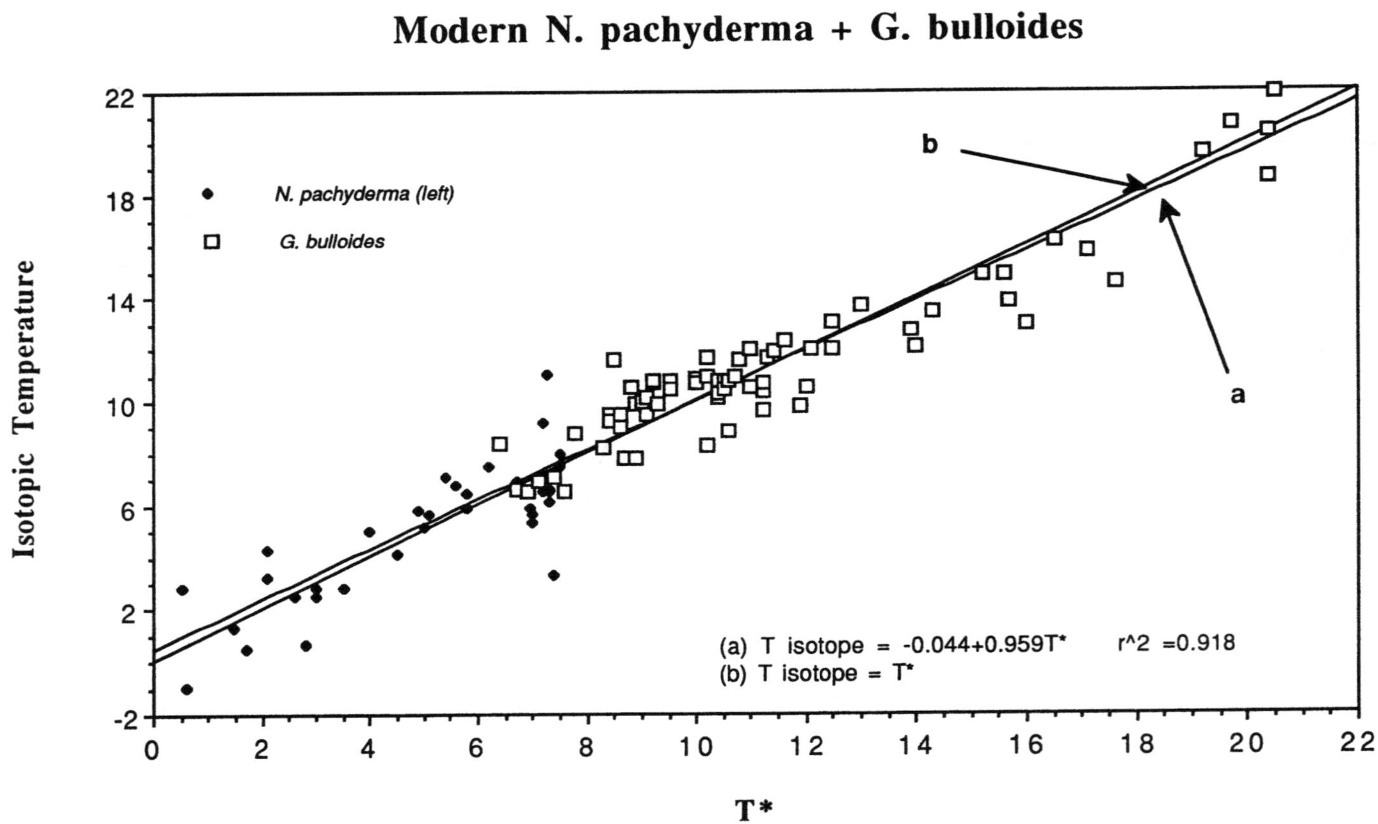

Figure 3

Plot of the isotopic temperature calculated from $\delta^{18}O$ values of *G. bulloides* and *N. pachyderma* (left coiling) which grew in their optimum temperature range against the effective temperature of calcification T*.

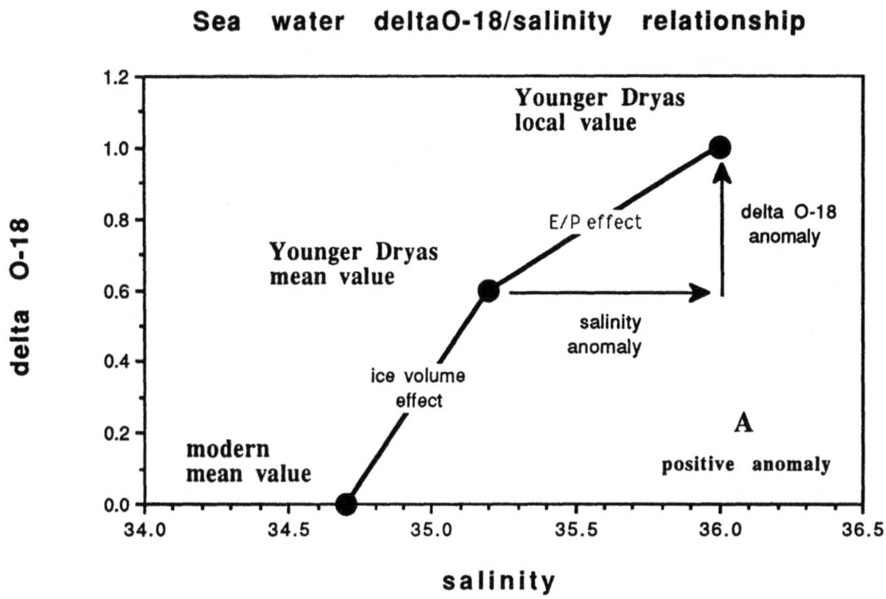

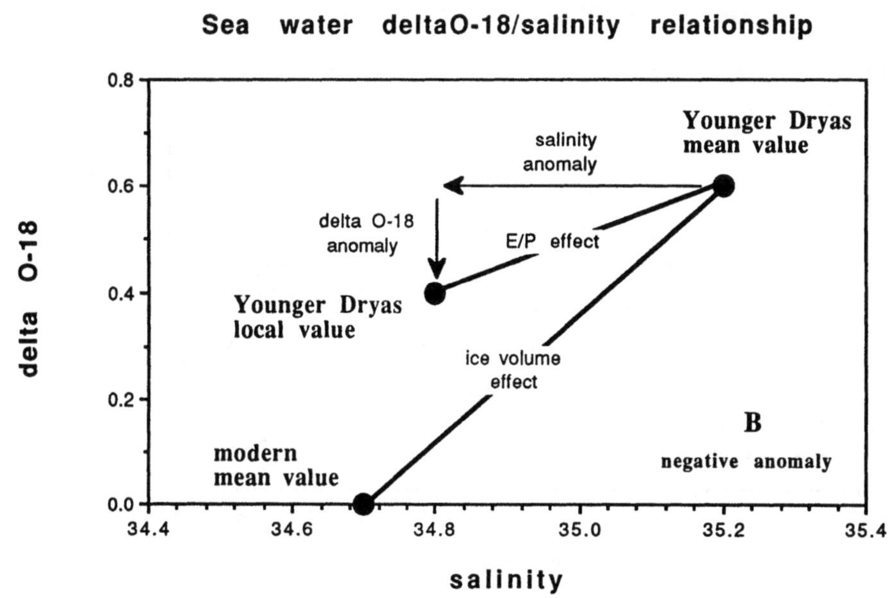

Figure 4
Surface water $\delta^{18}O$/salinity variations between the Younger Dryes cold event and today. Two hypothetical cases are considered : a positive anomaly due to a local increase of the E/P ratio (A) and a negative anomaly due to a local decrease of the E/P ratio (B).

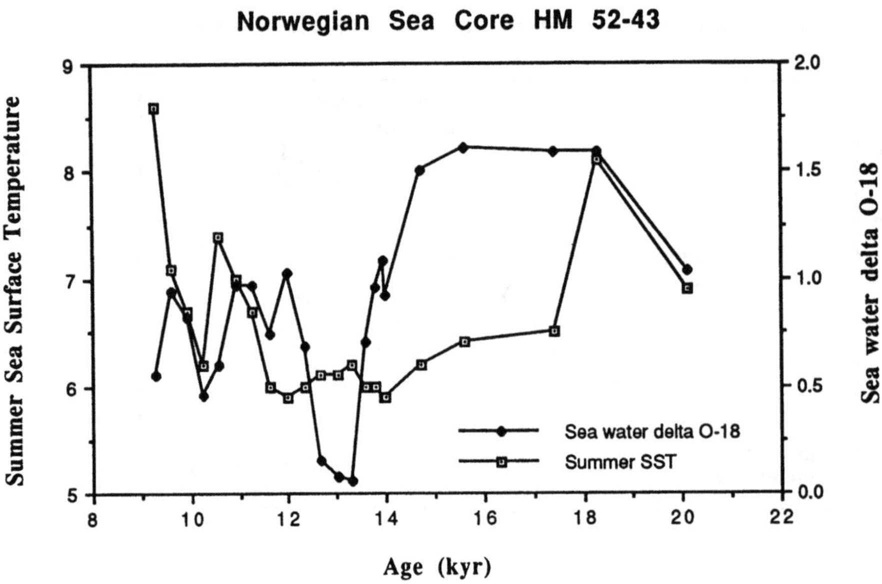

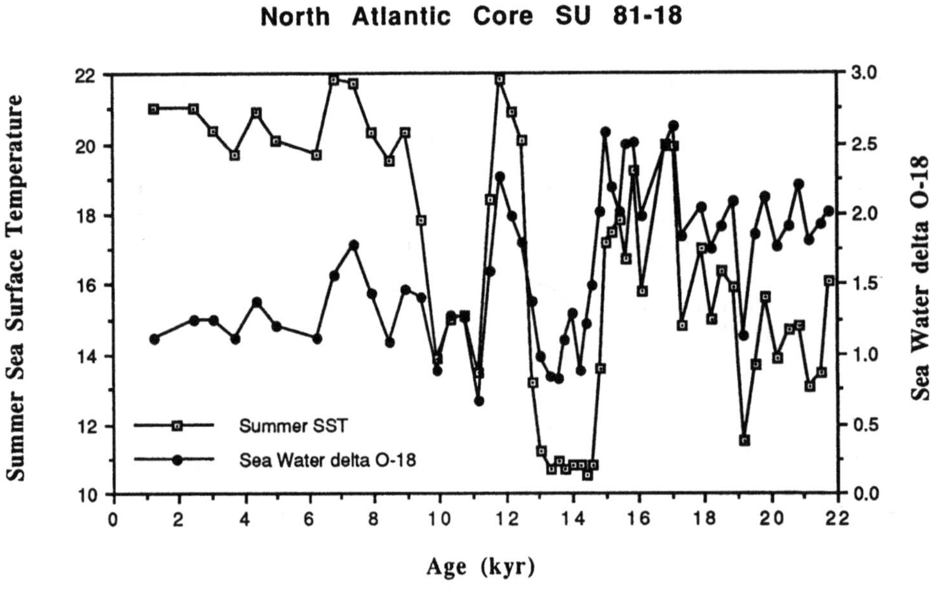

Figure 5

Plots of the summer sea surface temperature and $\delta^{18}O$ variations during the last deglaciation in Norwegian Sea core HM 52-43 and in North Atlantic cores SU 81-18 (off Portugal).

Continued next page

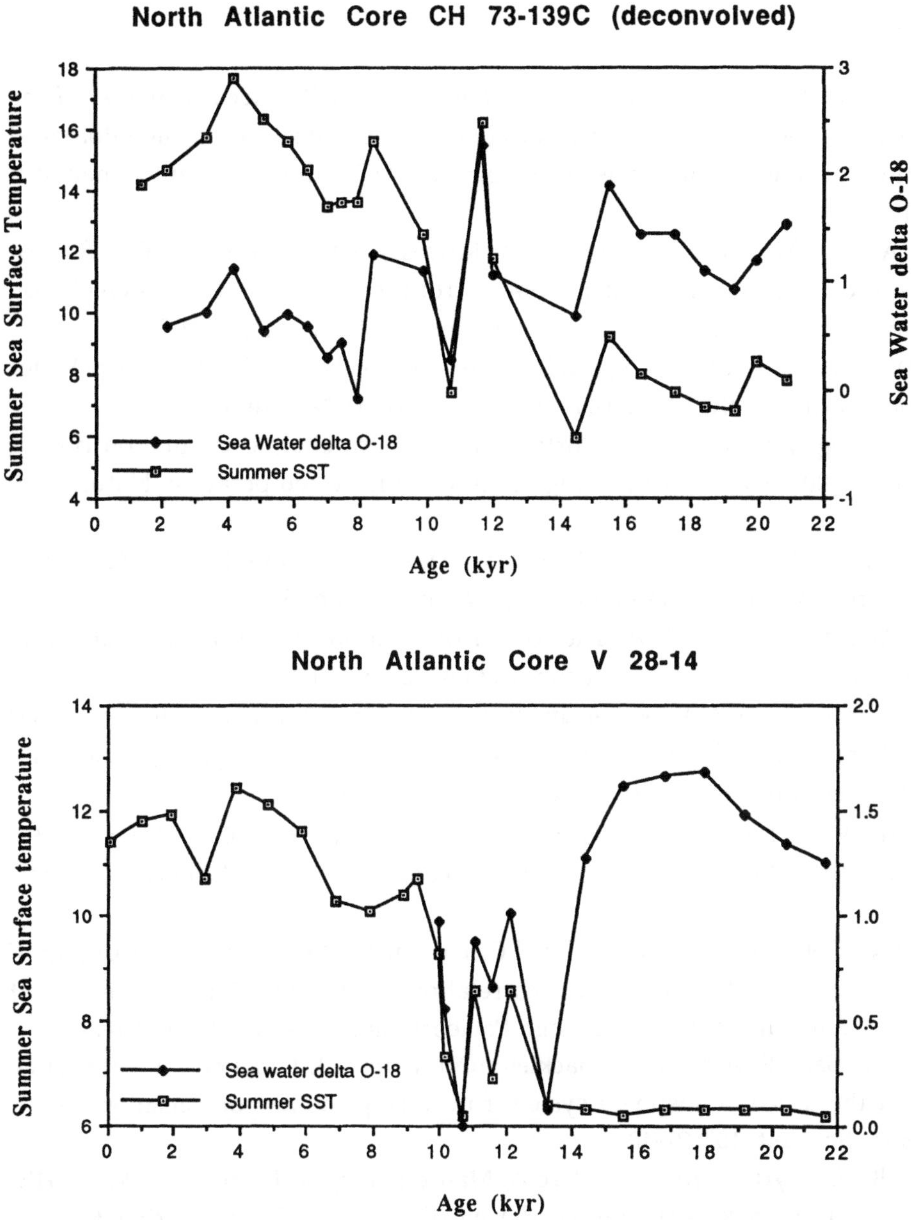

Figure 5

Plots of the summer sea surface temperature and $\delta^{18}O$ variations during the last deglaciation in North Atlantic cores CH 73-139C (off Ireland), deconvolved according to Bard et al. (1987a) and V 28-14 (west of Iceland).

REFERENCES

Bard E, Arnold M, Duprat J, Moyes J, Duplessy JC (1987a) Reconstruction of the last deglaciation : deconvolved records of $\delta^{18}O$ profiles, micropaleontological variations and accelerator mass spectrometric ^{14}C dating. Climate Dynamic 1: 101-112.

Bard E, Arnold M, Maurice P, Duprat J, Moyes J, Duplessy JC (1987b) Retreat velocity of the North Atlantic polar front during the last deglaciation determined by ^{14}C accelerator mass spectrometry. Nature 328: 791-794.

Berger WH, Vincent E (1986) Sporadic shutdown of North Atlantic deep Water production during the Glacial-Holocene transition? Nature 324: 53-55.

Boyle EA., Keigwin LD (1987) North Atlantic thermohaline circulation during the past 20,000 years linked to high-latitude surface temperature. Nature 330: 35-40.

Broecker WS, Peteet DM, Rind DM (1985) Does the ocean-atmosphere system have more than one stable mode of operation? Nature 315: 21-26.

Broecker WS, Denton G (1989) The role of ocean-atmosphere reorganizations in glacial cycles. Geochim. Cosmochim. Acta 53: 2465-2501.

Craig H, Gordon LI (1965) Deuterium and oxygen 18 variations in the ocean and the marine atmosphere. In Stable isotopes in oceanographic studies and paleotemperatures, Ed. E. Tongiorgi, 9-130, CNR, Pisa.

Dansgaard W, Clausen HB, Gundestrup N, Hammer CU, Johnsen SJ, Kristindottir P, Reeh N (1982) A new Greenland ice core. Science 218: 1273-1278.

Duplessy JC, Delibrias G, Turon JL, Pujol C, Duprat J (1981) Deglacial warming of the northeastern Atlantic Ocean ; correlation with paleoclimate evolution of the european continent. Paleaogeogr. Palaeoclimat., Palaeoecol. 35: 121-144.

Duplessy JC, Bard E, Arnold M, Shackleton NJ, Duprat J, Labeyrie L (1991a) How fast did the ocean-atmosphere system run during the last deglaciation? Earth Planetary Sci. Lett. 103: 27-40.

Duplessy JC, Labeyrie L, Juillet-Leclerc A, Maitre F, Duprat J, Sarnthein M (1991b) Surface salinity reconstruction of the North Atlantic Ocean during the last glacial maximum. Oceanologica Acta, in press.

Epstein S, Buchsbaum R, Lowenstam HA, Urey HC (1953) Revised carbonate-water isotopic temperature scale. Geol. Soc. Am. Bull. 64: 1315-1325.

Fairbanks RG (1989) A 17,000-year glacio-eustatic sea level record : influence of glacial melting rates on the Younger Dryas event and deep-ocean circulation. Nature 342: 637-642.

GEOSECS Atlantic, Pacific and Indian ocean expeditions, Vol. 7, (1987) Shorebased data and graphics, GEOSECS Executive Committee HG Ostlund, H Craig, WS Broecker, D Spencer, I. D. O. E., National Science Foundation.

Imbrie J, Kipp NG (1971) A new micropaleontological method for quantitative paleoclimatology: Application to a late pleistocene Caribbean core. In The Late Cenozoic Glacial Ages, K. K. Turekian ed. 71-181, Yale University Press.

Jansen E, Veum T (1990) Evidence for two-step deglaciation and its impact on North Atlantic deep-water circulation. Nature 343: 612-616.

Kellogg TB (1973) Late pleistocene climatic record in Norwegian and Greenland Sea deep sea cores. Dissertation, Columbia University, New York, 545 pages.

Kellogg TB, Duplessy JC, Shackleton NJ (1978) Planktonic foraminiferal and oxygen isotopic stratigraphy and paleoclimatology of Norwegian Sea deep-sea cores. Boreas 7: 61-73.

Labeyrie LD, Duplessy JC, Blanc PL (1987) Variations in mode of formation and temperature of oceanic deep waters over the past 125,000 years. Nature 327: 477–482.

Levitus S (1982) Climatological atlas of the world ocean. NOAA Professional Paper N°13, Rockville, Maryland (USA).

Mangerud J, Andersen ST, Berglund BE, Donner JJ (1974)Quaternary stratigraphy of norden, a proposal for terminology and classification. Boreas 3: 110-127.

Pujol C (1980) Les foraminifères planctoniques de l'Atlantique Nord au Quaternaire : Ecologie - Stratigraphie - Environnement. Thesis, University of Bordeaux 1, Bordeaux (France).

Ruddiman WF, McIntyre A (1973) Time-transgressive deglacial retreat of polar waters from the North Atlantic. Quat. Res. 3: 117-130.

Ruddiman WF, McIntyre A (1981) The North Atlantic Ocean during the last deglaciation. Paleaogeogr. Palaeoclimat. Palaeoecol. 35: 145-215.

Shackleton NJ (1974) Attainment of isotopic equilibrium between ocean water and the benthonic foraminifera genus *Uvigerina* : Isotopic changes in the ocean during the last glacia. Colloque CNRS N° 219, 203-210, CNRS, Paris.

Shackleton NJ (1987) Oxygen isotopes, ice volume and sea-level. Quaternary Sci Rev. 6: 183-190.

Veum T (1990) Sen Weichsel Paleoseanografi og paleoklima i Norsehavet og Nord Atlanteren. Thesis, 171 pages, Univ. of Bergen (Norway).

THE DETERMINATION OF PAST OCEAN-ATMOSPHERE RADIOCARBON DIFFERENCES

J.R.Southon*, D.E.Nelson and J.S.Vogel*
Archaeology Department
Simon Fraser University
Burnaby, B.C.
Canada V5A1S6

INTRODUCTION

In studying the course of the last deglaciation in detail, it is important to correctly relate radiocarbon chronologies for the land and different regions of the ocean. The global thermohaline circulation sequesters carbon in the deep ocean for periods of 1000-2000 years, and this is reflected in the reduced ^{14}C content or "age" of the dissolved inorganic carbon in deep ocean water. The ocean mixed layer exchanges carbon with both the deep ocean and the atmosphere and hence has an intermediate ^{14}C content. Globally, the preindustrial surface reservoir age (that is, the difference in ^{14}C age between atmospheric and surface marine carbon) is about 400 years, but regional values range from 300 years in parts of the Atlantic to 1300 years off Antarctica [Stuiver et al., 1986]. Since the age of the deep water has varied over time with changes in the thermohaline circulation [Shackleton et al., 1988; Broecker et al., 1988], past surface reservoir ages in some areas, particularly in upwelling regions at high latitudes, may have differed from those of today. These past reservoir ages must be known in order to correctly tie together terrestrial and ocean radiocarbon chronologies. Also, any changes would be of interest for the oceanographic information they would provide.

Mixed layer ^{14}C reservoir ages for recent samples have been determined previously by radiocarbon dating known-age shells [Stuiver et al., 1986], but the data for many critical areas is sparse, and few systematic studies of past reservoir ages have been carried out. However, a record of surface reservoir ages is preserved in the ^{14}C ages of associated marine and terrestrial

*Present address:
Center for Accelerator Mass spectrometry
Lawrence Livermore National Laboratory
P.O.Box 808, Livermore, CA94550

NATO ASI Series, Vol. I 2
The Last Deglaciation: Absolute and Radiocarbon Chronologies
Edited by E. Bard and W. S. Broecker

material. Possible samples include wood and shells from ancient shorelines buried in near-shore sediments or exposed on land in areas of post-glacial uplift, and material from coastal archaeological sites. Also, it has been shown that pollen can be radiocarbon dated by accelerator mass spectrometry (AMS) [Brown et al., 1989], and sufficient pollen is available in some ocean cores [van Campo et al., 1990] that comparisons of ^{14}C in pollen and foraminifera appear feasible.

In a recent study [Southon et al., 1990], we measured ^{14}C age differences in shell-wood pairs taken from coastal sediments in British Columbia, Canada, and compared them with an estimate of the pre-industrial reservoir age for the subpolar North Pacific gyre. Here we discuss the sampling requirements for such studies and present the late Pleistocene and early Holocene data from that investigation. The results suggest that reconstruction of past surface reservoir ages may indeed be possible using this technique.

RECENT RESERVOIR AGES AND THE FOSSIL FUEL EFFECT

The regional reservoir age for the recent past (Table 1) was calculated from ^{14}C data for pre-bomb known-age shells collected between Oregon and the Aleutians [Broecker, 1963; Robinson and Thompson, 1983], after correcting for ^{14}C depletion from fossil fuel CO_2 (the Suess effect). Global values for the marine Suess effect [Stuiver et al., 1986, Druffel and Suess, 1983] were reduced by one-third for the North Pacific to account for the effect of upwelling old deep Pacific water which is uninfluenced by short-term atmospheric ^{14}C changes. Any changes in the production rate of ^{14}C between 1850 and the mid 20th century were probably small [Stuiver and Quay, 1980,1981]; and there is little evidence of any atmospheric Suess effect before 1880 [Stuiver et al, 1984; Peng and Freyer, 1986]. We therefore used the post-1850 $\Delta^{14}C$ value (^{14}C depletion) of $-4 \pm 2‰$ as the atmospheric baseline for comparison with the 20th century marine data.

The scatter in the Alaskan reservoir ages in Table 1 is high, and may represent sampling problems or real variability. However, the mean of the Alaskan data is close to those for the other samples, and to estimated pre-nuclear $\Delta^{14}C$ values for the Central Pacific region between 45° and 55°N [Broecker et al., 1985]. Thus, the preindustrial reservoir age over large parts of the subpolar gyre was probably close to the value of 788 ± 33 ^{14}C years calculated here.

TABLE 1. Recent Northeast Pacific Reservoir Ages

Location, collection date	Source	Shell ^{14}C Age[c] years BP	$\Delta^{14}C$[d] per mil	Suess effect per mil	Reservoir Age[e] ^{14}C years
Adak, Aleutians,1950	a		-127±9	-6	1004±75
Alaska Peninsular,~1950	a		-106±6	-6	814±55
Pavlov Harbor, 1937	b	700±50	-82±6	-2	638±55
Orcas Is., Wash, ~1915	b	805±50	-92±6	-1	730±55
Orcas Is., Wash, ~1915	b	950±30	-108±4	-1	875±35
Sooke, B.C., 1916	b	850±50	-97±6	-1	776±55
Esquimalt, B.C., 1930	b	750±50	-87±6	-2	681±55
Yaquina Bay, Or., 1916	b	840±35	-96±4	-1	766±40
Yaquina Bay, Or., 1916	b	835±50	-95±6	-1	761±55
Sunset Bay, Or., 1936	b	895±50	-104±6	-2	831±55
				Mean	788±33

a: Broecker [1963].

b: Robinson and Thompson [1981].

c: Conventional ^{14}C ages calculated according to the conventions of Stuiver and Polach [1977].

d: Sample ^{14}C activities as per mil deviations from the standard after correcting for isotopic fractionation and for ^{14}C decay between the calender year of sample formation and AD 1950 [Stuiver and Polach, 1977].

e: The mixed layer $\Delta^{14}C$ values are corrected for the marine Suess effect, converted to conventional ^{14}C ages, and compared with an equivalent preindustrial atmospheric age of 32±16 radiocarbon years ($\Delta^{14}C$ = -4±2 per mil).

SAMPLE SELECTION AND MEASUREMENT

Samples of shell and associated wood and other organic material were collected from a sediment core and from several coastal sites. The Endacko 87A-23 core was recovered from Cook Bank, 20km north of Cape Scott on Vancouver Island, in 95m of water [Luternauer et al., 1989]. In this core, shallow-marine sediments containing shells and wood fragments overlie a middle unit of lagoonal or shallow pond sediments, which in turn overlie a Pleistocene paleosol containing in-situ rooted plant remains. Other samples were collected from a coastal bluff at Mary Point, which lies at the entrance to Naden Harbour on the north coast of Graham Island in the Queen Charlotte group. Sediments are exposed as the bluff erodes through periodic undercutting by wave action. A single shell-wood pair was taken from another eroding coastal bluff at Cape Ball on the east coast.

The ^{14}C concentrations in carbonate from coastal shellfish accurately mirror those in the surface ocean [Broecker, 1963; Mook and Vogel,1968], while wood provides a record of atmospheric ^{14}C. However, for these studies, we must also establish that the shell and associated wood sampled the ocean and the atmosphere at the same time. Careful selection of short-lived samples and avoidance of obviously disturbed sites can minimise the likelihood of errors from the deposition of old wood or from sediment reworking. Bias from other mixing processes, such as the penetration of old shell beds by roots from overlying plants, or the association of deep-burrowing shellfish with older buried wood, can be reduced by choosing subaerial material such as bark and small specimens of shallow-dwelling mollusc species. In practice, "noise" in the data is probably inevitable, and measurement of many pairs is extremely important for assessing dispersion in the data set and for identifying intrusive material.

In the present study, the richness of the sediment deposits and the use of AMS dating allowed the rapid measurement of many carefully selected small samples. We considered only well-preserved pairs of wood and shell samples in very close proximity (typically a few cm), and measured multiple pairs wherever possible. The presence of undisturbed statigraphic horizons in the sediment core, and of many intact bivalves in life position at the coastal sites, indicated that post-depositional reworking was minimal. The shells were all small or medium-sized specimens of intertidal and subtidal bivalve species. Most of the wood samples consisted of obviously short-lived materials such as bark, twigs, and small branches; or included the outer surface of a branch or stem. The small and fragile nature of some of the organic samples, and the presence of intact bark, suggest that the time between cessation of growth and deposition was probably rather short.

TABLE 2. Radiocarbon Measurements

Sample (RIDDL-)	Shell Sample [a]	∂13C ‰	14C Age, years B.P.	Sample (RIDDL-)	Wood Sample	∂13C ‰	14C age, years B.P.	Shell-Wood Difference
					Endacko 87A-23 Core			
978	25-32 cm: *S. giganteus*	0.4	10,770 ± 100	979	25-32 cm: shell contents	-26.6	9,940 ± 75	833 ± 123
980	40 cm: *C. nuttalli*	2.2	10,960 ± 85	981	41 cm: wood fragment	-25	10,485 ± 70	473 ± 113
982	50-53 cm: *C. nuttalli*	1.9	11,020 ± 80	983	54-55 cm: conifer twig or root	-27.4	10,290 ± 80	729 ± 111
				1433	80-86 cm: wood fragment	-27	10,370 ± 160	
				984	83-86 cm: wood fragment	-27.1	10,650 ± 350	
				985	93 cm: root	-27.6	10,470 ± 75	
							Mean (Endacko)	678 ± 107
					Cape Ball			
1191	*M. nasuta*	1.3	9,560 ± 140	1192	wood (small branch, incl. bark)	-24.1	9,090 ± 160	467 ± 209
					Mary Point Location B			
879	*S. giganteus*	0.4	9,930 ± 120	880	wood (twig, including bark)	-26.1	9,070 ± 110	870 ± 152
881	*P. staminea*	-0.8	9,710 ± 100	882	wood (small branch, incl. bark)	-24.8	8,930 ± 90	779 ± 127
883	*S. giganteus*	-0.1	9,790 ± 120	884	wood (including outer surface)	-23.3	9,070 ± 120	720 ± 165
1193	*S. nuttalli*	-0.4	9,780 ± 110	1194	bark fragments	-27.4	9,070 ± 90	711 ± 141
1195	*S. giganteus*	-0.4	9,870 ± 110	1196	wood (small branch)	-25.8	9,130 ± 120	741 ± 163
1205	*S. giganteus*	0	9,850 ± 130	1206	wood (including outer surface)	-26.4	9,290 ± 160	550 ± 202
1207	*S. giganteus*	-2.1	9,690 ± 110	1208	wood (small branch)	-26.2	8,800 ± 120	898 ± 158
1211	*S. giganteus*	0.2	9,560 ± 110	1212	wood (small branch)	-28.7	8,850 ± 120	710 ± 161
1213	*P. staminea*	0.2	9,760 ± 120	1214	wood (twig)	-25.4	8,860 ± 150	895 ± 187
1215	*S. giganteus*	-0.4	9,650 ± 170	1216	wood (small branch)	-25.2	8,950 ± 150	698 ± 222
1217	*S. giganteus*	-0.3	9,730 ± 100	1218	wood (twig)	-23.8	9,030 ± 100	694 ± 135
1219	*S. giganteus*	-0.4	9,670 ± 130	1220	wood (small branch, incl. bark)	-26.0	8,970 ± 130	704 ± 181
							Mean (Mary Point B and Cape Ball)	726 ± 48

a: Shell species: *Saxidomus giganteus*: Butter clam; *Macoma nasuta:* Bent-nose clam; *Schizothoerus nuttalli*: Horse clam; *Clinocardium nuttalli*: Cockle; *Protothaca staminea*: Native little-neck clam.

The shells were cleaned by washing and scraping, and small aliquots were strongly etched with 0.1N HCl and subsequently converted to CO_2 by hydroysis with 85% phosphoric acid. The wood samples were cleaned by washing, and small portions were subjected to acid-base-acid treatment with ultrasonic agitation, followed by sealed tube combustion with CuO. The techniques used for AMS sample preparation and measurement are described in detail in Vogel et al. [1987] and Nelson et al. [1986].

RADIOCARBON DATES

The full study involved a total of 58 ^{14}C dates on samples from these locations. To illustrate the quality of the data, results for two horizons from the late Pleistocene and early Holocene are shown in Table 2. We emphasize that no "outliers" have been discarded from this data set. Three of 17 measurements from a more recent horizon (not shown in Table 2) were discarded as anomalous, but two of these were from samples collected earlier to date the deposits, rather than specifically as a shell-wood pair. The other 55 dates all appeared valid. Thus, it does appear that consistent data can be obtained with this method if sufficient care is taken in sample selection.

Three dates on wood and in-situ root material from below 80 cm in the Endacko core give a weighted mean of 10440±60 BP. The lowest shell-containing layer is dated to 10390±100 BP by the wood dates from the lowest 2 shell-wood pairs. These results indicate that the marine transgression occurred close to 10400 BP, during the latter part of the Younger Dryas stadial. The 9940±75 BP date on detrital material in a shell found closer to the surface suggests that deposition may have continued at this site for some time. The data give a mean shell-wood difference of 678±107 years over this period.

The dates from the single shell-wood pair from Cape Ball are similar to those from Mary Point B (Table 2), and we assume that they represent the same post-glacial marine transgression. The combined data give mean wood and shell dates of 9010±40 BP and 9730±30 BP respectively, and the mean shell-wood difference is 726 ± 48 years.

Previous data on wood and shell ages from British Columbia sediments (eg., Fulton [1971]; Linden and Schurer [1988]; J.J.Clague, pers commun.) generally support high reservoir ages, but show shell-wood age differences ranging from 350 to 1130 years. Much of this variation is probably spurious, as the degree of association between wood and shell was in some of these measurements was rather uncertain. A comparison of shell and other archaeological material from four locations in the Gulf of Georgia ca. 400-1100

BP [Robinson and Thompson, 1980] showed low scatter and good agreement with the present-day reservoir age, highlighting the potential of archaeological sites in these studies.

Earlier measurements on a single shell-wood pair by another laboratory gave a reservoir age of 390±120 years for Cape Ball (9060±90 BP; GSC2867, 2884); (J.J.Clague, pers commun). This result and the reservoir age from our single Cape Ball pair are both low compared with those from the same time period at Mary Point. We place much more weight on the Mary Point B data because of the large number of pairs which gave concordant results, and because smaller reservoir ages may be artifacts of old wood effects, but the Cape Ball data nevertheless raise important questions about local variability.

DISCUSSION

We have derived past reservoir ages for the northeast Pacific by measuring ^{14}C age differences of shell-wood pairs from coastal sediments. The data indicate that some of the scatter in previous shell-wood comparisons can be avoided by careful sample selection and by carrying out multiple measurements. AMS dating provides the necessary high measurement throughput and small sample size capability.

The reservoir ages of 678±107 and 726±48 years for terrestrial ^{14}C ages of 10200 and 9000 BP, respectively, are very similar to the recent pre-industrial value of 788±33 years, and may be identical if some small bias from old wood effects is allowed for. If our results are representative of the entire region, it would appear that the strength of upwelling into the North Pacific subpolar gyre during the latter part of the Younger Dryas and the early Holocene was similar to that of the recent past. This provides a constraint on possible modes of North Pacific thermohaline circulation at the end of the Younger Dryas and supports the use of present-day reservoir ages from the region to link terrestrial and surface ocean chronologies in the period around the final Pleistocene/Holocene transition.

The data contain hints of local spatial variations of reservoir ages within this ocean gyre and further measurements are clearly needed to assess this possibility. Such variations would be important for establishing accurate regional and local chronologies, and also for the detailed information on ocean mixing and upwelling they would provide. We stress that reservoir age data for many areas of the ocean are very sparse, and emphasize the need for caution in linking terrestrial and marine chronologies until this issue has been explored further.

REFERENCES.

Broecker WS (1963) $^{14}C/^{12}C$ ratios in surface ocean water, National Academy of Science, Natural Resources Council Publication no. 1075, 138-149.
Broecker WS, Peng T-H, Ostlund G and Stuiver M (1985) The distribution of bomb radiocarbon in the ocean. J. Geophys. Res., 90, 6953-6970.
Broecker WS, Andree M, Bonani G, Wolfli W, Oeschger H, Klas M, Mix A and Curry W (1988) Preliminary estimates for the radiocarbon age of deep water in the glacial ocean. Paleoceanography, 3, 659-669.
Brown T, Nelson DE, Mathewes RW, Vogel JS and Southon JR (1989) Radiocarbon dating of pollen by accelerator mass spectrometry. Quat. Res., 32, 205-212.
Druffel ERM and Suess HA (1983) On the radicarbon record in banded corals: exchange parameters and net transport of $^{14}CO_2$ between atmosphere and surface ocean. J. Geophys. Res, 88, 1271-1280.
Fulton RJ (1971) Radiocarbon geochronology of southern British Columbia, Geological Survey of Canada paper GSC 71-37, Ottawa.
Linden RH and Schurer PJ (1988) The sediment characteristics and sea-level history of the Royal Roads Anchorage, Victoria, British Columbia. Can. J. Earth Sci, 25, 1800-1810.
Luternauer JL, Clague JJ, Conway KW, Barrie JV, Blaise B and Mathewes RW (1989) Late Pleistocene terrestrial deposits on the continental shelf of western Canada: evidence for rapid sea-level change at the end of the last glaciation. Geology, 17, 357-360.
Mook WG and Vogel JC (1968) Isotopic equilibrium between shells and their environment. Science, 159, 874-875.
Nelson DE, Vogel JS, Southon JR and Brown TA (1986) Accelerator radiocarbon dating at SFU, Radiocarbon, 28, 215-222.
Peng, T-H., and Freyer HD (1986) Revised estimates of Atmospheric CO_2 variations based on the tree-ring ^{13}C record, in: The changing carbon cycle: A global analysis, edited by J.R.Trabalka and D.E.Reichle, pp 151-159, Springer-Verlag, New York.
Robinson SW, and Thompson G (1981) Radiocarbon corrections for marine shell dates with application to southern Pacific Northwest Coast prehistory. Syesis, 14, 45-57.
Shackleton, NJ, Duplessy J-C, Arnold M, Maurice P, Hall Mand Cartlidge J (1988) Radiocarbon age of the last glacial Pacific deep water. Nature, 335, 708-711.
Southon JR, Nelson DE and Vogel JS (1990) A record of past ocean-atmosphere radiocarbon differences from the northeast Pacific. Paleoceanography, 5,

197-206.

Stuiver M, and Polach H (1977) Reporting of ^{14}C dat. Radiocarbon, 19, 355-363.

Stuiver M, and Quay PD (1980) Changes in atmospheric ^{14}C attributed to variable sun. Science, 207, 11-19.

Stuiver M, and Quay PD (1981). Atmospheric ^{14}C changes resulting from fossil fuel release and cosmic ray flux variability. Earth Plan. Sci. Lett., 53, 340-362.

Stuiver M, Burk RL and Quay PD (1984) $^{13}C/^{12}C$ ratios in tree rings and the transfer of biospheric carbon to the atmosphere. J. Geophys. Res., 89, 11731-11748.

Stuiver M, Pearson GW, and Brazunias T (1986) Radiocarbon age calibration of marine samples back to 9000 cal yr BP. Radiocarbon, 28, 980-1021.

van Campo E, DuplessyJ-C, Prell WL, Barratt N, and Sabatier R (1990), Comparison of terrestrial and marine temperature estimates for the past 135kyr off southeast Africa. Nature, 348, 209-212.

Vogel JS, Nelson DE and Southon JR (1987) ^{14}C background levels in an accelerator mass spectrometry system. Radiocarbon, 29, 323-333.

THE LAST DEGLACIATION IN ANTARCTICA : FURTHER EVIDENCE OF A "YOUNGER DRYAS" TYPE CLIMATIC EVENT

J. Jouzel[1,2], J. R. Petit[2,1], N. I. Barkov[3], J. M. Barnola[2], J. Chappellaz [2], P. Ciais[1], V. M. Kotkyakov[4], C. Lorius[2], V. N. Petrov[3], D. Raynaud[2] and C. Ritz[2]

1/Laboratoire de Géochimie Isotopique, CEA/DSM CEN Saclay
91 191 Gif-sur-Yvette, France
2/Laboratoire de Glaciologie et Géophysique de l'Environnement, B.P. 96
38 402 St Martin d'Hères, France
3/Arctic and Antarctic Research Institute, Beringa Street 38
199226 Leningrad, USSR
4/Institute of Geography, Staromonetny per 29
109017 Moscow, USSR

ABSTRACT

Dome C results have shown that the last climatic transition was, in this Antarctic ice core, a two-step process with two warming trend periods interrupted by a slightly colder period estimated to have taken place from about 12.8 to 13.6 kyr B.P. New results show that this cooling event is also well recorded in the Vostok record and probably in a new East Antarctic core drilled at Komsomolskaïa. Although apparently preceding the Younger Dryas dated from 11.0 to 10.2 kyr B.P. (^{14}C ages) it is possible that the Antarctic cold reversal is a Southern Hemisphere counterpart of the well documented Northern Hemisphere cold event 1) because the accuracy of ice core absolute dating is no better than 10 % and 2) because the dating of the Younger Dryas, itself, may require a correction of up to 2 kyr. Recent measurements of methane concentrations in the Vostok core should provide useful information for linking Northern and Southern Hemisphere observations. Additional information concerning this period also includes the CO_2 record and the changes in dust concentration, with data now available both for Dome C and Vostok cores.

NATO ASI Series, Vol. I 2
The Last Deglaciation: Absolute and Radiocarbon Chronologies
Edited by E. Bard and W. S. Broecker

I - INTRODUCTION

The Younger-Dryas is a climatic stage which took place during the second half of the last deglaciation and was originally defined for a pollen zone in Europe (Jensen, 1938 ; Iversen, 1954). It corresponds to a cold event spanning approximately a millennium from ~11 to 10 kyr ^{14}C ages B.P. (Before Present) which followed the warming trend of the Bölling and Alleröd interstadials.

Available data on the Younger Dryas were recently compiled by Rind et al. (1986) both for Europe, where this event is best recognized, and for the whole world. They comprise terrestrial pollen records which indicate that trees which started growing in response to the climatic warming of the deglaciation were suddenly replaced by shrubs and herbs characteristic of a glacial regime, isotope records from the Camp Century and Dye 3 Greenland ice cores (Dansgaard et al., 1982 ; Oeschger et al., 1984) and isotope records from lake sediments in France (Eicher et al., 1981) and Switzerland (Oeschger et al., 1980).

This is illustrated in figure 1 which shows the remarkable correlation between lake Gerzensee (Switzerland) and Dye 3 (Greenland) climatic records, as deduced from oxygen18 profiles over this period ($\delta^{18}O$ expressed in per mil with respect to SMOW, the Standard Mean Ocean Water). At Dye 3 the cooling corresponding to the Younger Dryas is estimated at ~ 7 °C and recent studies (Dansgaard et al., 1989) have shown that this event ended very abruptly and possibly in less than 20 years with a transition characterized by a sudden shift in nearly all parameters studied in this core (deuterium-excess, chemical trace elements, acidity and continental dust).

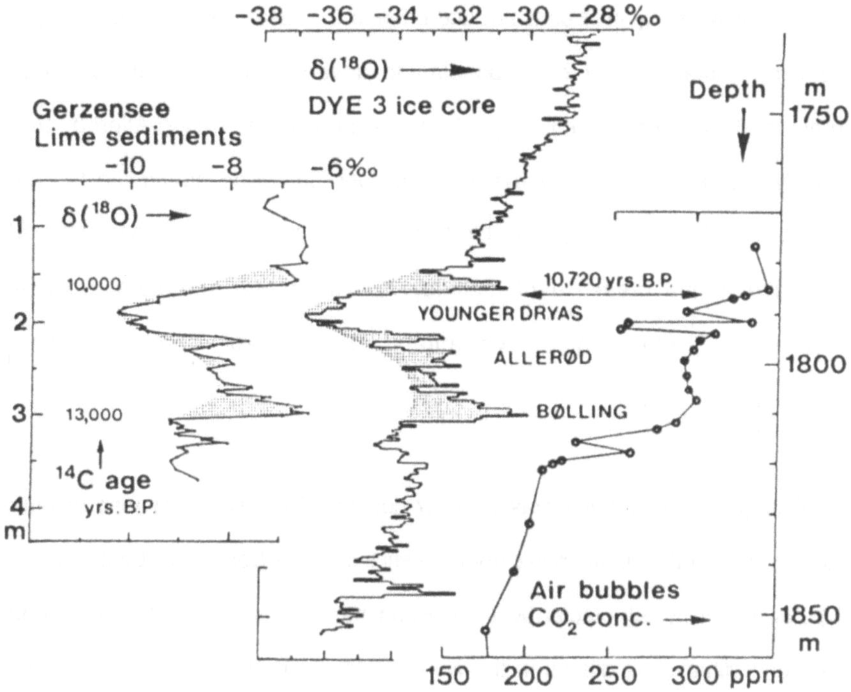

Figure 1 : Detailed ^{18}O profile along a 120 m increment of the Dye 3 deep core containing ice deposited during the entire Pleistocene to Holocene transition (Middle curve). CO_2 concentration in air bubbles and ^{18}O for lime sediments in the Swiss lake Gerzensee are shown on the right and left curves respectively (adapted from Dansgaard and Oeschger, 1989).

Observed clearly in Greenland, and often in European records, the Younger Dryas is a well defined climatic event corresponding to a very significant change in climatic and environmental conditions with respect to the preceding Bölling-Alleröd period, with a particularly abrupt transition towards the conditions prevailing during the Holocene. On the other hand, the change in CO_2 concentration shown in figure 1 has to be confirmed because it may be due to the presence of melt layers. This profile therefore probably does not represent true atmospheric variations (Dansgaard and Oeschger, 1989 ; Stauffer et al., 1985).

During the Younger Dryas a major cooling of the North Atlantic ocean took place, resulting in the North Atlantic polar front advancing to the South and East to a position about 5 °C poleward of its full glacial position (Ruddiman and McIntyre, 1981). Indeed, the close link between climatic conditions in the North Atlantic and the Younger Dryas cooling is well recognized (Rind et al., 1986 ; Broecker et al., 1985 ; Broecker and Denton,1989 ; Fairbanks, 1989 ; Jensen and Veum, 1990) although the mechanisms involved and their relation with the complex melting history of the Northern Hemisphere ice sheet are largely debated (Shackleton, 1989 ; Fairbanks, 1990). Another indication of the specific role of the North Atlantic is that, for North America, only regions adjacent to the Atlantic report uniequivocal evidence of a climatic oscillation (Rind et al., 1986 ; Peteet et al., 1990). This pattern of a more intense cooling over Greenland, Europe and the eastern part of North America generally agrees with the results of a climate simulation in which the sensitivity of global climate to a colder North Atlantic sea surface temperature was investigated (Rind et al., 1986).

These authors noted that outside the regions under the influence of the North Atlantic, such evidence for a Younger Dryas cooling was not convincing (Figure 2). However, recent results such as those obtained in the Caribbean and Gulf of Mexico (Overpeck et al., 1989

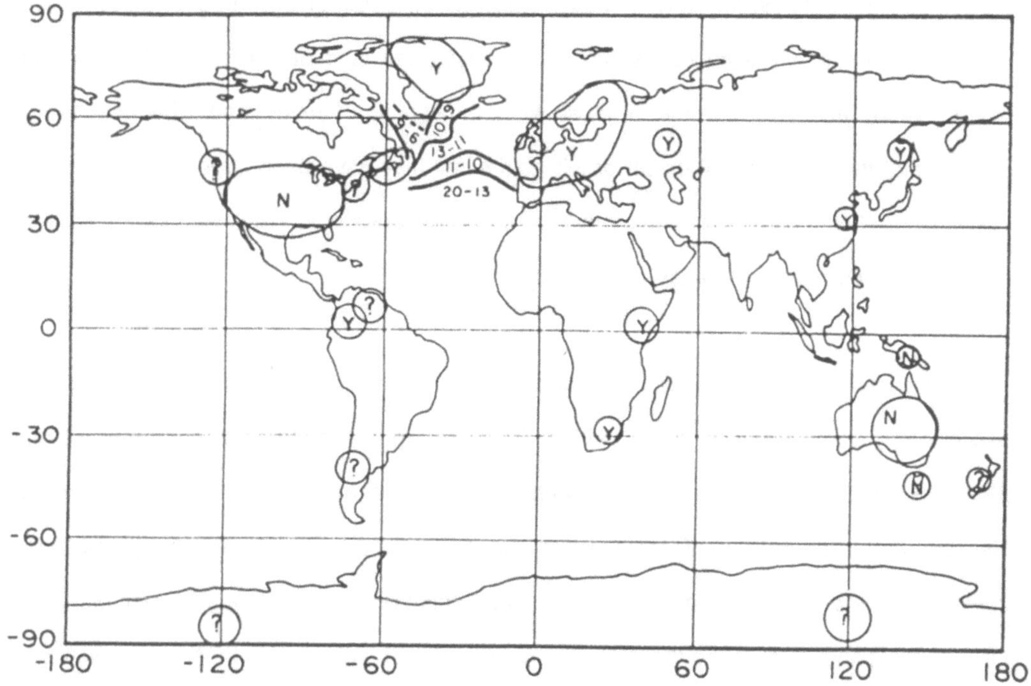

Figure 2 : This map adapted from Rind et al. (1986) shows the worldwide assessment of the published results indicating the presence (Y), absence (N) or possible (?) paleoclimatic indication of a late-glacial oscillation which may or may not be correlative with the Alleröd-Younger Dryas. Also included is the location of the North Atlantic polar front for different time periods (Ruddiman and McIntyre, 1981). Deglacial retreat was interrupted by the readvance from 11-10 kyr B.P. This map does not take into account the results obtained since the review paper of Rind et al. (See text).

Flower and Kennet, 1990), North Africa (Street-Perrott and Perrott, 1990 ; Gasse et al., 1990), Southeastern Alaska (Engstrom et al., 1990), the Gulf of California (Keigwin and Jones, 1990) and even in the tropical western Atlantic (Linsley and Thunnel, 1990) suggest that the geographical extent of this cold event may have been larger in the Northern Hemisphere. This has been confirmed by recent investigations in the Sulu Sea in the Eastern Indonesian Archipelago (Kudrass et al., 1991).

For the Southern Hemisphere, there is some indication of a late-glacial climatic oscillation in South America, South Africa, South Georgia and New Zealand (Rind et al., 1986) as recently confirmed for New-Zealand by Newnham (unpublished manuscript). Heusser and Rabassa (1987) established the correspondence between the late-glacial cooling recorded in southernmost South America and the Younger Dryas but for Markgraf (1989), the concept of a Younger Dryas equivalent in southern South America was too simplistic to correctly describe the late-glacial climatic changes in this region. In a recent study, Labracherie et al. (1989) also showed that the last deglaciation in the southern ocean was interrupted by a cooling phase similar to the Younger Dryas. However, detailed accelerator mass spectrometry (AMS) dating shows that this cooling took place between 12 and 11 kyr B.P. (^{14}C ages), therefore preceding the Younger Dryas by about 1 kyr.

Antarctic ice cores offer a unique opportunity to obtain high resolution records of the last deglaciation providing many parameters relevant to climate and environmental changes over this period. In an article dealing with the Antarctic ice record during the late Pleistocene, Jouzel et al. (1987a) reviewed the isotopic results available for Byrd and Dome C, indicating that the warming associated with the last climatic transition was a two-step process with two warming trend periods interrupted by a cold reversal. These authors noted that this cold reversal may have preceded the Younger Dryas by 1 kyr or more. The objective of the present article is to extend this study of the last deglaciaton in Antarctica, first by examining the isotopic results now available for the Vostok core, and for a new core drilled at Komsomolskaïa in East Antarctica, and second, by discussing two other types of relevant parameters namely the dust fallout and the concentration in trace gases (in particular in methane).

II - THE ISOTOPE RECORDS

Figure 3 gives the locations of the four Antarctic sites discussed in this article. Figure 4 shows the deuterium data (expressed in per mil versus SMOW) with respect to depth for the three East Antarctic sites discussed in this article together with the oxygen 18 profile measured along the Byrd ice core in West Antarctica (Johnsen et al., 1972) for the period covering the Holocene and the last deglaciation.

The 905 m long ice core obtained at Dome C, an ice divide region, (annual temperature : -53.5 °C and mean accumulation : 3.4 g cm^{-2} yr^{-1}) has previously been studied both for the full record (Lorius et al., 1979 ; Jouzel et al., 1982, 1989) and over the last two millenia (Benoist et al., 1982). The excellent core recovery has made isotope measurements possible on a quasi continuous basis.

At Vostok station the present mean temperature is -55.5 °C with a snow accumulation rate of about 2.3 g cm^{-2} yr^{-1}. A drilling program initiated two decades ago reached a depth of 2546 m in 1989. The climatic record for the top 2083 m, which spans a whole climatic cycle, has been presented elsewhere (Lorius et al., 1985 ; Jouzel et al., 1987b ; Jouzel et al., 1989). Four cores cover at least the Holocene period and the climatic transition corresponding to depths from the surface to ~ 350 m. The deuterium profile published by Jouzel et al. (1987b) was obtained on core 3Γ, below 138 m, and on a new adjacent core (4Γ) from the surface down to 279 m using 1-m ice increments. We used the new 4Γ drilling to obtain a more detailed (0.5 m) profile for the last deglaciation. It was also noted that at the time of the cold reversal which occurred during the transition, there

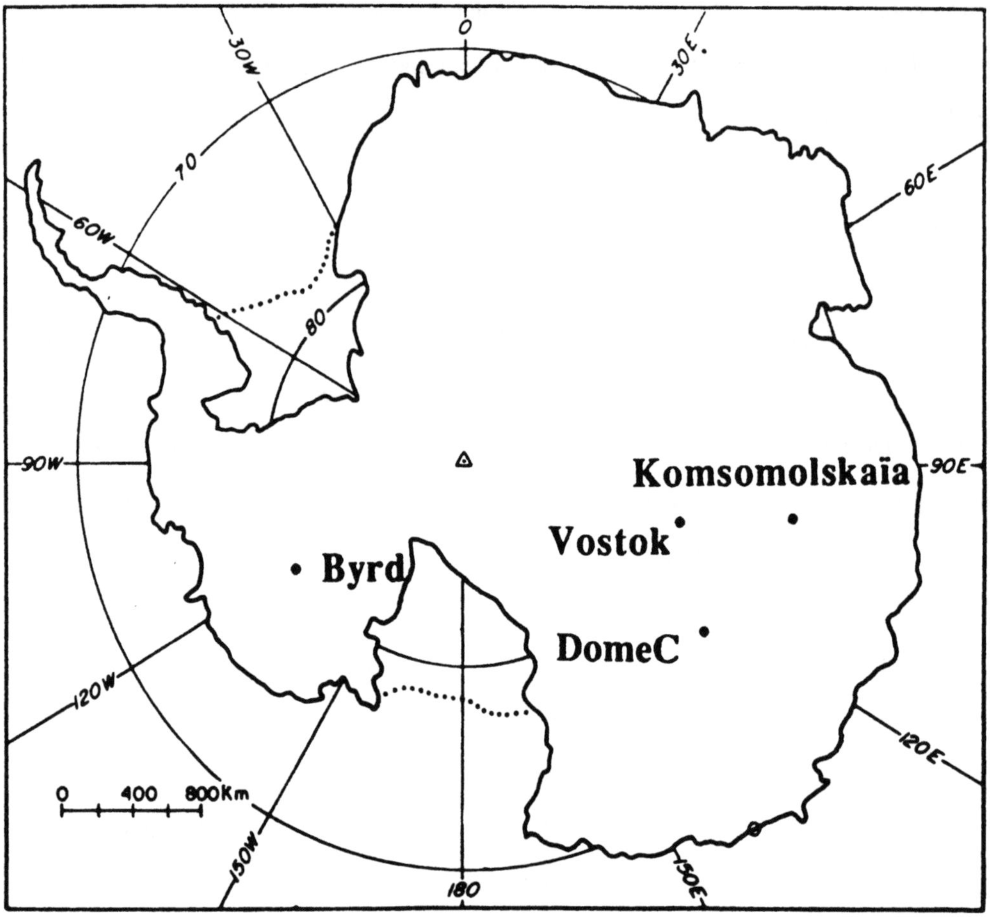

Figure 3 : Map of Antarctica showing the location of the drilling sites

was some missing ice (8 m) in the 3Γ core. The new dataset now available from 280 to 400 m which extends the 4Γ record down to 400 m allowed us to fill this gap and to obtain a continuous record for the last deglaciation. The record shown in figure 4 is the one obtained along 4Γ by combining these new data (from 280 to 400 m) and previously published data for the same core (Jouzel et al., 1987b). The two isotope records (3Γ and 4Γ) are shown separately in Figure 5.

The new deep ice core presented here was extracted by the Soviet Antarctic Expeditions at summer camp of Komsomolskaia in 1983 and extends down to 850 meters. The site location is 74°05' S, 97°29' E with an annual temperature of -52.6 °C and an average accumulation rate at least a factor of two higher than at Vostok station (about 5 g cm^{-2} yr^{-1} after Young et al., 1982). We have carried out deuterium measurements on two meter long sections, discontinuously selected every five meters along the core. Ice is missing between 77 m and 120 m and the resolution is reduced to 1 sample every 25 meters below 725 m. Note that an oxygen 18 profile was obtained in parallel at the Institute of Geography in Moscow (Nikolaïev et al., 1988). A detailed comparison of the two isotopic profiles (δ^2H and $\delta^{18}O$) and of the variation in deuterium-excess (δ^2H - 8 $\delta^{18}O$) will be reported elsewhere.

Lastly, we have for Byrd (mean annual temperature : - 28 °C ; accumulation : 16 g cm^{-2} yr^{-1}) reported the upper part (down to 1600 m) of the $\delta^{18}O$ record of Johnsen et al. (1972).

III - ICE CORE CHRONOLOGY

Owing to the very low accumulation values over the central part of the East Antarctic Plateau, both removal processes at the surface and diffusion of isotopic species in firn may smooth out the seasonal variations of the isotopic signal (Johnsen, 1977 ; Whillans and

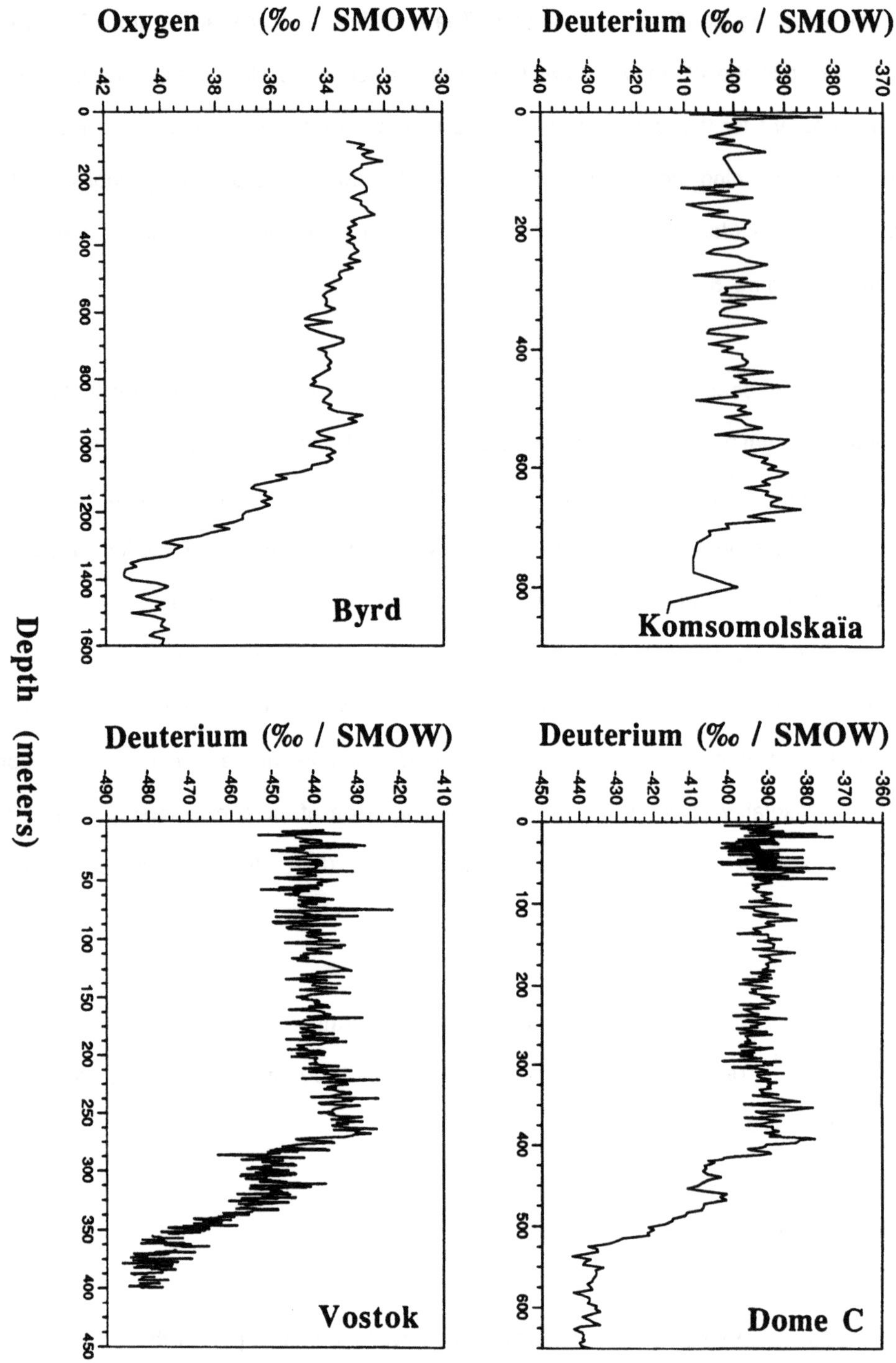

Figure 4 : Isotopic ice versus depth of the four Antarctic deep ice down to the depth corresponding to ~ 18 kyr BP. Deuterium data are reported for Vostok (core 4Γ), Dome C and Komsomolskaia and Oxygen 18 data (from Johnsen et al.;1972) for Byrd.

Grootes, 1982) which can no longer be used for dating purposes. The method that was employed is thus based on specific ice flow models. In central East Antarctica from where the three cores originate, the ice sheet velocities do not exceed a few meters per year and the Holocene period covers depths sufficiently distant from the bedrock so that the sequences are quasi undisturbed by ice flow conditions. We review here how the Dome C and Vostok time scales have been established.

For the Dome C core, located on an ice dome region, a simple one-dimensional vertical thinning model, assuming constant vertical strain rate, was used (Lorius et al., 1979). It was proposed to account for the fact that the accumulation was lower during the glacial than during the Holocene period because the amount of precipitation over the Antarctic plateau is essentially controlled by the mass of water vapour available in the inversion layer and thus by the temperature of this layer (Robin, 1977). Lorius et al. (1979) simply assumed that the accumulation during the glacial part of the record was 75% of that during the Holocene, with a linear increase toward the Holocene value during the climatic transition. From ^{10}Be measurements, Raisbeck et al. (1981) suggested that this decrease was larger (around 50%) implying that the bottom of the core was older than 32 kyr B.P. This variable accumulation approach was further refined by Benoist (1986), who assumed that the accumulation rate was directly proportional to the derivative of the saturation vapour pressure, derived itself from the temperature record. As fully discussed in Jouzel et al. (1989), such an approach of estimating accumulation change although having some limitations (see below) is fairly well supported by several independent arguments. This improvement significantly extended the Dome C chronology at the bottom of the core from 32 to 39 kyr B.P. (Benoist, 1986 ; Jouzel et al., 1989). For cold reversal part of the record, which we consider here, the difference between the two chronologies never exceeds a few hundred years.

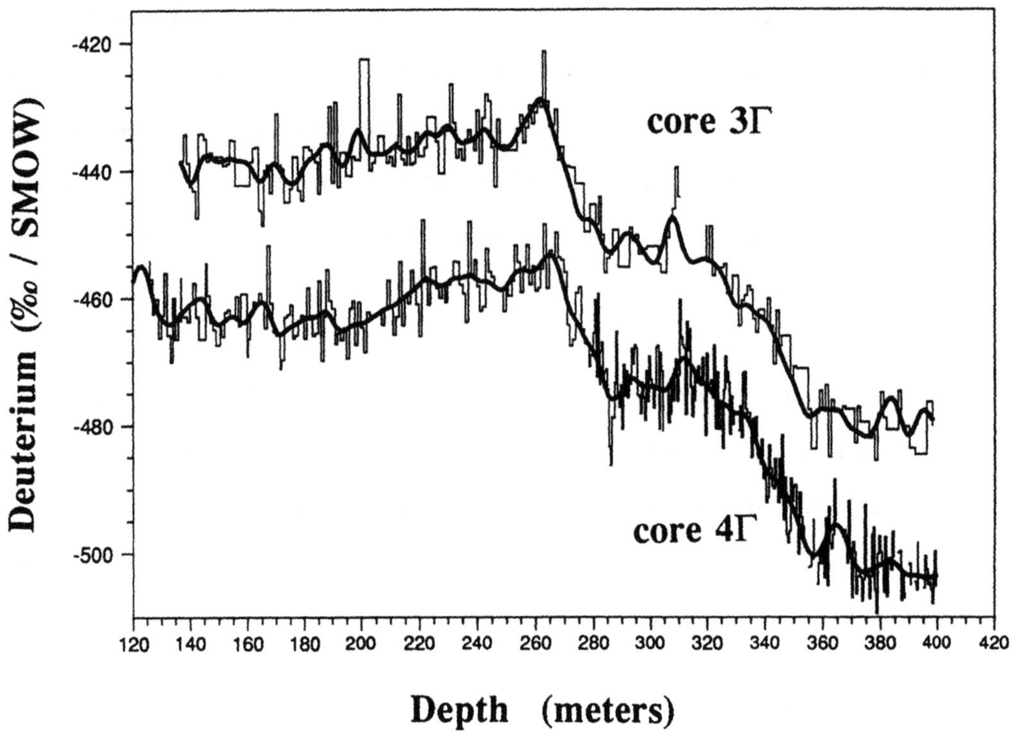

Figure 5 : Isotopic profiles versus depth from adjacent cores 3Γ and 4Γ at Vostok station between 138 and 400m.

The chronology of Vostok is based on a two-dimensional steady state glaciological model in which are introduced the known bedrock and surface elevations along the flow line (Lorius et al., 1985 ; see Ritz, 1991 for a full description). In this dating the saturation vapour pressure approach was used, based on the discontinuous oxygen 18 profile available in 1985. This dating was further slightly modified using the continuous deuterium profile (Jouzel et al., 1989) with no noticeable change in the Holocene part of the record. Remarkably, these Dome C and Vostok datings with common use of the saturation vapour pressure approach put the ^{10}Be peaks discovered at Vostok around 600m and at Dome C around 820m (Raisbeck et al., 1987), practically in phase, so that these datings with a further minor adjustment (3 % for Dome C) were adopted by Jouzel et al. (1989). We have, in Figure 6, reported the 18 to 8 kyr BP part of the Dome C and Vostok deuterium records using these time scales

As previously discussed (Jouzel et al., 1987a), the transition is at Dome C a two-step process with two warming trend periods interrupted by a cold oscillation. The two warming trends took place from 17.2 to 13.7 and to 11.7 to 10.2 kyr B.P. respectively with the colder period lasting ~1.5 kyr from 13.2 to 11.7 kyr B.P. Because of the Dome C redating, these figures are different from those of Jouzel et al., (1987a), i.e 15.5 to 13.2, 11.6 to 10.2 kyr B.P., for the two warming trends respectively and 12.8 to 11.6 kyr BP for the colder period. This cold reversal is now well marked in the Vostok 4Γ isotope record but starts about 1.5 kyr later than at Dome C. This phase lag is probably not real but due to dating inaccuracy. This is well supported by available dust data that we will now discuss briefly.

The interest of examining how dust concentration varied during the last deglaciation in ice cores is twofold : first, large changes in the atmospheric dust loading (which are

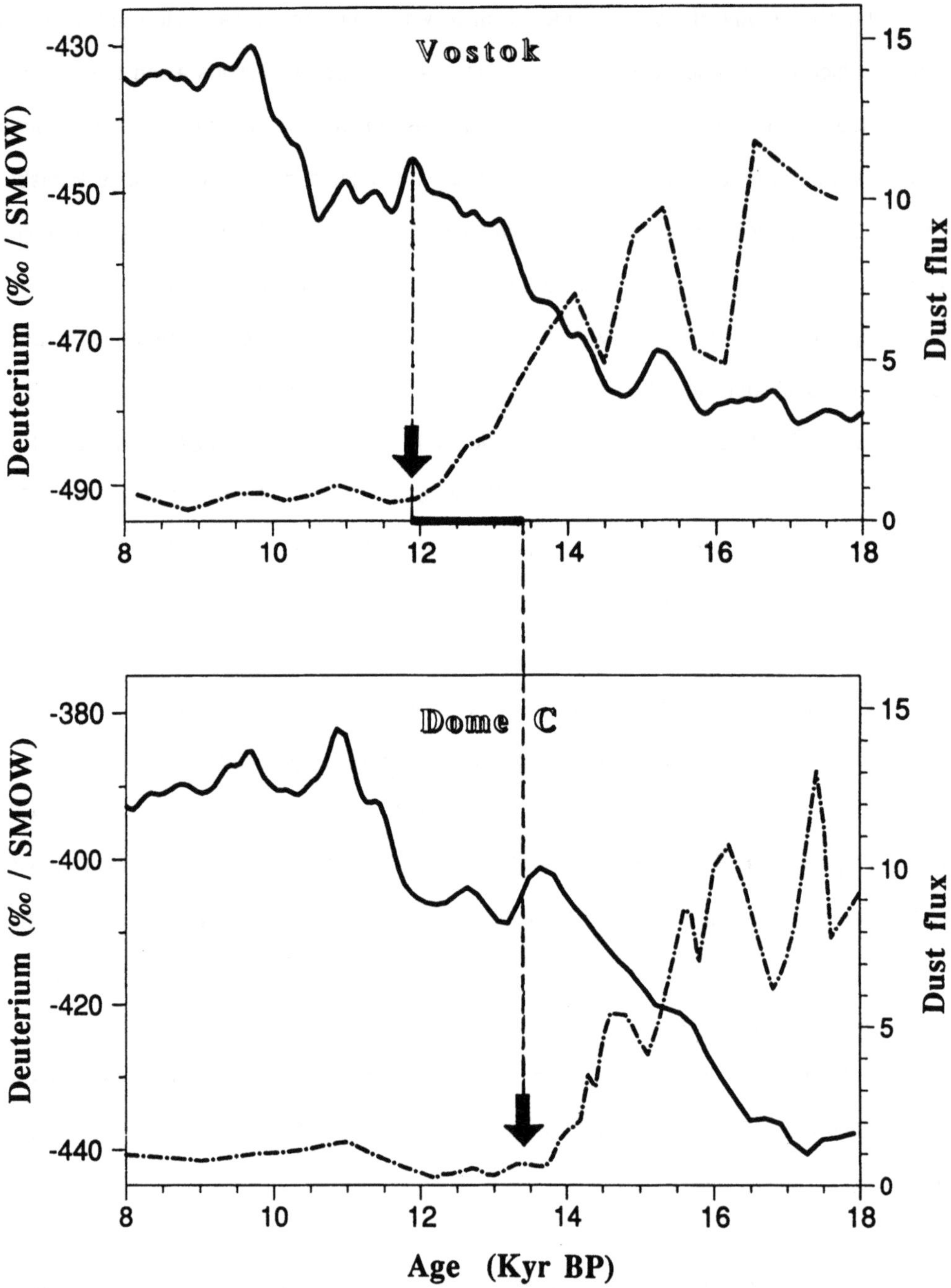

Figure 6 : Comparison of the Dome C and Vostok dust records over the 8 to 18 kyr B.P. period along with the smoothed deuterium records with datings from Jouzel et al.(1989). The dust fallout is expressed in mg m^{-2} yr-1 assuming a particle density of 2. The two arrows indicate the point for which "Holocene" dust concentrations are reached.

recorded in ice) may have a potential climatic role (Petit et al., 1981 ; Royer et al., 1983 ; Harvey, 1988 ; Broecker and Denton, 1989) ; second, dust events may serve as stratigraphic markers between ice core and other dust records (Petit et al., 1990). One common and striking feature of the Dome C and Vostok dust records is the large dust fallout increase recorded in both cores during the Last Glacial Maximum with respect to Holocene values. Available data including part of this dust peak are reported in Figure 6. This figure shows that dust fallout values close to those corresponding to the Holocene are reached at the end of the first warming trend and appear to prevail, to the extent of available data, during the cold reversal (as illustrated on Figure 9 the situation is different for the Younger Dryas in the Northern Hemisphere with a well documented dust peak at Dye 3). As far as the relevance of the dust records as stratigraphic markers is concerned, we note that the end of the dust peak, although occurring at the same isotopic level in the two cores, is referring to the time scales of figure 6, dated 1.5 kyr earlier at Dome C (13.4 kyr B.P.) than at Vostok (11.9 kyr B.P.). This suggests, but would have to be confirmed through higher resolution dust records, that at least one of the Dome C and Vostok time scales has to be revised for this deglaciation part of the records.

Provisionally, we have chosen to put the Dome C and Vostok cold reversals in phase by simply decreasing the Vostok modern accumulation rate by 12 % and keeping the Dome C chronology unchanged (Figure 7). This choice is quite arbitrary but is adopted here because it has the advantage of putting the beginning of the cold reversal in phase with that also seen in the oxygen 18 record of the Byrd ice core (Figure 7) dated independently, as proposed by H.B. Clausen and C.U. Hammer (personal communication, 1988). This Byrd time scale is based on the existence of pronounced seasonal variations in the detailed acidity record that allowed estimates to be made of past accumulation changes. The smoothed Byrd $\delta^{18}O$ record is reported in Figure 7. The cold reversal took place at Byrd from ~ 13.2 to 12.5 kyr B.P. with an amplitude comparable to that of Dome C but on a shorter period (however its relative importance with respect to the deglaciation signal is less at Byrd than at Dome C)

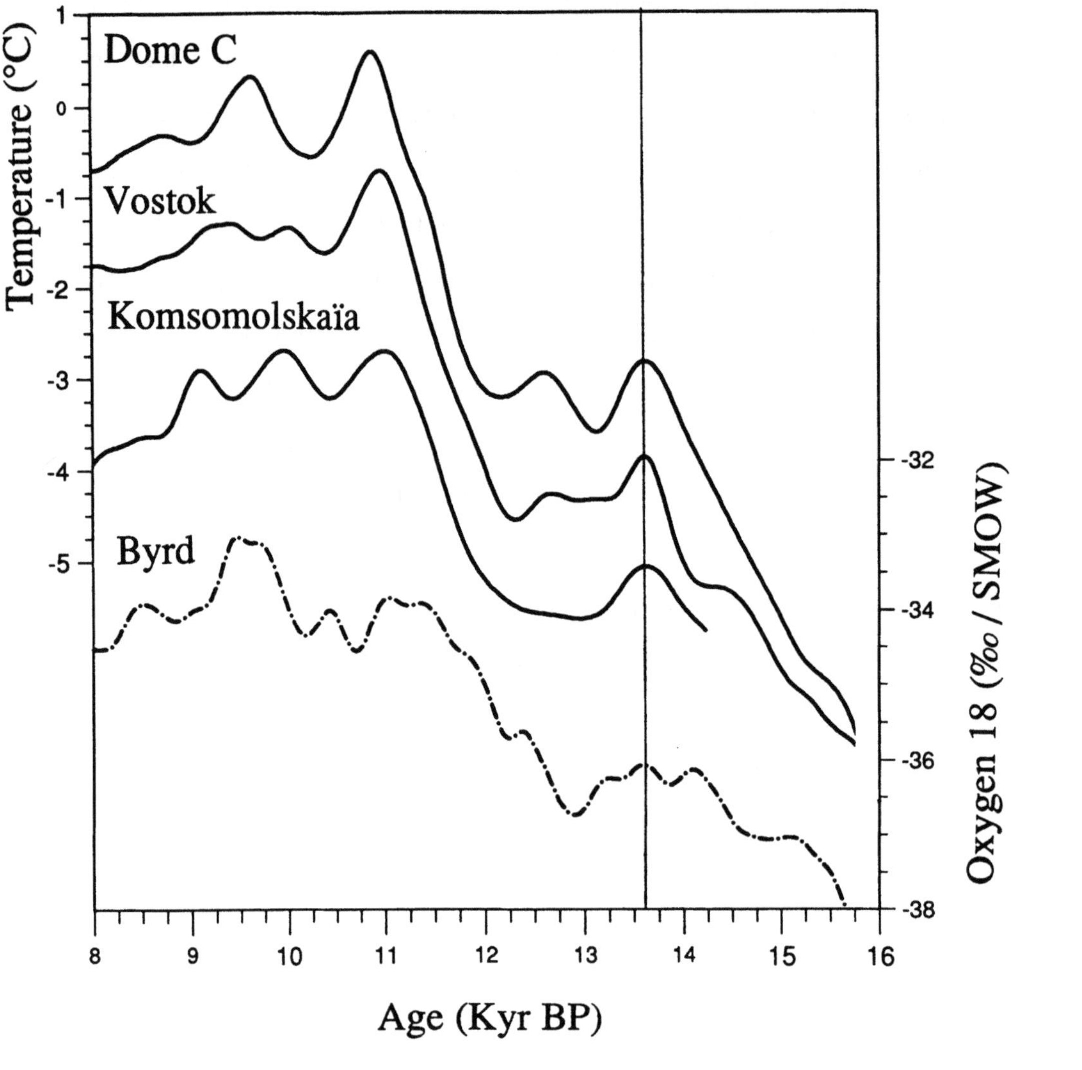

Figure 7 : Comparison of the temperature profiles for the 3 East Antarctic cores with the Vostok time scale modified as described in the text along with the $\delta^{18}O$ profile along the Byrd West Antarctic core (lower curve)

This strategy has thereby the advantage to keep a consistent chronology for the Dome C, Vostok and Byrd ice cores for the last deglaciation. Although the dust fallout at Byrd (Thompson, 1977) is not sufficiently documented in this part of the record to confirm this stratigraphic correlation, this appears well justified. On the other hand, keeping the Byrd and Dome C time scales unchanged rather than the Vostok time scale is a quite arbitrary choice. Indeed, the Byrd time scale used by Beer et al. (this volume) and based on ice flow and accumulation change modeling is for the cold reversal ~ 1 kyr younger than that of Clausen and Hammer (Johnsen, personal communication). This points out the difficulty of deriving absolute time scales for deep antarctic ice cores and must be kept in mind in the remaining part of the discussion.

This difficulty is further illustrated when comparing the Dome C and Vostok chronologies reported in Figure 7 down to the bottom of the Dome C core. A consequence of the adopted strategy is that the ^{10}Be peaks around 35 kyr B.P. are put out of phase of 12 % which is clearly unsatisfactory and has to be further discussed. This unability of the water vapour pressure approach to put in phase both the end of the dust peaks during the deglaciation and the ^{10}Be peaks around 35 kyr B.P. certainly poses a lower limit to the accuracy of this approach.

For Komsomolskaïa, the glaciological model applied at Dome C may be used as a first approximation but we face the problem of a very poor estimate of the modern accumulation, the representativeness of available data, 4.6 to 5.3 g cm^{-2} yr^{-1}, being limited to 4 years of accumulation (Young et al., 1982). Using 5 g cm^{-2} yr^{-1} as the modern accumulation leads to an age of 18 kyr B.P. at the bottom of the core which is quite unrealistic given that the isotope profile indicates that this bottom ice was deposited during the climatic transition. Indeed, this profile suggests that the ice between 700 and 765m corresponds to the cold reversal and we have used the assumption that this event is in phase with that recorded at

Vostok and Dome C to obtain the time scale of Figure 7. This corresponds to taking a value of 6.6 g cm^{-2} yr^{-1} for the modern accumulation rate.

IV - TEMPERATURE RECONSTRUCTION

With this provisionally adopted chronology, the three East Antarctic records (Dome C, Vostok and Komsomolskaïa) are reported in figure 7 on a temperature scale. We only briefly recall here the generally well accepted climatic interpretation of these ice core isotope records. Isotopic fractionation processes which occur during the atmospheric water cycle lead to a general decrease of the δ values with temperature. For modern polar precipitation, there is a well explained linear relationship between the annual averages of the δ values and the surface temperature T_S (Dansgaard et al., 1973 ; Lorius and Merlivat, 1977). The existence of such a relationship is well understood through various theoretical approaches (Jouzel and Merlivat, 1984 ; Jouzel et al., 1987c ; Fisher, 1990 ; Joussaume and Jouzel, submitted). The observed isotope-temperature slope is of 6 ‰/°C for the deuterium in East Antarctica (Lorius and Merlivat, 1977). Correspondingly, it is assumed that, at a given site, snow which fell under glacial conditions is isotopically impoverished with respect to modern snow (Figure 4). Before being translated into a temperature scale, isotope profiles must be corrected for the isotopic changes of sea water (Labeyrie et al., 1987) leading to an estimate of ΔT_S of ~ 9 °C between glacial and interglacial conditions at Vostok and Dome C and ΔT_S ~ 12 °C at Byrd. This interpretation based on the use of the observed deuterium-temperature slope for modern precipitation, refers to the temperature change at the precipitation site. It does not account for the possible changes of elevation either due to the movement of the ice from its origin, or the general changes in the thickness of the ice sheet, which could explain the higher Byrd ΔT_S value (although the difference between East and West Antarctica may be real, climatic interpretation of the Byrd profile must be done very cautiously and this is why it is preferable to keep the $\delta^{18}O$ record). Note lastly that due to the presence of a strong inversion over Antarctica, the temperature change in the atmosphere above the inversion, ΔT_a, is lower than at the surface with $\Delta T_a \sim 2/3\ \Delta T_s$ (Lorius et al., 1979 ; Jouzel et al., 1987b) i.e. around 6 °C at Dome C and Vostok.

The amplitude of the cold reversal defined peak to peak is of ~10 ‰ in δ^2H at Dome C (Figure 4) corresponding to a ΔT_s of ~1.5 °C. At Vostok, this amplitude again defined peak to peak, is greater than 20 ‰ in δ^2H for the 4Γ core (see Figure 5) corresponding to a ΔT_s of 3 to 4 °C. One very likely reason for this large amplitude is indeed the higher time resolution of the Vostok record (1 sample represents about 25 yr at Vostok instead of 100 yr at Dome C). A significant part of this amplitude is related to the presence of two spikes which may be missing at Dome C due to the coarser time resolution. As this part of the Komsomolskaïa record is essentially discontinuous (4 samples for the cold reversal) the temperature amplitude cannot be estimated at this site.

V - CO_2 and CH_4 RECORDS

Measurements of contents in CO_2 and CH_4 in the air bubbles trapped in Greenland and Antarctic ice cores have revealed substantial concentration variations during the last climatic transition. The CO_2 level increased from a glacial value of ~200 ppmv (parts per million) to an interglacial value of 280 ppmv (Delmas et al., 1980 ; Berner et al., 1980 ; Neftel et al.,1982 ; Stauffer et al., 1985 ; Barnola et al., 1987) and the CH_4 concentration almost doubled from 350 to 650 ppbv (Stauffer et al., 1988 ; Chappellaz et al., 1990). Detailed knowledge of how these changes occurred and how they are temporally related to climate change and, in particular, to an event such as the Younger Dryas, is crucial. Because the two greenhouse gases are globally well mixed, at the timescales considered here, the ice core records of CO_2 and CH_4 may, in addition, be helpful in examining the timing between the Northern Hemisphere Younger Dryas oscillation and the cold Antarctic reversal.

In addition to the obtention of a detailed set of measurements, an accurate determination of the air-ice difference in age is required. This is highly dependent on the accumulation rate at the site. For the Antarctic site of Byrd, and for the Greenland sites of Dye 3 and Camp Century (all with a present-day accumulation higher than 20 g cm^2 yr^{-1}) the present-day difference of age between the enclosed air and the surrounding ice lies between 100 and 250 years and the temporal representativiteness of a sample is around a few tens of years (Schwander and Stauffer, 1984). Thus, any phase relationship between the climate and the atmospheric composition changes can be determined with a relatively good accuracy. In the case of Dome C and Vostok low accumulation sites, (3.4 and 2.3 g cm^{-2} yr^{-1} respectively) these temporal characteristics increase by one order of magnitude with respective values of 1700 and 2600 years for the air-ice age difference and in the 300-600 year range for the mixing time of the air (Schwander and Stauffer, 1984 ; Barnola et al., 1991). Thus for these two sites, the phase relationship strongly depends on the accuracy of the age difference between the air and the ice.

Barnola et al. (1991) reevaluated the air-ice difference in age at Vostok, based on assumptions different from those previously considered (Barnola et al., 1987 ; Chappellaz et al., 1990). This does not change significantly the estimate for the present-day conditions but results in younger ages than previously assumed for glacial conditions (by about 2000 years during the coldest stages). For the period of interest in this paper - the Antarctic cold reversal -, the new estimate of the air dating reduces the age of the air by 300 to 600 years.

Values in CO_2 concentrations obtained from the Greenland ice cores of Camp Century (Neftel et al., 1982) and Dye 3 (Stauffer et al., 1985) for the last deglaciaton both reveal the 30 % increase typical of the major climatic transition. In the case of Dye 3 (figure 1) values close to 300 ppmv are obtained at the beginning of the Bölling period (1812 m in depth scale) but Stauffer et al. (1985) consider that all the CO_2 values from samples above 1815m could be affected by melt layers. Thus the CO_2 profile of figure 1 most probably do not represent the true atmospheric variations and a good CO_2 record from the Northern Hemisphere is still needed.

In the Southern Hemisphere, CO_2 data are available from three cores : Byrd (Neftel et al., 1988), Dome C (Raynaud and Barnola, 1985) and Vostok (Barnola et al., 1987 ; Barnola et al., 1991). In Figure 8, we present the CO_2 results corresponding to the last deglaciation from Dome C (shaded circles) and Vostok (black circles) cores along with the Vostok climatic record. Three levels around 11 kyr B.P. of the Dome C core showing CO_2 values higher than present have been excluded because of their very low gas content. The air dating was obtained using the ice chronologies of figure 7 and calculating the air ice age difference following the procedure of Barnola et al. (1991). The profile resulting from this compilation suggests a CO_2 increase in two steps, separated by a relatively constant level of about 230 ppmv during the cold reversal, but we remain cautious in the interpretation of the duration of this CO_2 plateau because of the air dating uncertainties for each of the two cores. In the Byrd core, Neftel et al. (1988) have obtained a very detailed CO_2 profile with unfortunately a data gap approximately at the time of the cold reversal. Nevertheless, the interpolation of the data suggests a slower CO_2 increase during this period, and thus does not contradicts the Dome C and Vostok data. Furthermore, even if more measurements are needed to determine the detailed CO_2 increase during the last deglaciation, it is important to note that none of these three Antarctic records shows either CO_2 values of about 300 ppmv, or a strong drop in CO_2 concentration during the entire deglaciation, as suggested by Dye 3.

Values in CH_4 concentration for the period of the last deglaciation have been published in two recent studies (Stauffer et al., 1988 ; Chappellaz et al., 1990). We show in Figure 8 the results obtained on the Vostok ice core (Chappellaz et al., 1990). They reveal a large CH_4 oscillation from about 650 ppbv and 480 ppbv, at the time of the cold

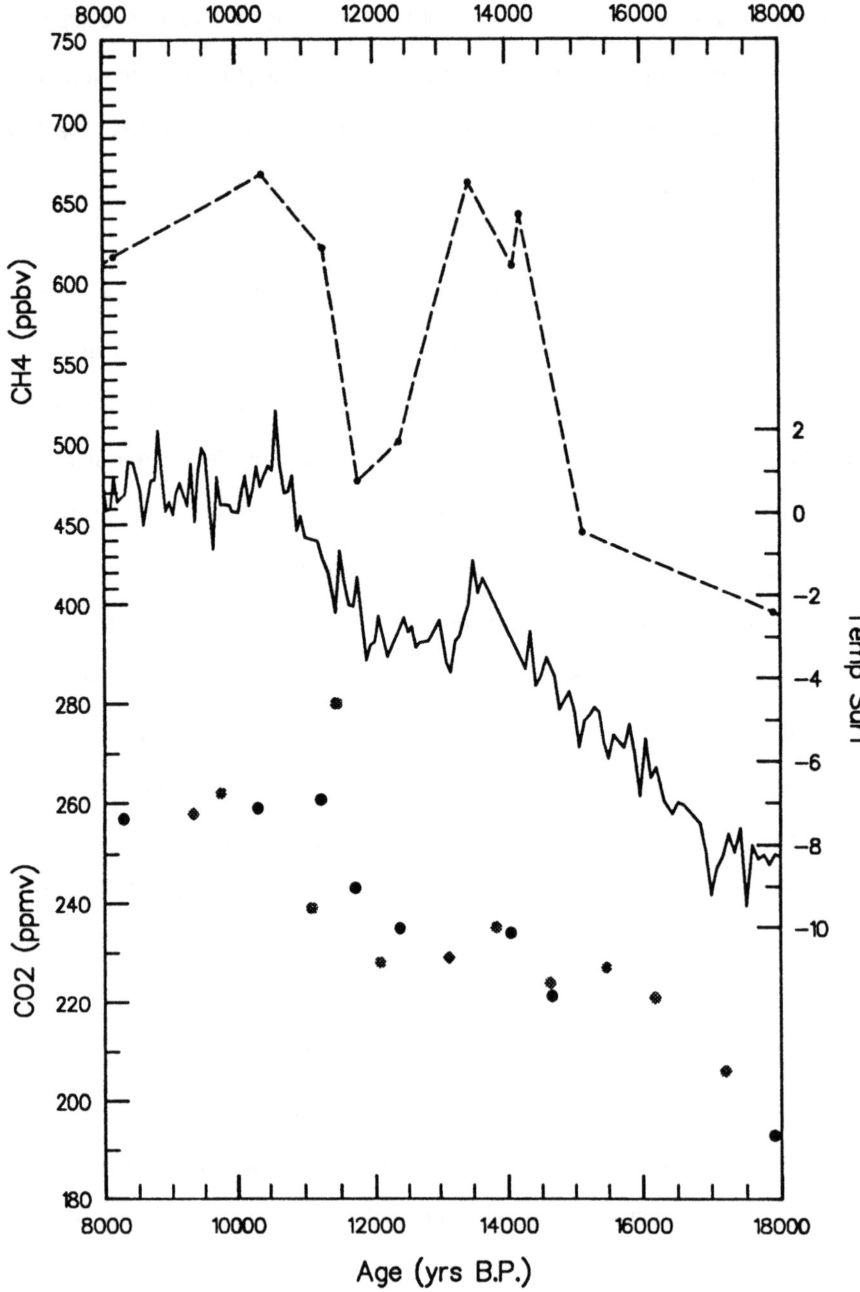

Figure 8 : Variation of CO_2 and CH_4 concentrations during the last deglaciation as deduced from gas measurements along the Vostok (CH_4) and the Dome C (shaded circles for CO_2) and Vostok cores (black circles for CO_2). The smoothed deuterium Vostok profile is shown in addition. The dating procedures used for dating the air in these 2 cores are described in the text.

reversal. We strongly believe that this oscillation is not due to experimental problems. This has been confirmed by replicate analyses at several depth levels (7 samples for the low values). Note also that the extreme concentrations are probably low estimates because of the long mixing time of the air in the Vostok core. Figure 8 suggests that CH_4 changes may have slightly lagged the Vostok temperature change (or at least that the lower CH_4 values occurred at the end of the cold reversal), but the determination of this phase relationship which strongly depends on the air ice age difference (calculated here as for CO_2) needs further measurements. In any case this CH_4 drop (possibly also recorded in the Dye 3 core but at only one level, see Stauffer et al., 1988) will provide, when fully documented, important information on the climatic coupling between the Northern and Southern Hemispheres.

There are indeed two reasons to expect that this drop was synchronous with the Northern Hemisphere Younger Dryas event through the link between natural CH_4 concentration in the atmosphere and the global wetland areas (discussed on the basis of the full CH_4 profile covering the last climatic cycle in Chappellaz et al., 1990). During the Younger Dryas, paleodata indicate a strong atmospheric cooling in the middle latitudes of the Northern Hemisphere (as much as 7 °C in Greenland and in Western Europe ; Rind et al., 1986 ; Dansgaard et al., 1989) which could have frozen potential bogs in these regions. In the low latitudes, lake levels (Street-Perrott and Perrott., 1990), organic-rich layers in the Mediterranean Sea (Rossignol-Strick et al., 1982) and the mineralogy of deposits in lakes (Gasse et al., 1990) reveal an episode of high aridity, mostly synchronous with the Younger Dryas, that could reflect a decreased area of wetlands in tropical regions. A good test for the hypothetic synchronousness of the oscillation and the Younger Dryas will be provided by from the future measurements performed on the GRIP and GISP II ice cores expected to reach the bedrock in Greenland, where both events may be well-recorded. Together with more detailed analyses from the Antarctic ice cores, the knowledge of the phase relationship of these events should then be largely improved.

VI - DISCUSSION

The results presented in this article fully confirm, as discussed for example in Jouzel et al. (1987a), that the last deglaciation in Antarctica was a two step process with two warming trends interrupted by a cold reversal. Evidence of this reversal previously based essentially on Dome C data (and to a lesser degree on the Byrd record), now includes the Vostok isotope profile based on a new, relatively high resolution (0.5 m i.e. ~ 25 years) record. Owing probably to this high resolution, the shape of this reversal is now better documented and Vostok data suggest the existence of two spikes (one at the beginning and the other at the end of this period). This would, if confirmed by even more detailed records from Vostok or from other Antarctic sites, correspond to a temperature amplitude of up to 3 to 4 °C. It is however only half of the amplitude of the Younger Dryas as recorded in Greenland cores (7 °C). Other features associated with this cold reversal that we have examined are a significant drop in CH_4 concentration, the possible existence of a CO_2 plateau and the absence of a dust signal (the dust fallout being quite similar to the low Holocene values).

A key question to be solved for determining if and how this Antarctic cold reversal is related with the Northern Hemisphere Younger Dryas is that of dating. Unfortunately the datings of neither the cold Antarctic reversal nor of the Northern Hemisphere Younger Dryas are firmly established. The ~1 kyr discrepancy between Dome C and Vostok datings for this part of the isotope profile was noted whereas the dust records suggest that they are in phase. We have provisionally and, somewhat arbitrarily, adopted the Dome C chronology and modified that of Vostok with a cold reversal centered for the three East Antarctic cores between 13 and 12 kyr B.P. This is older than the currently accepted ^{14}C dates for the Younger Dryas i.e. from 11 to 10 kyr B.P.

Bard et al. (1990) recently questioned the dating of the Younger Dryas, proposing an age of ~11.5 kyr B.P. for the Preboreal/Younger Dryas boundary instead of 10 ^{14}C kyr. This is based on a new calibration of the ^{14}C time scale using U-Th ages from Barbados corals. Bard et al. (1990) noted the agreement then obtained with the dendrochronological estimates (> 11.3 kyr B.P.). The age for the beginning of the Younger Dryas with this new calibration curve would be around 13 kyr B.P., which would practically cancel the time lag between the cold Antarctic reversal and the Younger Dryas with both events lasting from around 13 to 11.5 kyr B.P. (except for Byrd where this event is shorter). Also it is pointed out (Bard et al., 1990) that the Younger Dryas climatic event may have been as long as 1.7 kyr in agreement with the qualitative estimates of Oeschger et al. (1980) and, thus except for Byrd, in agreement with the duration of the cold reversal as recorded in East Antarctic ice (1.5, 1.3 and 1.7 kyr at Dome C, Vostok and Komsomolskaïa respectively).

This new calibration of the ^{14}C time scale indicates that the Younger Dryas and the Antarctic reversal may be in phase. However, as extensively discussed at the NATO symposium on the Last Deglaciation, this new dating of the Younger Dryas is not at the moment fully accepted by the scientific community involved in the dating of this climatic event. There is still disagreement between these new dates and those deduced independently from Greenland ice cores (Hammer et al., 1985a) or varves chronology (Cato, 1985). On the other hand one must keep in mind that dating of our ice cores is arbitrary within the limit of ~ 1 kyr for this part of the records.

To obtain independent information on the timing of the last deglaciation in Antarctica, one cannot rely on a direct correlation with the cold period observed in some of the Southern Hemisphere continental records. First, this event is far from being as clearly defined as in Western Europe and regions adjacent to the North Atlantic and the climatic interpretation of these records is, for example for South America, still debated (Markgraf, 1989). Second, when recognized, the chronology of this cold event covers a relatively large range of dates between 12 and 10 ^{14}C kyr B.P. for South America (as compiled by Heusser and Rabassa, 1987) and between 13 and 14 ^{14}C kyr B.P. for New-Zealand (Newnham, unpublished). In any case, there is, for none of the records a guarantee of a large geographical significance.

It is only recently (Labracherie et al., 1989) that the existence of a cold event during the last deglaciation was depicted in the Southern Hemisphere oceanic record. From the study of AMS dated deep sea cores in the Indian sector of the southern ocean, (between 44°S and 55°S), these authors showed that the deglacial warming was abruptly interrupted by a cool oscillation dated between 12 and 11 ^{14}C kyr B.P. therefore preceding the Northern Hemisphere Younger Dryas by about 1 kyr. They noted the correspondence with the cooling evidenced by Heusser and Rabassa (1987) for southernmost South America. As a relatively close link may be expected between the East Antarctic climate and the adjacent Indian ocean, a correlation with marine records may offer, to the extent that a major geographical significance of the cold marine event can be demonstrated and its chronology confirmed on other cores, a way to indirectly date the Antarctic cold reversal.

Another potentially interesting approach consists in using the well marked drop in dust concentration at mid-deglaciation (Figure 6), thought to be of large geographical significance (see, for example, the discussion in Petit et al., 1990) as a stratigraphic marker. In this line, we have reported on Figure 9 the deglaciation part of the well documented dust profile in Dye 3 (Hammer et al., 1985b). A noticeable feature is the sudden drop, around 1810m, which occurred at the beginning and in the warmest part of the Alleröd period preceding the Younger Dryas cooling. Assuming that this drop and the one observed at Dome C and Vostok, just before the cold reversal (figure 6), correspond to a global event, (which has to be further documented), the end of the first warming trend observed in the Antarctic cores and in Dye 3 (beginning of the Alleröd) are then synchronous On the other hand, one can cite the study of marine sediments located east of New-Zealand by Stewart and Neal (1984) showing a rapid decline of accumulation rates of quartz ^{14}C dated

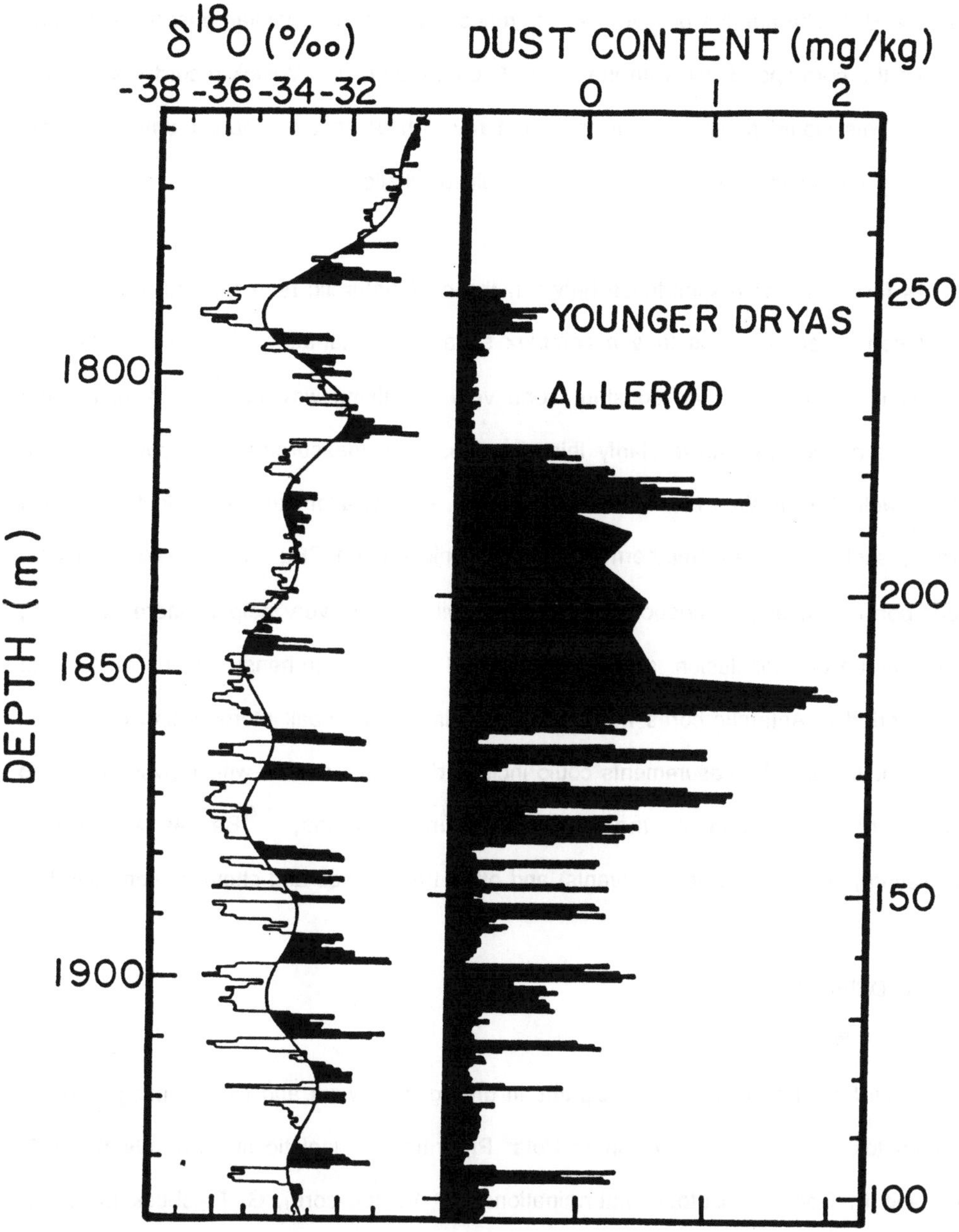

Figure 9 : Deglaciation part of the Dye 3 dust profile (Hammer et al.,1985). This figure is adapted from Broecker (This volume)

at 14.7 kyr B.P. The authors believe that this reflects a change in aeolian dust transport and they noted the correspondence with the Dome C dust record (the Vostok record was not yet available). This would place the Antarctic cold reversal at an even earlier date than that proposed by Labracherie et al. (1989) for the cold oscillation.

As far as the correspondence between the cold Antarctic reversal and the Younger Dryas is concerned, we thus face a complex situation. Fortunately, CH_4 measurements performed in the enclosed air constitute one very promising way to link Greenland and Antarctic ice cores and thus to clarify this situation. More measurements in both polar ice sheets in well chosen sites (i.e. with relatively high accumulation) are required to establish accurately such a Northern/Southern Hemisphere link through CH_4 well marked variations. Radiocarbon dating of air trapped in these cores will also be very helpful. More generally, and this will be our conclusion, a very detailed and more comprehensive study of the cold reversal period in Antarctic cores would be beneficial for a global interpretation of the last deglaciation. Additional measurements could include deuterium-excess which gives access to the conditions prevailing in the moisture source regions, ^{10}Be (as a proxy of accumulation), acidity (for volcanic events) and other trace gases and chemical compounds.

ACKNOWLEDGEMENTS

We thank all Soviet and French participants in drilling, field work and ice sampling at Dome C and Vostok and the U.S. Division of Polar Programs for logistic support. We thank R. Chiron and P. Doira for isotope determinations (4Г Vostok core), G. Raisbeck for useful discussions and C. Hammer and H. Clausen for providing us with their unpublished Byrd chronology. This work is supported in France by Programme National d'Etudes de la Dynamique du Climat (PNEDC), Institut de Protection et de Sûreté Nucléaires (CEA/IPSN), Terres Australes et Antarctiques Françaises (TAAF), and the Commission of the European Communities and in the USSR by the Soviet Antarctic Expeditions.

REFERENCES

Bard E, Hamelin B, Fairbanks RG and Zindler A (1990) Calibration of the ^{14}C time scale over the past 30 000 years using mass spectrometric U-Th ages from Barbados Corals. Nature 345 : 405-410.

Barnola JM, Raynaud D, Korotkevitch YS and Lorius C (1987) Vostok ice core provides 160000 year record of atmospheric CO_2. Nature (London) 329 : 408-414.

Barnola JM, Pimienta P, Raynaud D and Korotkevich YS (1991) CO_2 climate relationship as deduced from the Vostok ice core : a reexamination based on new measurements and on a reevaluation of the air dating. Tellus, in press.

Benoist JP, Jouzel J, Lorius C, Merlivat L and Pourchet M (1982) Isotope climatic record over the last 2.5 Ka from Dome C, Antarctica, ice core. Annals of Glaciology 3 : 17-21

Benoist JP (1986) Analyse spectrale de signaux glaciologiques: Etude des glaces sédimentaires dèposées à Dome C, Morphologie du lit d'un glacier. Thèse USMT Grenoble France.

Berner W, Oeschger H and Stauffer B (1980) Information on the CO_2 cycle from ice core studies. Radiocarbon 22 : 237-245.

Broecker WS, Peteet DM ad Rind D (1985) Does the ocean-atmosphere system have more than one stable mode of operation? Nature 315 : 21-26.

Broecker WS and Denton GH (1989) The rôle of ocean-atmospheric reorganisation in glacial cycles. Geochemica Cosmochimica Acta 53 : 2465-2501.

Broecker WS (1991) The strength of the Nordic heat pump 13,500 to 9500 BP. This volume

Cato I (1985). The definitive connection of the Swedish geochronological time scale with the present, and the new date of the zero year in Döviken, northern Sweden. Boreas 14 : 117-122.

Chappellaz J, Barnola JM, Raynaud D, Korotkevich YS and Lorius C (1990) Ice core record of atmospheric methane over the past 160,000 years. Nature 345 : 127-131.

Dansgaard W, Clausen HB, Gundestrup N, Hammer CU, Johnsen SJ, Kristinsdottir P and Reeh N (1973) Stable isotope glaciology. Meddeleser om Greenland 2 : 1-53.

Dansgaard W, Clausen HB, Gundestrup N, Hammer CU, Johnsen SJ, Krinstindottir P and Reeh N (1982) A new Greenland deep ice core. Science 218 : 1273-1277.

Dansgaard W, White JW and Johnsen SJ (1989) The abrupt termination of the Younger Dryas climate event. Nature 339 : 532-534.

Dansgaard W and Oeschger H (1989) Past environmental long-term records from the Arctic. In : H Oeschger and CC Langway (eds) The environmental record in glaciers and ice sheets. S. Bernhard, Dahlem Konferenzen, John Wiley and Sons, p287

Delmas RJ, Ascencio JM and Legrand M (1980) Polar ice evidence that atmospheric CO_2 20000 yr BP was 50 % of the present. Nature 284 : 155-157.

Eicher U, Siegenthaler U, Wegmuller S (1981) Pollen and isotope analysis on late and post-glacial sediments of the Tourbière de Chirens (Dauphiné, France). Quat Res 15 : 160-170.

Engstrom DR, Hansen BCS and Wright HE (1990) A possible Younger Dryas record in Southeastern Alaska. Science 250 : 1383-1385.

Fairbanks RG (1989) A 17 000 year glacio-eustatic sea level record : influence of glacial meltig rates on the Younger Dryas event and deep ocean circulation. Nature 342 : 637-642.

Fairbanks RG (1990) The age and origin of the "Younger Dryas climate event" in Greenland ice cores. Paleoceanography 5-6 : 937-948.

Fisher DA (1990) A zonally averaged stable-isotope model coupled to a regional variable-elevation stable-isotope model. Annals of Glaciology 14 : 65-71.

Flower BP and Kennet JP (1990) The Younger Dryas cool episode in the Gulf of Mexico. Paleoceanography 5-6 : 949-961.

Gasse F, Tehet R, Durand A, Gilbert E and Fontes JCH (1990) The arid-humid transition in the Sahara and the Sahel during the last deglaciation. Nature 346 : 141-146.

Hammer CU, Clausen HB and Langway CC (1985a) The Byrd ice core : continuous acidity measurements and solid electrical measurements. Annals of Glaciology 7 : 214.

Hammer CU, Clausen HB, Dansgaard W, Neftel A, Krinstindottir P and Johnson E (1985b) Continuous impurity analysis along the Dye 3 deep core. Geophysical Monograph 33 : 90-94.

Harvey LDD (1988) Climatic impact of ice age aerosols. Nature 334 : 333-335.

Heusser CJ and Rabassa J (1987) Cold climatic episode of Younger Dryas age in Tierra Del Fuego. Nature 328 : 609-611.

Iversen J (1954), The late glacial flore of Denmark and its relation to climate and soils. Danm Geol Unders 80, Serie II : 87-119.

Jensen E and Veum T (1990) Evidence for two-step degalciation and its impact on North-Atlantic deep-water circulation. Nature 343 : 612-616.

Jensen K (1938) Some West Baltic pollen diagrams. Quater 1 : 124-139.

Johnsen SJ (1977) Stable isotope homogeneisation of polar firn and ice. In : IAHS (ed) IAHS Publication n°118, Isotopes and impurities in snow and ice, Proceedings of the Grenoble Symposium Aug/Sep 1975, p210

Johnsen SJ, Dansgaard W, Clausen HB and Langway CC (1972) Oxygen isotope profiles through the Antarctic and Greenland ice sheets. Nature 235 : 429-434.

Joussaume S and Jouzel J (submitted) Paleoclimatic tracers : an investigation using an atmospheric general circulation model under ice age conditions. Water isotopes, J of Geophys Res

Jouzel J, Merlivat L and Lorius C (1982) Deuterium excess in an East Antarctic ice core suggests higher relative humidity at the oceanic surface during the Last Glacial Maximum Nature 299 : 688-691.

Jouzel J and Merlivat L (1984) Deuterium and oxygen 18 in precipitation : modeling of the isotopic effects during snow formation. J of Geophys Res 89 : 11749-11757.

Jouzel J, Lorius C, Merlivat L and Petit JR (1987a) Abrupt climatic changes : The Antarctic ice record during the late Pleistocene. In : WH Berger, LD Labeyrie (eds) Abrupt climatic change : evidence and implications. NATO ASI Series, D Reidel Publishing Company, Dordrecht Boston Lancaster Tokyo, p235.

Jouzel J, Lorius C, Petit JR, Genthon C, Barkov NI, Kotlyakov VM and Petrov VN (1987b), Vostok ice core : a continuous isotope temperature record over the last climatic cycle (160000 years). Nature 329 : 403-408.

Jouzel J, Russell GL, Suozzo RJ, Koster RD, White JWC and Broecker WS (1987c) , Simulations of the HDO and $H_2{}^{18}O$ cycles using the NASA/GISS general circulation model : the seasonal cycle for present day conditions. J of Geophys Res 92 : 14739-14760.

Jouzel J, Raisbeck G, Benoist JP, Yiou F, Lorius C, Raynaud D, Petit JR, Barkov NI, Korotkevitch YS and Kotlyakov VM (1989) A comparison of deep Antarctic ice cores and their implications for climate between 65000 and 15000 years ago. Quat Res 31 : 135-150.

Keigwin LD and Jones GA (1990) Deglacial climatic oscillations in the Gulf of California. Paleoceanography 5-6 : 1009-1023.

Kudrass HR, Erlenkeuser H, Vollbretch R and Weiss W (1991) Global nature of the Younger Dryas cooling event inferred from oxygen isotope data from Sulu sea cores. Nature 349 : 406-409.

Labeyrie LD, Duplessy JC and Blanc PL (1987) Variation of mode of formation and temperature of oceanic deep waters over the past 125 000 years. Nature 327 : 477-482.

Labracherie M, Labeyrie LD, Duprat J, Bard E, Arnold M, Pichon JJ and Duplessy JC (1989) The last deglaciation in the southern ocean. Paleoceanography 4 : 629-638.

Linsley BK and Thunell RC (1990) The record of deglaciation in the Sulu Sea : Evidence for the Younger Dryas event in the tropical western Pacific. Paleoceanography 5-6 : 1025-1039.

Lorius C and Merlivat L (1977) Distribution of mean surface stable isotope values in East Antarctica. Observed changes with depth in a coastal area. In : IAHS (ed) IAHS Publication n°118, Isotopes and impurities in snow and ice, Proceedings of the Grenoble Symposium Aug/Sep 1975, p127

Lorius C, Merlivat L, Jouzel J. and Pourchet M (1979) A 30000 yr isotope climatic record from Antarctic ice. Nature 280 : 644-648.

Lorius C, Jouzel J, Ritz C, Merlivat L, Barkov NI, Korotkevich YS and Kotlyakov VM (1985) A 150000 year climatic record form Antarctica. Nature 316 : 591-596.

Markgraf V (1989) Paleoclimates in central and South America since 18000 BP based on pollen and lake level records. Quat Sci Res 8 : 1-24.

Neftel A, Oeschger H, Schwander J, Stauffer B and Zumbrunn R (1982) Ice core sample measurements give atmospheric CO_2 content during the past 40000 years. Nature 295 : 220-223.

Neftel A, Oeschger H, Staffelbach T and Stauffer B (1988) CO_2 record in the Byrd core 50000-5000 years B.P. Nature 331 : 609-611.

Nikolaïev VI, Kotlyakov VM and Smirnov KE (in Russian) (1988) Isotope studies of the ice core from the Komsolmolska¨la station, Antarctica. Data of Glaciological studies of the USSR academy of Sciences 63 : 97-102.

Oeschger H, Welten M, Eicher V, Moll M, Reisen T, Siegenthaler V, Wegmuller S (1980) ^{14}C and other parameters during the Younger Dryas cold phase. Radiocarbon 22 : 299-310.

Oeschger H, Beer J, Siegenthaler U, Stauffer B, Dansgaard W and Langway Jr CC (1984) Late glacial climate history from ice cores. Geophysical Monograph 29 : 299-306.

Overpeck JT, Peterson LC, Kipp N, Imbrie J and Rind D (1990) Climate change in the Circum-North-Atlantic region during the last deglaciation. Nature 338 : 553-557.

Peteet D, Vogel J S, Nelson DE, Southon JR, Nicknam RJ and Heusser LE (1990) Younger Dryas climatic reversal in Northeastern USA? AMS ages for an old problem. Quat Res 33 : 219-230.

Petit JR, Briat M, Royer A (1981) Ice age aerosol content from East Antarctic ice core samples and past wind strength. Nature 293 : 391-394.

Petit JR, Mounier L, Jouzel J, Korotkevitch YS, Kotlyakov VI and Lorius C (1990) Paleoclimatological and chronological implications of the Vostok dust record. Nature 343 : 56-58.

Raisbeck GM, Yiou F, Fruneau M, Loiseaux JM, Lieuvin M, Ravel JC and Lorius C (1981) Cosmogenic ^{10}Be in Antarctic ice during the past 30000 years. Nature 292 : 825-826.

Raisbeck GM, Yiou F, Bourles D, Lorius C, Jouzel J and Barkov NI (1987) Evidence for two intervals of enhanced ^{10}Be deposition in Antarctic ice during the last glacial period. Nature 326 : 273-277.

Raynaud D and Barnola JM (1985) CO_2 and climate : information from Antarctic ice core studies. In : A Ghazi and R Fantechi (eds) Current issues in Climate Research. D. Reidel Publishing Compagny, Dordrecht Boston Lancaster Tokyo, p240

Rind D, Peteet D, Broecker W, McIntyre A and Ruddiman W (1986) Impact of cold North Atlantic sea surface temperatures on climate implications for the Younger Dryas cooling. Climate Dynamics 1 : 3-33.

Ritz C (1991) Chronology of the Vostok ice core based on precipitation and ice flow modelling. Journal of Glaciology : Submitted

Robin G de Q (1977) Ice cores and climatic changes. Philos Trans Roy Soc London 280, serie B :143-160.

Rossignol-Strick M, Nesteroff W, Olive P and Vergnaud-Grazzini C (1982) After the deluge : Mediterranean stagnation and sapropel formation. Nature 295 :105-110.

Royer A, De Angelis M and Petit JR (1983) A 30000 year record of physical and optical properties of microparticles from an East Antarctic core and implications for paleoclimate reconstruction models. Climatic Change 5 : 381-412.

Ruddiman WF and McIntyre A (1981) The North Atlantic ocean during the last glaciation, Palaeogeogr Palaeoclimatol Palaeoecol 35 :145-214.

Shackleton NJ (1989) Deep trouble for climate change. Nature 324 : 616-617.

Schwander J and Stauffer B (1984) Age difference between polar ice and the air trapped in its bubbles. Nature 311 : 45-47.

Stauffer B, Neftel A, Oeschger H and Schwander J (1985) CO_2 concentration in air extracted from greenland ice samples. CC Langway et al. (eds) Geophysical Monograph 33, American Geophysical Union Washington, p85

Stauffer B, Lochbronner E, Oeschger H and Schwander J (1988) Methane concentration in the glacial atmosphere was only half that of the preindustrial value. Nature 332 : 812-814.

Staffelbach T, Stauffer B, Sigg A and Oeschger H (1991) CO_2 measurements from polar ice cores : more data from different sites. Tellus 43 B(2) : 91-96.

Stewart RB and Neal VE (1984) Chronology of paleoclimatic change at the end of the last glaciation. Nature 311 : 47-48.

Street-Perrott FA and Perrott RA (1990) Abrupt climate fluctuations in the tropics : the influence of Atlantic Ocean circulation. Nature 343 : 607-612.

Thompson LG (1977) Variations in microparticle concentration, size distribution and elemental composition found in Camp Century Greenland and Byrd station, Antarctica, deep cores. In : IAHS (ed) IAHS Publication n°118, Isotopes and impurities in snow and ice, Proceedings of the Grenoble Symposium Aug/Sep 1975, p351

Whillans IM and Grootes PM (1985) Isotopic diffusion in cold snow and firn. J of Geophys Res 90 : 3910-3918.

Young NW, Pourchet M, Kotlyakov VM, Korolev PA and Duygerov MB (1982) Accumulation distribution in the IAGP area, Antarctica 90°E-150°E. Annals of Glaciology 3 : 333-338.

Possible Ice-Core Evidence for a Fresh Melt Water Cap

Over the Atlantic Ocean in The Early Holocene

David Andrew Fisher
Terrain Sciences, Glaciology section
Geological Survey of Canada
601 Booth St
Ottawa Ontario K1A 0E8
Canada

Abstract

During the last deglaciation there is ocean evidence (Fairbanks, 1989) for two large infusions of melt water from ice sheet wastage, MWP-1A and MWP-1B. These should have spiked the Atlantic surface waters with fresh water that would evaporate and find their way into the Arctic ice cores. The first part of this paper argues against Fairbanks candidate for an ice core MWP-1B. Stable isotope and summer melt layer data is then presented from the Agassiz A84 ice core from Arctic Canada that suggests the O-18 signature of MWP-1B is present.

A stable isotope/precipitation model is run to verify that fresh water "spiking" of the surface waters can produce the O-18 changes suggested for A84. Lower latitude injections of melt water seem to produce a bigger O-18 effect in the High Arctic.

The models sensitivity is assessed to changes in humidity, windiness, sea ice, meridional mixing lengths and cold cloud vapour characteristics. The model deuterium excess **d** is very sensitive to each of; sea ice, humidity and windiness separately. When humidity-windiness are crudely coupled the sensitivity increases but when sea ice- humidity-windiness are coupled the model **d** sensitivity is gone. O-18 is rather stable in all cases. The importance of near site variables connected to cold clouds is also stressed.

Introduction

The main theme of the workshop is the reconciliation of the timing and length of various ocean and continental Younger Dyras signatures. Since the Canadian ice cores' time scales (Fisher et al. 1983) are based on spot measurements of annual layer thickness measurements and fine tuned by volcanic layer tie-ins to the Greenland ice cores these time scales can only be as accurate as those from Greenland cores, (Hammer et al., 1986),(Hammer and Johnsen this meeting).

NATO ASI Series, Vol. I 2
The Last Deglaciation: Absolute and Radiocarbon Chronologies
Edited by E. Bard and W. S. Broecker

However one of the unifying new facts in the discussion of the Younger Dryas is the detailed Barbados sea level curve (Fairbanks 1989). If as Fairbanks notes, it can be shown that "Barbados is without significant episodic uplift or subsidence" then this curve's slope is a detailed record of ice sheet wastage and shows two major maxima in melt water additions to the worlds oceans. The older MWP-1A peaks at 13.2 ka(cal) [12.1 ka C-14] and the younger MWP-1B at 10.2 ka (cal) [9.5 ka C-14]. MWP-1A is about 2 ka wide and -1B is about 3 ka wide. Between them the minimum centered at 11.8 ka(cal) is about 1 ka wide. The whole interval from 14 ka to about 7 ka(cal) had elevated sea level increase.

In Fairbanks (1989) there is the suggestion that MWP-1B and 1A are echoed in the Arctic ice core stable isotope records. MWP-1B was said to have caused or significantly contributed to the last big negative excursion of $\delta(^{18}O)$. This interpretation of the ice cores contradicted a well documented empirical annual layer dating by Hammer et al. (1986), and furthermore could not explain why the ice cores should have sharply defined impurity boundaries (Dansgaard et al. 1989) coincident with this negative δ excursion. These difficulties combined with the varve dated $\delta(^{18}O)$ record from Gosciaz Lake (Rozanski and Pazdur, this meeting) strongly suggested the Fairbanks' interpretation of the late Wisconsinan δ negative excursion was wrong. The question still remains then ;" Is there any evidence in ice cores for very low δ meltwater moisture sources during Fairbanks' MWP- 1A or MWP-1B ?"

This contribution gives a tentative yes to the question. The first part presents the experimental evidence from an ice core obtained at the top of a flow-line of the Agassiz ice cap, Ellesmere Island. The second part reviews numerical experiments done with a stable isotope/precipitation model. The model is run for Arctic ice cap sites using present condi tions but changing $\delta(^{18}O)$ and $\delta(D)$ of source water over a wide range and as function of latitude.Finally the model is run using present conditions but altering sea ice coverage, win diness and relative humidity.

Empirical Evidence for Melt Water Cap on Source Ocean

Agassiz Ice Cap Ice Cores

The Agassiz Ice Cap is on Ellesemere Island, figure 1. The ice core discussed A84 was taken at the top of a local high 1760 masl where the mean annual temperature is about -24 C, the accumulation 9.8 cm/a (ice equivalent), and the average summer melt percent about 5 % by mass.

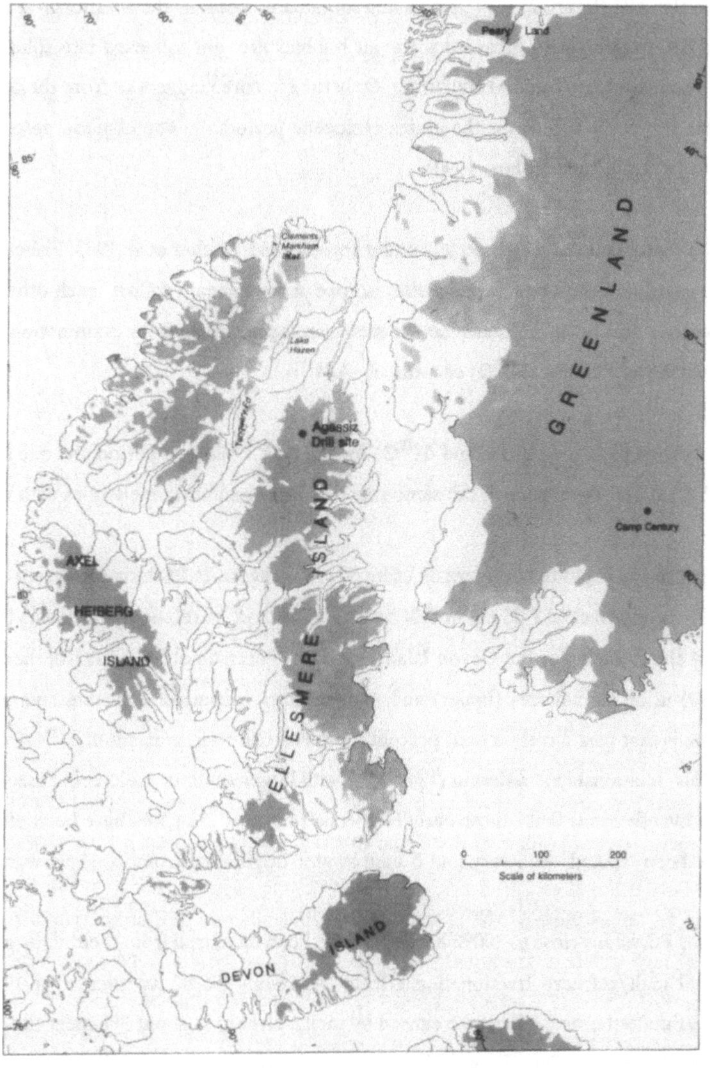

Figure 1. Location map for the ice coring sites mentioned.

The time scale and stable records for this core and others nearby are described in (Fisher et al., 1983; Koerner and Fisher, 1990) Most summers there are periods of a few weeks when some snow is melted at the surface soaks into the firn and refreezes usually within that years' layer. The refrozen layers have a higher density and fewer bubbles then the surrounding firn even after the firn has been itself compacted into ice at depth (Koerner, 1977). The melt layer percent PC is defined;

$$PC = [\text{melt layer thickness }]/[\text{total accumulation}]$$

where both the amounts are in cm of ice equivalent/year.

The PC record in the A84 (figure 2) is preserved and obtainable because the ice though spanning about 100 ka is only about 130 m thick. This means that the air bubbles are not squeezed into solid solution nor are they masked by stress-release internal fracturing. Only the ice core sequences from the top of the Agassiz have a discernible ice layer record for the entire Holocene period. The climatic interpretation of this record is given in (Koerner and Fisher 1990).

Since the winter snow at A84 is greatly under represented (Fisher et al.,1983; Fisher,submitted) both the melt layer record and the undisturbed stable isotope record should follow each other. The evidence for an early Holocene MWP-1B influence on the stable isotopes comes from examination of the changing relationship between the PC and $\delta(^{18}O)$ records of A84.

Figure 2 shows melt percent PC and $\delta(^{18}O)$ for the A84 Holocene period and one can see that from 0 to 6.5 ka BP, $\delta(^{18}O)$ and PC change in the same sense; ie higher summer melt goes with more positive δs.

This relationship is plotted in figure 3a using 50 yr averages. As PC increases beyond 35% a constant maximum $\delta(^{18}O)$ is approached (-26 ‰ in this case). This sort of relatio nship seems to be typical. For example, figure 4 shows melt for the Devon Island ice core (Paterson et al. 1977; Koerner, 1970; Koerner and Russell, 1979) using annual data (upper) and non-summer accumulation versus summer melt (lower). The point to note is that past a certain melt percent there is a con stant maximium $\delta(^{18}O)$ value. Older than 6.5 ka however this relationship is different (figure 3b) with larger summer melt being associated with more negative δs. The hypotheses is that these early Holocene (6.5 to 10.2 ka) δs have been effected by source oceans that have been "spiked" with very low δ melt water originating from ice sheet wastage.

The core δs for PC values close to 100% have ceratinly been increased from their undisturbed (unmelted) values by several melt/refreeze fractionation effects (Arnason, 1969; Koerner et al.,1973). Hooke and Clausen (1982) estimate the net δ increase caused by melt/refreeze is about 3‰ near the equilibrium line (PC=100%) of the Barnes Ice Cap. From NW Greenland δ data (Reeh et al.,1990) one can see the increase is about 2.5 ‰ Thus the early Holocene (9.2 to 10.2 ka) δs with very high PC values in figures 2 and 3 should probably be shifted about 3 ‰ to more negative values because of melt/refreeze effects. This suggests the undisturbed δs for this interval are 3.5 ‰ below the hypothesized -26 ‰ constant value. This 3.5 ‰ could be due to the influence of the MWP-1B. Of course, if this is true then it means other Northern Hemisphere ice core records must also have similar stable isotope traces of this event. Since this introduces an extra variable in ice core interpretation, the weaknesses in the hypothesis should be addressed.

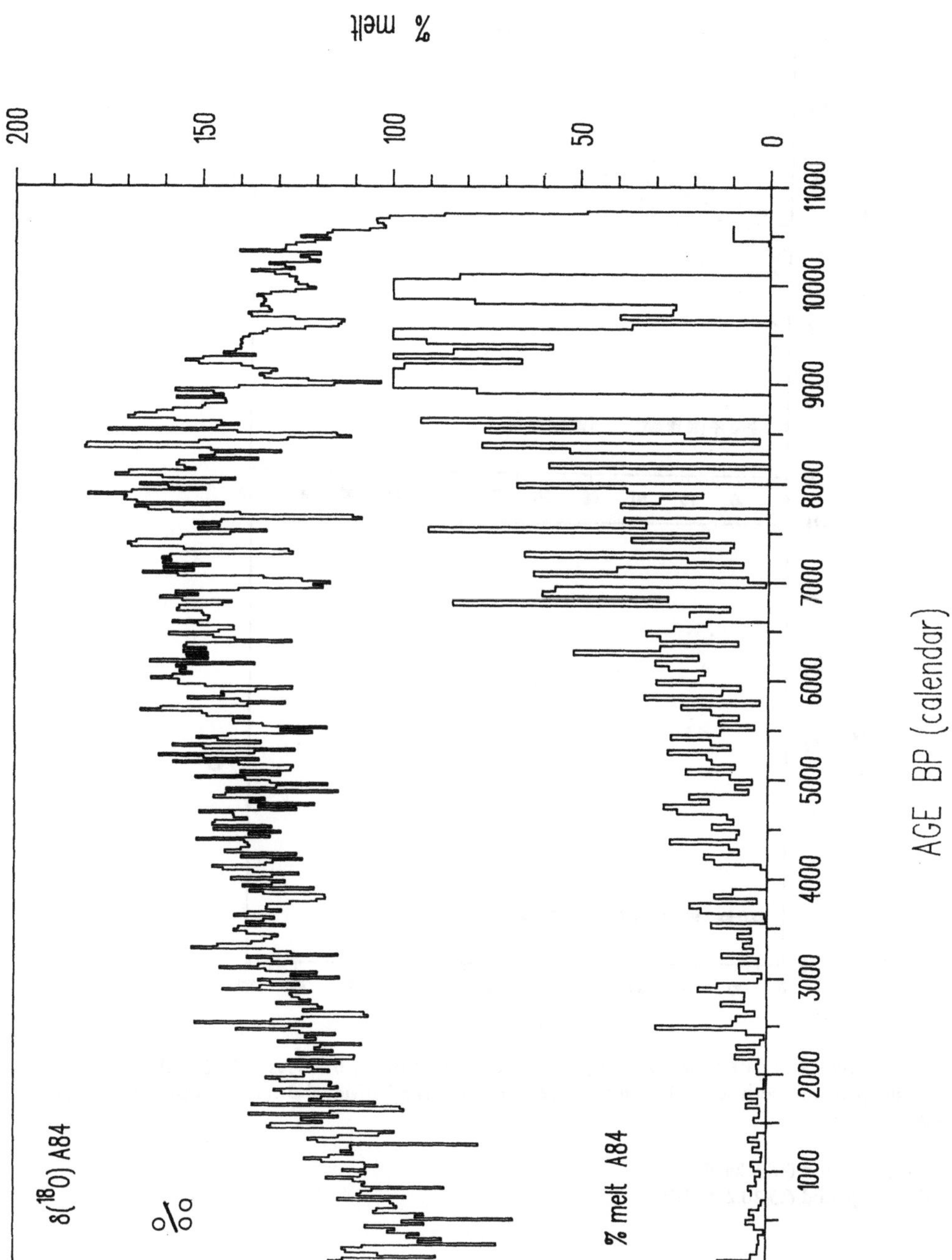

Figure 2. Holocene oxygen-18 and percent melt, PC, for the Agassiz A84 ice core. The time scale to 7ka BP is based on measured layer thickness and tuned by cross referencing volcanic layers to the Greenland ice cores. The sharp $\delta(^{18}O)$ shift at the end of the Wisconsin is given the Dye 3 date of 10.7 ka (cal) (Hammer et al., 1986). The time scale here can at best only be as good as the Greenland ice core dates.

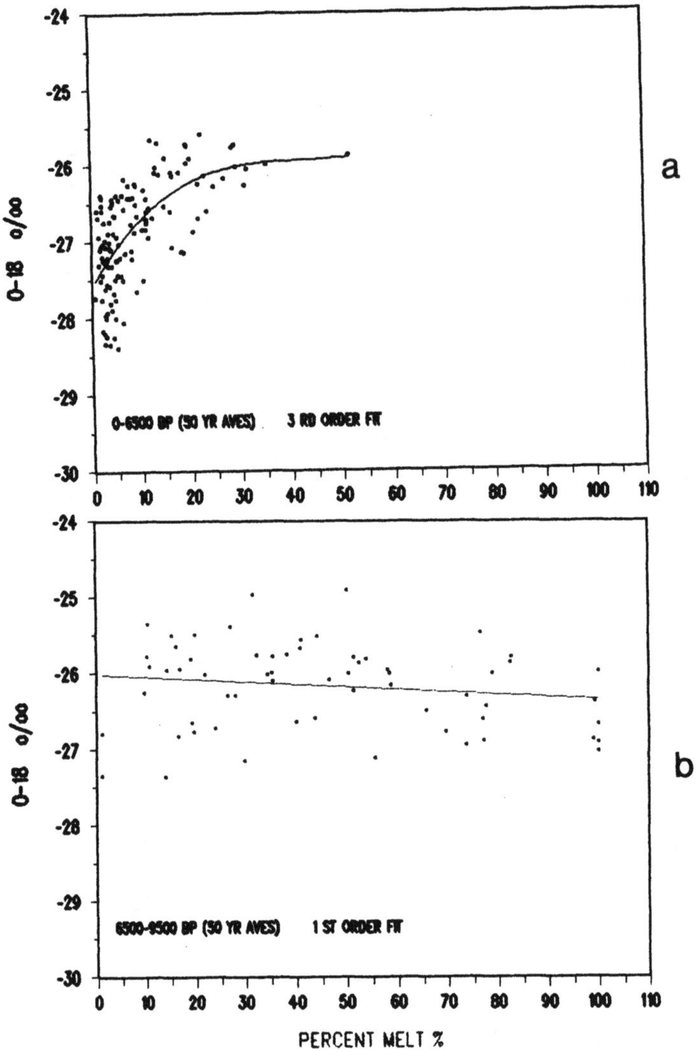

Figure 3. Time series data from figure 2 plotted against each other, δ versus PC, % melt. The lines that fit the data pairs are derived by the least squares method. The individual δ and PC values used represent 50 year averages.

3a. For the period 0-6.5 ka BP.
3b. For the period 6.5-10.2 ka BP.

Since the δ s are naturally weighted to seasonal snow fall distribution the argument assumes this distribution is constant. The present under representation of winter snow at A84 is largely caused by wind erosion of the loose cold winter snow (Fisher et al, 1983). Could the winters of the early Holocene have been so warm that the snow was too wet to scour away by wind? This seems unlikely because the present three coldest months at A84 have an average temperature of -37°C (Alt and Labine, personal communication). Koerner and Fisher (1990) have estimated that the A84 summer temperatures needed to

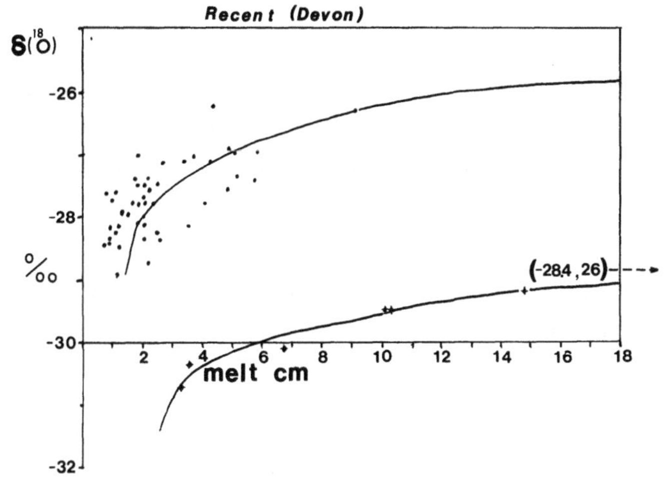

Figure 4. $\delta(^{18}O)$ versus melt amount from the Devon Island Ice Cap. The upper data uses annual averages from the Devon 1971 ice core. The lower data gives total melt versus δ of non-summer snow along a traverse up the ice cap.

produce 100 % melt at that site would have to be about 2.5 C warmer than present. Even if the winter temperatures were warmer than present by three times that amount it would still not be warm enough inhibit wind erosion of the winter snow. The other complicating effects of isostatic depression/rebound of the region in the deglacial period have been discussed in (Koerner and Fisher, 1990). There is at present no δ-elevation effect for this part of the Queen Elizabeth Islands so there should be no δ correction for depression/rebound. In any case,if a standard elevation correction is made for the early Holocene the arguement for source ocean spiking is made stronger.

Another possible explaination of the "abnormally" low δs in the early Holocene could be found in a 'precipitation shadow' effect caused by the still large ice masses of the retreating Laurentide (S. Johnsen, personnal communication). This of course could be offset by evaporation from glacial lakes and from the ablation zones of the ice sheet.

One would have preferred a sharp δ minimum in the Agassiz record centered at 10.2 ka (cal) to relate to MWP-1B instead of an anomaly in the PC-δ relationship between 10.2 and 9 ka (cal). Nevertheless, the hypothesis that there is evidence of MWP-1B in the early Holocene Agassiz core can be supported by the PC-δ anomaly and later by the modelling results. However it is far from proven by either and only is suggested as a possiblity.

Model Sensitivity and the Melt Water Pulses

It is hypothesized that spiking the ocean surface water with ice sheet dervived melt water caused significant shifts in δ in northern ice core records during MWP-1B. The response of an existing S-precipitation model to changes in the amount and latitude distribution of melt water spiking is now examined.

Since other changes in source waters and cold cloud charcterists can also produce discernable stable isotope effects at high elevation Arctic sites, model sensitivity to these parameters is also reviewed.

The Model

The model described in (Fisher 1990) is empirical using zonal averages of (figure 5):

E evapouration,**T** near surface air temperature, T_c sea temperature, meridional water vapour flux, **h** near surface relative humidity, wind speed, sea ice cover, **Si** supersaturation in the clouds and T_s the temperature for shifting from vapour-water phase changes to vapour-ice ones.

The model like its progenitor (Fisher and Alt 1985) is designed to calculate zonal averages of sea level $\delta(^{18}O)$, $\delta(D)$ the precipitation rate **P** and **W** the precipitable water. The model is a multi-source one and sites at high latitudes receive moisture in significant amounts from all "up wind" latitude zones. Input and output are functions of x the distance from the equator.

Global solutions are input to regional solutions along assumed vapour trajectories, up ice caps,over ice shelves etc. The elevation, air temperature and percipitation are inputs along such trajectories and $\delta(^{18}O)$ and **d** are outputs.

The main sources of the global empirical data used as input are:

Oort (1983) for (T,T_c,vapour flux,wind speeds, **h**), IAEA/WMO (1969,70,71,73,75,79,81,83,86)for (**T**, **h**, $\delta(^{18}O)$, $\delta(D)$). There are supplimentary sources for **E** from (Sellers 1967),(Peixoto and Oort 1983), (Oberhuber 1988), for Sea ice cover from (Jacka and others 1985), (Claire and others 1987), for zonal precipitation from (Jaeger 1983). Atlantic ocean surface water δ's (Redfield and Friedman 1965) are used as input.

The model is fully discussed in (Fisher 1990). Most of the input variables, figure 5, are zonal averages of well known meteorological quantities. However some are not so obvious and are briefly discussed below.

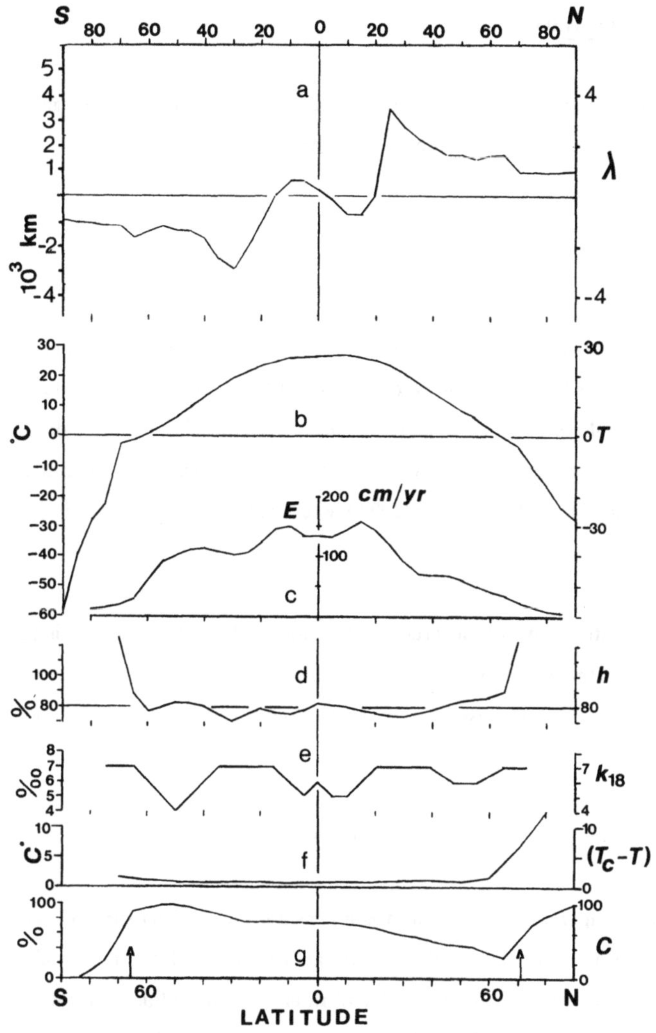

Figure 5. Zonal averages of annual input data fields versus latitude, averaged in 5 degree strips.

5a. λ survival distance for water vapour. Positive means that there is a northward vapour transport, and negative values mean southward vapour flux. $\lambda = 0$ implies that all the water evapourated in that zone falls out in that zone.

5b. T near surface air temperature.

5c. E evapouration rate.

5d. h near surface relative humidity.

5e. k_{18} kinetic evapouration fractionation factor for O^{18}. k=7 ‰ is for for smooth oceans and 4 ‰ is for rough oceans.

5f. T_c- T difference between sea temperature T_c and air temperature.

5g. C percent of the zone that is ocean. The arrows show mean annual sea ice fronts.

The Empirical Survival Function λ

The water vapour survival distance λ is the meridional distance overwhich a given vapour slug travels before it is reduced in mass by a factor 1/e. It is a function of latitude.

Supersaturation Si, and T_s

For air temperatures $T > T_s$ equilibrium fractionation coeffecients (vapour to water) are used in the model (Fisher 1990). For $T < T_s$ however (Jouzel and Merlivat 1984) the water vapour in the clouds is assumed to sublimate direc tly to ice crystals instead of condensing to water droplets. For $T<T_s$ the clouds are thought to be supersaturated wrt. vapour-ice phase changes. Jouzel and Merlivat (1984) developed a supersaturated model and Fisher (1990) made slight improvements to it. Fisher's version is used here.

Si is the supersaturation ratio in precipitating clouds $Si = e/e_s$, where e_s is the saturation vapour pressure over ice and **e** is the actual vapour pressure at the cloud tempera ture. **Si** functions are usually close to being linear functions of the air masses temperature (Jouzel and Merlivat 1984) and here **Si** for $T < Ts$ is given by ;

$$Si = 1.0 - mx10^{-3}\ T$$

Where **T** is the cloud temperature in C° and **m** is a constant for a given run and ranges from 0 for air masses in equilibrium with ice crystals to about 12 for air saturated wrt water. Any cloud where $T<Ts$ and $m > 0$ is supersatuared wrt ice crystals. The value of **m** has a major influence on **d**, the deute rium excess and a much lesser effect on the δ's. Modelling experience and limited meteorological data suggests that single function representations of **Si** best simulate **d** data when $4 < m < 7$. This paper and others (Jouzel and Merlivat 1984), (Fisher 1990), (Fisher in press) find $m = 5$ or 6 produce the best simulations of the **d** data. This is possibly because the mixing of air masses with many different **Si** functions (Fisher in press) produces an effective average **m**. The temperature **Ts** at which the clouds switch from water droplets to ice crystals is also very important in determining **d**. Suggested values of **Ts** range from -5 (Johnsen et al., 1989) to as low as -26 C° for clean air (Mason 1975), (Curry et al., 1990). Jayaweera and Ohtake (1973) measured temperatures of -20 in Arctic stratus that contained no ice crystals. Mason (1975) points out that cleaner (less dusty) air can reach lower temperatures before ice crystals form. **Ts** is an "event" temperature of a particular air mass whereas most input data temperatures are temporal means over a year or a month. Air temperatures should ideally be weighted to actual storm events before being used with a given **Ts**.

Since for most of the global sea level model runs **T > -10** the **Si** function is not too important but once a coast is crossed in a regional solution and **T << -10**, **Si** becomes very important in determining **d**. The model **δ**'s are stable over quite a wide range of **Si** functions but the model **d**'s are extremely sensitive to which **Si** function is used.

Temperature Inversion and Precipitation Weighting

For model runs over high latitude non-marine regions it is often the case that the surface temperatures are lower then the condensation temperatures, ie there is a temperature inversion. This is particularly important in cold areas where the model must chose temperatures to feed to the **Si(T)** function if **T < T_s**. The Antarctic inversion function of Jouzel and Merlivat (1984) is used.

For annual average runs the annual average temperatures over cold non-marine areas is not likely to be equal to the year's precipitation weighted temperature. Usually the latter is several degrees warmer than the temporal annual average. Annual regional solutions should use the preciptation weighted temperatures. These are obtained by "correcting" temporal annual averages using nearby met station data.

Global Solutions

Using the empirical source input functions shown in figure 5, cold cloud constants **T_s** = -10 C and **m**=6, the global model produces results given in figure 6a for **δ(^{18}O)** and figure 6b for **d**. The dots are measured values from IAEA reports (1969,1970,1971,1973,1975,1979,1981,1983,1986) for coastal and island sites. A zonal model of course can only be expected to run through the swarm of data points. The model reproduces quite well the zonal averages of **δ(^{18}O)** and **d** including the low **δ**'s around the equator in the tropical convergence zone, see (Fisher 1990) for a full desciption of these results.

Example of a Regional Solution, Camp Century, NW Greenland

The elevations and precipitation from the coast up to Camp Century is available in detail from (Mock 1965). The air temperature at the Camp Century site (referred to as CC) is - 24.5 C (Mock 1965) and at sea level (Thule) -11.6 C (IAEA, 1975). A linear variation is assumed in annual temperature between sea level and CC at 1885 m asl. Here various values of **Ts** and **m** are tried to assess the effects on simulated **d**'s. The off shore precipitation from (Jaeger 1983) and nearby met station data is put at 12.3 (g cm^{-2} a^{-1}).

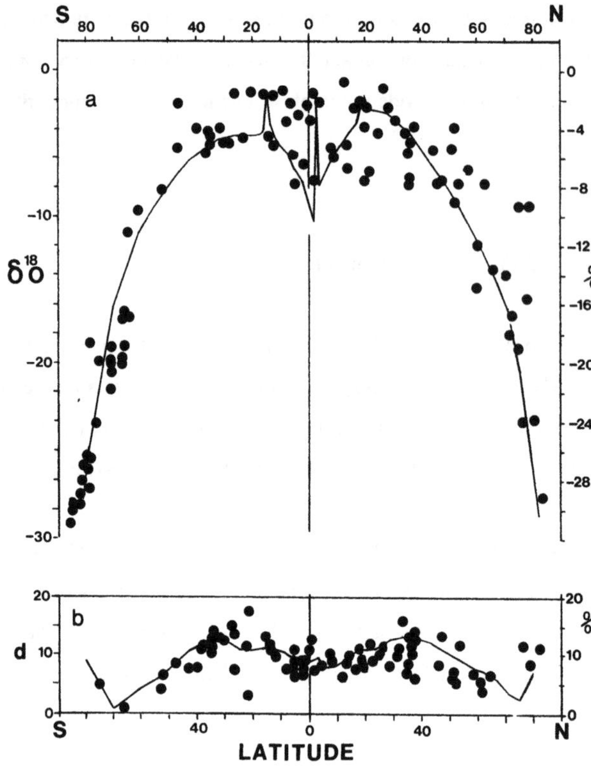

Figure 6. Annual average δ's and d'd .

6a. Annual averages of δ(^{18}O) versus latitude. The dots are measured values and the line is from the model. The model gives values near -2 ‰ where λ = 0 .

6b. Annual averages of the deuterium excess
d = δ(D) - 8*δ(^{18}O). The dots are from measured stable isotopes and the line is given by the model.

The moisture trajectory is assumed to follow the line perpendicular to the precipitation isopleths. With **Ts**=-10 C the calculated δ(^{18}O)'s appear as a line in figure 7a and predicted **d**'s are shown below it for a range of **m**'s. The effect of changing **Ts** while holding **m** at 5 is displayed in figure 7b. The δ data for NW Greenland (squares) has been assembled by (Reeh and others 1989). The data points in figure 7a between elevations 1200 and 500 m asl are from sites that have undergone extensive summer melting and thus the δs to not represent the undisturbed annual value (Arnason 1969).

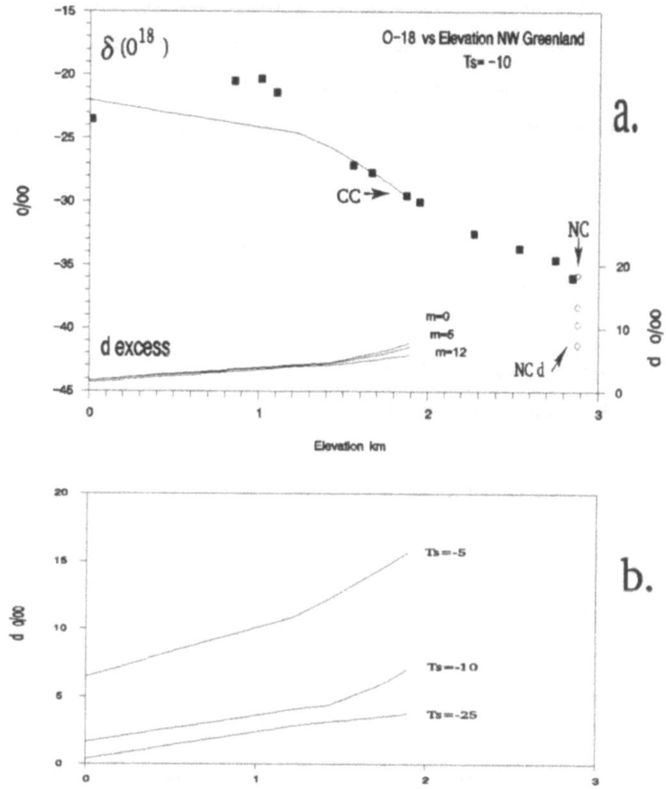

Figure 7a. The solid line shows the $\delta(^{18}O)$ simulation up a line from the coast to Camp Century (CC). This line stops at CC (1885 m asl). The "o" marked NC is the predicted δ for the North Central site using another trajectory see (Fisher, in press). The squares are data from (Reeh and others, 1989). Simulated **d** appears under the δ's for **Ts** = -10 C and the full range of supersaturation functions $0 < \mathbf{m} < 12$. The diamonds are the predicted **d**'s that go along with NC.

7b. The deuterium excess **d** predicted along the CC trajectory using **m**=5 and a range of temperatures **Ts** at which clouds shift from being predominantly water droplets to ice crystals.

$\delta(^{18}O)$ and $\delta(D)$ of Source Ocean Surface Waters

Each source strip contributes water vapor with an initial δ given by;

$$\delta_v^o = \left\{ \left(\frac{1-k}{1-kh} \right) \frac{1}{\alpha(T_c)} - 1 \right\} \{1 + \delta_{ocean}\} + \delta_{ocean}$$

where **k** is the kinetic evaporation fractionation factor, **h** the relative humidity over the ocean boundary layer and **Tc** the ocean temperature. In earlier models, Jouzel and Merlivat 1984) and (Fisher 1990) δ_{ocean} has been taken as everywhere equal to zero. This is not really true for surface water however. δ_{ocean} from measured δ(D)'s from the Atlantic surface waters (Redfield and Friedman 1965) and $\delta(^{18}O)$ from the empirical co-isotope relation for north Atlantic surface waters; $\delta(D) = 6.5\ \delta(^{18}O)$ (Craig 1966),(Ferronsky and Brezgunov 1989) are used for modern simulations. $\delta(^{18}O)$ averages used as input for source ocean strips is shown in figure 8.

In the melt water spiking tests the δ_{ocean} values are assigned using various surface mixtures of ice sheet melt water with 18ka BP bulk (fully mixed) ocean water with:

$\delta(^{18}O)$ = +1.2 ‰ (Dansgaard and Tauber,1969)
and $\delta(D)$ = +9.32 ‰ (assumes d = -0.3 ‰).

The pure ice sheet melt water is assumed to have :

$\delta(^{18}O)$ = -35 ‰
and $\delta(D)$ = -272 ‰, (d=8.0 ‰).

For example surface water which is 1 part fresh ice sheet melt and 6 parts 18 ka bulk ocean water has:

$$\delta_{ocean}(^{18}O) = 1/7[-35 + 6(1.2)] = -3.97\ ‰.$$

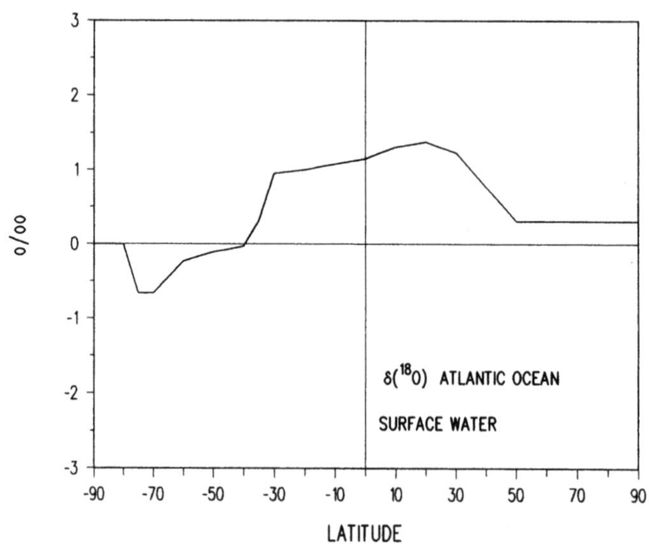

Figure 8. The inferred $\delta(^{18}O)$ of Atlantic surface waters used as input to the present global model, (Redfield and Friedman 1965), (Craig 1966).

The resulting surface δ s are given some assumed latitude distribution so as to crudely simulate fresh water coming into the Gulf of Mexico, out the St Lawrence or evaporating directly from fresh lakes on and around the Laurentide ice sheet.

The melt spike experiments for Camp Century were done using modern empirical input for the coldest northern month (Fisher, 1990) so as to roughly simulate colder average conditions. Other runs using annual average data resulted in roughly the same relative changes. Also these runs assume modern surface δs are 0, and the melt spiked surface waters have not had their assigned δs altered by evaporation. The melt spike experiments are designated as follows:

EXPERIMENT NAME	DESCRIPTION	FIGURE
SMOW1	δ_{oceans} = 0‰	
SMOW2	18 ka bulk ocean thoroughly mixed and δ constant with latitude	9a
SMOW3	1 part fresh (out the St Lawrence) and 6 parts 18 ka mixed ocean	9b
SMOW4	1 part fresh (out Gulf of Mexico) and 6 parts 18 ka mixed ocean + 1 part fresh (out St Lawrence) and 20 parts 18 ka mixed ocean.	9c
SMOW5	1 part fresh (out St Lawrence + proglacial lakes + wet surface of ice sheet) and 1 part 18 ka mixed ocean.	9d
SMOW6	1 part fresh (out Gulf of Mexico) and 4 parts 18 ka mixed ocean +1 part fresh (out St Lawrence) and 6 parts 18 ka mixed ocean.	9e

These various trials are meant to show how sensitive the model is to changing the surface waters over a wide range. The results for $\delta^{18}O$) and **d** are given in figure 10.

The SMOW1 represents the CC δs one would expect from a modern well mixed source ocean in March and SMOW2 from a well mixed 18 ka ocean of the late Wisconsin time otherwise using the modern March input fields. The other melt cap experiments should be compared to the SMOW2 results. The SMOW3 and SMOW4 experiments are the more conservative fresh water caps. The SMOW4 (out the Gulf of Mexico) produces a δ shift comparable to the +3.5 ‰ claimed to be present in A84 due to MWP-1B. The SMOW3 (out the St Lawrence) experiment which uses a similar mixing ratio for the surface waters produces only about a 1.5 ‰ shift. Of course the surface area of the ocean co vered by a melt cap introduced between 45 to 70°N (figure 9c) is smaller than for one between 20 to 50°N (figure 9c) so for an equal volume of fresh water input the mixing ratio for the northern fresh cap should possibly be higher. Never theless, because of the shape of the source ocean contribution function for CC in March (see figure 12a) one gets a bigger effect from a lower latitude fresh water cap.

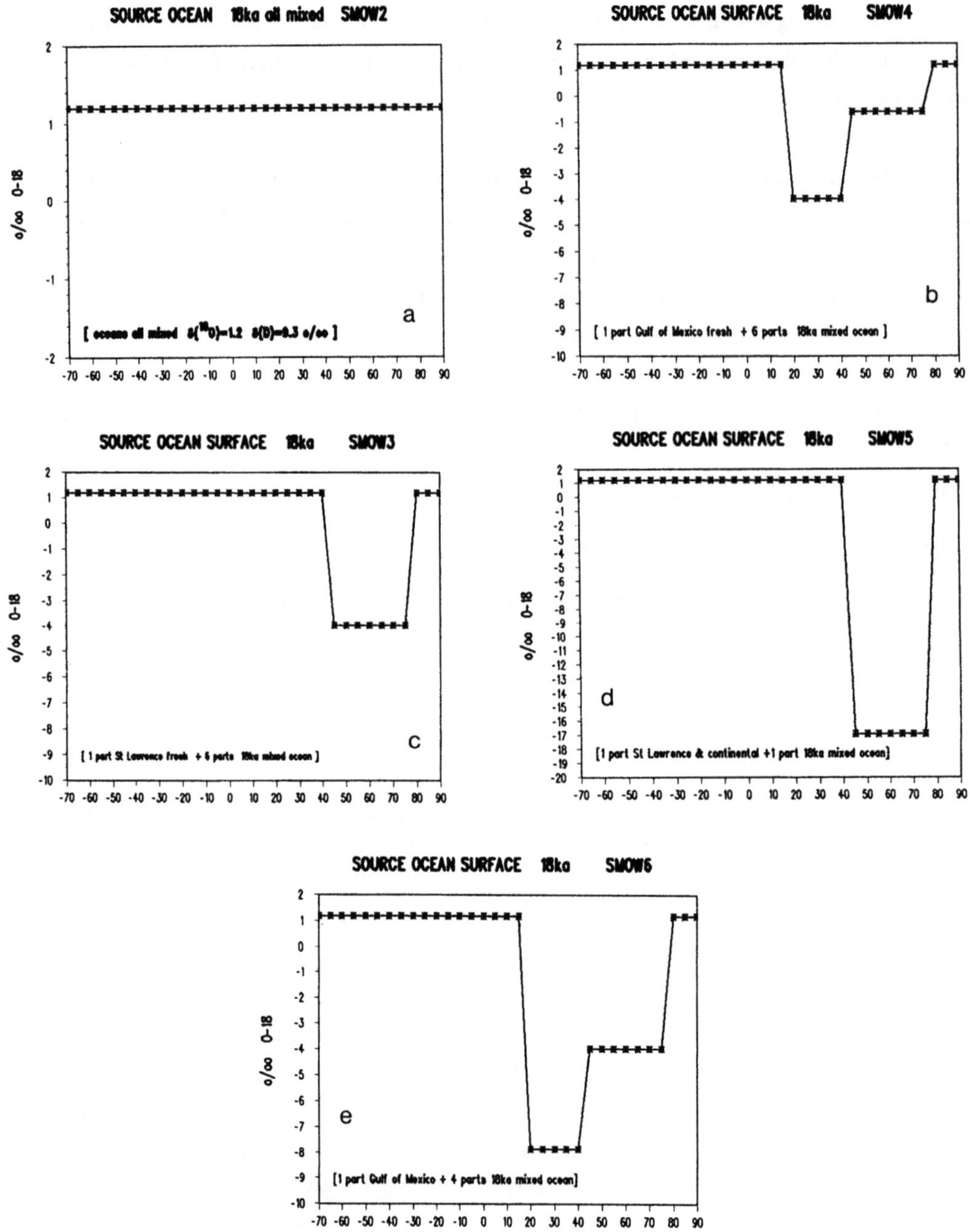

Figure 9. Surface water $\delta(^{18}O)$ for the source ocean assuming various various fresh water melt caps.

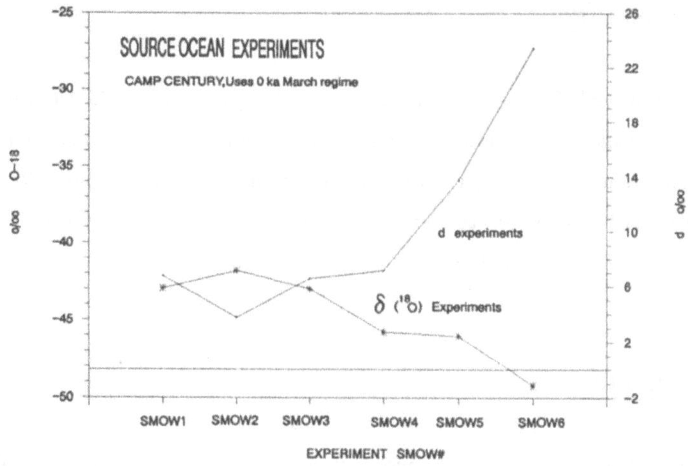

Figure 10. $\delta(^{18}O)$ and **d** at Camp Century using present March conditions and the various δ_{ocean} functions of figure 9 in order to investigate different fresh water melt caps on the source oceans.

The SMOW5 experiment might be worth consideration in that during the warm seasons of the deglaciation there must have been a very large wet evaporating area at higher latitudes on and around the ice sheets. These areas would have contributed pure very negative unmixed source water.

The SMOW6 experiment is extreme, producing a large δ shift and huge **d** shift.

Other sensitive variables that cause stable isotope changes

There follows a number of experiments in which the stable isotopes are calculated at Camp Century using modern annual input fields (figure 5) but altering one or more variable at a time. These then are sensitivity tests rather than simu lations. The variables altered are :

a. **h** source oceans relative humidity alone.
b. Wind speed **u** (k_{18}) of source oceans alone.
c. Source oceans relative humidity **h** coupled with wind speed **u** (k_{18}).
d. Source oceans sea ice front alone.
e. Source oceans sea ice front. Assuming a windy band 15 degrees south of the front. Wind speed coupled with humidity as in experiment c.
f. Horizontal meridional mixing length.
g. Temperature **Ts** at which clouds shift from predominantly vapor->water droplets to vapor->ice crystals.

h relative humidity, figure 11a

As has been shown before by Merlivat and Jouzel (1979) and Fisher and Alt (1985) the deuterium excess at any site can be altered considerably by changing the average relative humidity of the source waters. This is also true for this model and figure 11a shows $\delta(^{18}O)$ and **d** at Camp Century for various average **h** values, (averaged over the whole source area for CC). Presently **h** for CC sources is 81 %, figure 5d. As shown in figure 11a the δ value is quite stable over a wide range of **h**.

Windiness u, k_{18}, figure 11b

Changing the wind speed **u** over the source oceans also can alter **d** without changing δ very much. The wind speed in this model is introduced through Merlivat and Jouzel's (1979) kine tic evaporation factor k_{18}, where roughly speaking $k_{18}=4$ ‰ for rough oceans and 7 ‰ for smooth oceans, see figure 5e. A rough ocean has $u > 7\ ms^{-1}$ 10 m over the surface of the ocean. Petit et al. (1991) have pointed out that k_{18} is not monotonic in **u** but increases again for **u** greater than about 12 m/s, (60 km/hr).

Just for these crude tests,and provided $u < 12$ m/s the simplest monotonic relation between **u** and k_{18} is used, namely;

$$u = (8 - k_{18})3.1 \tag{1}$$

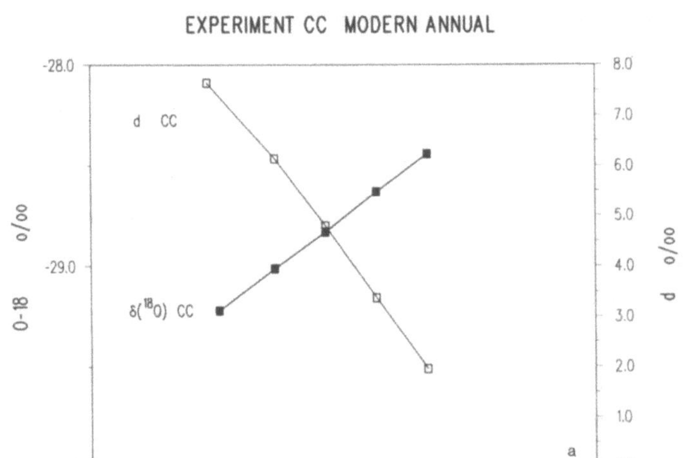

Figure 11a. Sensitivity of the model $\delta(^{18}O)$ and **d** at Camp Century using modern annual inputs and varying the source area's relative humidity **h**.

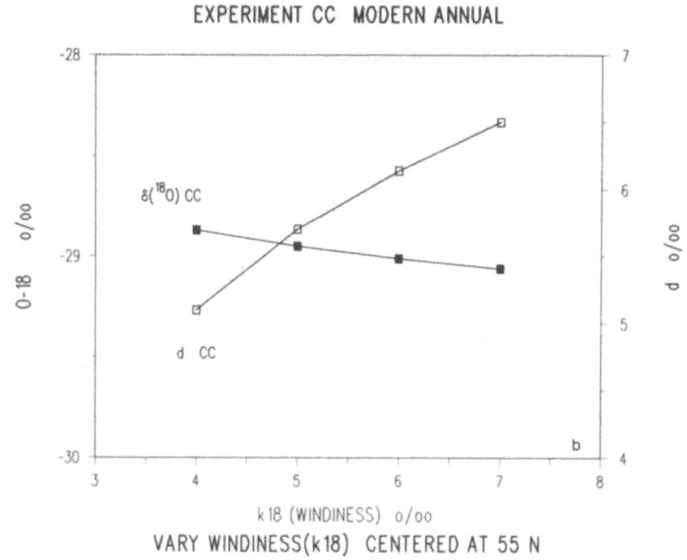

Figure 11b. Sensitivity of the model **δ(¹⁸O)** and **d** at Camp Century using modern annual inputs and varying the windiness of the band centered at 55 degrees N. Windiness is expressed through $\mathbf{k_{18}}$ as explained in the text.

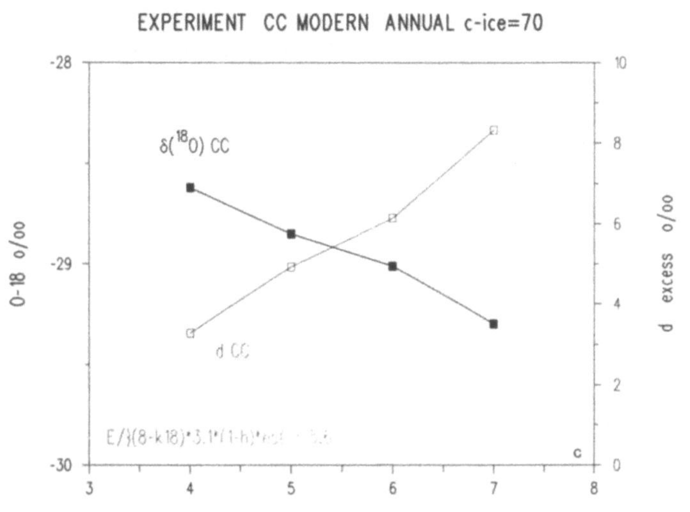

Figure 11c. Sensitivity of the model **δ(¹⁸O)** and **d** at Camp Century using modern annual inputs and varying the humidity and windiness of the band at 55 degrees N. The **h** and $\mathbf{k_{18}}$ variables have been coupled as described in the text.

Figure 5e shows that there is presently a windy band in the northern hemisphere centered at 55 degrees, (there is a much stronger one in the southern hemishere). This windiness expe riment consists of changing the magnitude and width of this band only. The windier it gets the wider it gets. The results at Camp Century appear in figure 11b. **δ** and **d** are plotted against the minimum k_{18} (highest **u**) for the windy band. The windiest experiment has k_{18} = 4‰ and the smoothest k_{18} = 7‰. $\delta(^{18}O)$ is stable wrt wind but **d** increases as the wind decreases, ie as k_{18} goes up.

Relative humidity **h** coupled with wind speed **u** (k_{18}), figure 11c

Wind speed and relative humidity are not independant. Ocean evaporation rates are given empirically by (Sellers, 1967);

$$E = K\,u(e_s - e) \qquad (2)$$

where **u** is wind speed, e_s saturation vapor pressure at the temperature of the ocean/air boundary layer, **e** the vapor pressure just above the boundary layer and K is a constant. The relative humidity is;

$h = e/e_s$. Using equations 1 and 2 the constant becomes;

$$K = E/[u(1-h)e_s] = E/[3.1(8-k_{18})(1-h)e_s] \text{....(3)}$$

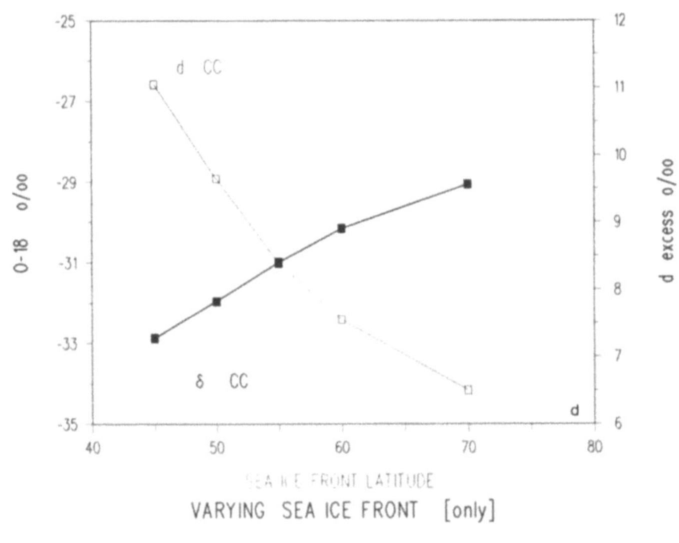

Figure 11d. Sensitivity of the model $\delta(^{18}O)$ and **d** at Camp Century using modern annual inputs and varying the sea ice front alone.

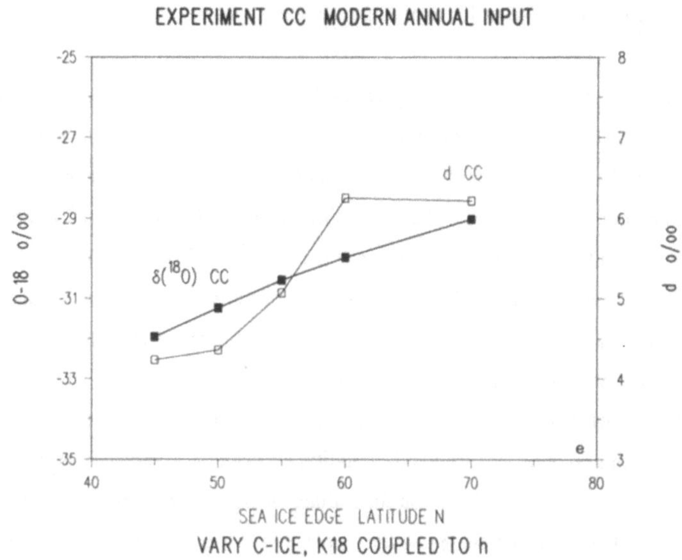

Figure 11e. Sensitivity of the model $\delta(^{18}O)$ and **d** at Camp Century using modern annual inputs and varying the sea ice front which has a windy band 15 degrees south of it in which **h** and k_{18} are coupled as in the 11c experiment.

With the modern annual zonal averages for **E**, k_{18},e_s(Tc) and **h**, (figure 5) **K** is 5.6 +/- 0.5, provided one stays outside the tropics ie poleward of 17 degrees latitude.

The experiment uses the values of **E** and e_s(Tc) for the source oceans, figure 5, and the windiness, k_{18} is varied while **h** is calculated using equation 3.

The results appear in figure 11c. Again δ at CC is not changed very much by altering the windiness coupled to humidity. The **d** values however are more variable when windiness and humidity are coupled because these two variables work together wrt **d** changes.

Source oceans sea ice front alone, figure 11d

As indicated by the arrows at the bottom of figure 5 the annual northern sea ice front is presently at about 70 degrees N. Using the modern zonal input of figure 5, the model sensi tivity to changing only the sea ice front is examined. The results for Camp Century, figure 11d, show considerable changes in δ and **d** when the sea ice is moved down to 45 degress. This is a bit too extreme because during the coldest parts of the last ice age the Atlantic sea ice probably went no further south than 50 to 55N.

Sea ice front plus a coupled windy zone 15° S of it, figure 11e

At present there is a windy band 15 degrees south of the mean annual sea ice front, see figure 5. As an experiment then, the sea ice runs above are repeated but keeping the windy band 15 degrees south of wherever the sea ice is. In addition humidity and the windiness are coupled assuming pre sent evaporation rates **E** and saturation vapor pressures $\mathbf{e_s}$. The results are shown in figure 11e and are surprising. The δ sensitivity is small and about the same as when the sea ice is changed by itself, but the deuterium excess sensitivity is now almost gone! One would think that water arriving at CC from more distant warmer sources would have larger **d** values as is in fact demonstrated for the figure 11d experiment. Here however one is counteracting this distant warm source effect by an equally strong offsetting effect namely moving the windy humid band southward with the sea ice front. The two effects at CC are about equal and thus the resulting **d** sensitivity to this coupled [sea ice-wind-humidity] experiment is very low. At face value this result contradicts Dansgaard et al.'s(1989) explaination for sudden changes in **d** at the Younger Dryas boundary. It should be remembered though that these model experiments are not real simulations as the variable couplings are not complete and at best empirical.

Meridional mixing length λ_h, figure 12

As discribed in (Fisher, 1990), water added at a given source takes a certain meridional distance λ_h to mix vertically to some defined height. The height chosen for the runs above is 2.5 km and corresponds to the top of the mid-troposhere (at about 800 mb). The Arctic high elevation runs, like Camp Century have

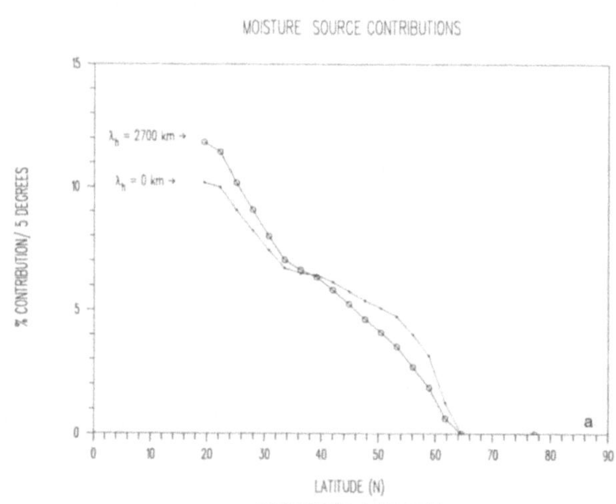

Figure 12a. Winter source water contribution percents at Camp Century using a typical horizontal meridional mixing length of 2700 km and one of 0 km.

all been done with λ_h = 2700 km. This means that water has to originate about 23 degrees south of a given site that is at 2.5 km elevation before it makes a large contribution to the sites precipitation. The smaller λ_h the larger the input from "local" water.

The percent contribution of source water to CC is given in figure 12a for λ_h = 2700 and 0 km. The $\delta(^{18}O)$ and **d** sensitivity to changing λ_h appears in figure 12b. As expected the smaller λ_h the more positive δ and smaller **d** at CC. The range of λ_h is probably $1500<\lambda_h<3500$ km so the δ changes possible are not large and the **d** changes are about 1‰. Figure 12a shows that for the winter regime not much local water gets up to CC. This increases for annual conditions but is still rather small, (Fisher 1990).

Temperature Ts cold cloud shift temperature, figure 13

The temperature at which to shift from vapor->water droplets to vapor->ice crystals has a large effect on **d** and a significant effect on δ, (Fisher, 1990). Using modern input data the CC δs and **ds** over a wide range of **Ts** values were calculated. This **Ts** sensitivity appears in figure 13. The probable range of **Ts** is -17 < **Ts** < -5 C°. It could be sensitive to air turbidity, with "dirty" clouds having more positive **Ts** values. The value of **Ts** becomes important for high cold sites but as figure 13 shows for a site like CC there is a large sensitivity of the model **ds** to **Ts** and even a significant sensitivity in δ.

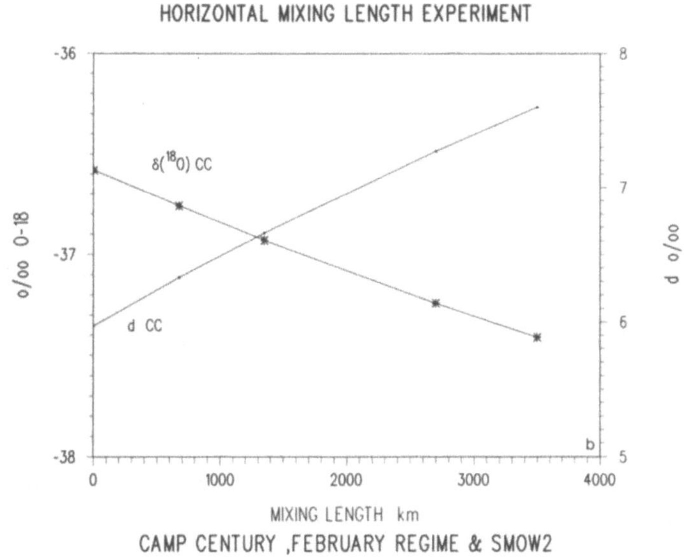

Figure 12b. Sensitivity of winter δs and **ds** at Camp Century to changes in horizontal meridional mixing lengths.

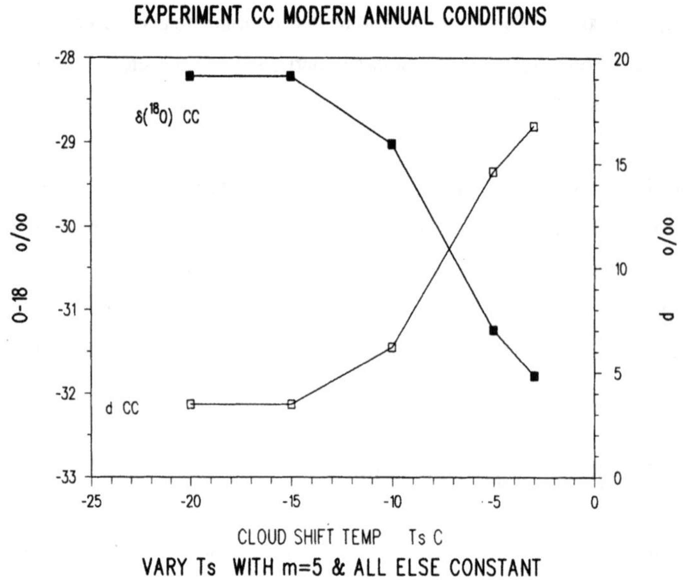

Figure 13. Sensitivity of annual δs and ds at Camp Century to changes in the cold cloud shift temperature Ts.

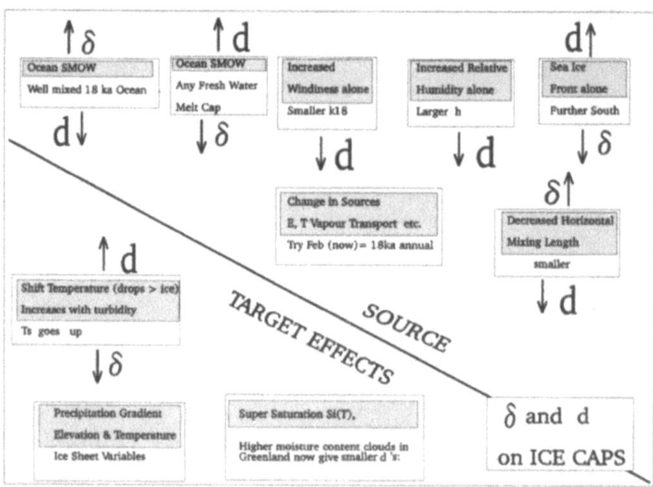

Figure 14. Summary of source and near site effects that can influence the **δ(^{18}O)** and **d** values of precipitation on high elevation ice caps.

Summary

Figure 14 gives a summary of source and near site effects that can change **δ(^{18}O)** and **d** for high ice cap sites.

At the beginning of this paper it is hypothesized that the early Holocene part of the Agassiz A84 δ(^{18}O) record shows evidence of a melt water cap, probably associated with MWP- 1B. The melt cap sensitivity experiments suggest that Atlantic surface water spiked with ice sheet melt could produce the required negative shifts if a sustained mixing ratio of at least 1 melt to 6 bulk ocean could be produced. However, as the last part of the paper and figure 14 emphasize there are host of effects in the source water zones and in the near site areas that produce large changes in ice core stable isotopes. Since modern input data fields are used,all the calculated results are sensitivity tests and not simulations.

References

Arnason, B (1969) The exchange of hydrogen isotopes between ice and water in temperate glaciers. **Earth and Planetary Science Letters**, 6(6), 423-430.

Claire, L.P., J.C. Comiso, H.J. Zwally, D.J. Cavalieri, P. Gloersen and W.J.Campbell (1987) **Arctic Sea Ice 1973-1976: Satellite Passive-microwave Observations**, NASA SP-489, 295. Scientific and Technical Information Branch NASA, Washington.

Curry, J.A., Meyer, F.G., Radke, L.F., Brock, C.A. and Ebert, E.E (1990) Occurrence and characteristics of lower tropospheric ice crystals in the Arctic. **International Journal of Climatology**, vol 10, 749-764.

Craig, H. (1966) Isotopic composition and origin of the Red Sea and Salton Sea geothermal brines. **Science**, 154. 1544-1548.

Dansgaard, W. and Tauber, H. (1969) Glacier oxygen-18 content and Pleistocene ocean temperatures. **Science**, vol 166, 499-502.

Dansgaard, W., White, J.W.C. and Johnsen, S.J. (1989) The abrupt termination of the Younger Dryas climate event. **NATURE**, vol 339, 532-534.

Fairbanks, R.G. (1989) A 17,000-year glacio-eustatic sea level record:influence of glacial melting rates on the Younger Dryas event and deep-ocean circulation. **NATURE**, vol 342, 637-642.

Ferronsky, V.I. and Brezgunov, V.S. (1989) Stable isotopes and ocean dynamics. **in** Fritz, P. and Fontes, J. Ch. (eds) **Handbook of Environmental Isotope Geochemistry**, vol 3, The Marine Invironment, A. Elsevier. 1-27.

Fisher, D.A., Koerner, R.M., Paterson, W.S.B., Dansgaard, W.,Gundestrup, N. and Reeh, N. (1983) Effect of wind scouring on climate records from ice-core oxygen-isotope profiles. **NATURE**, vol 301 no 5897, 205-209.

Fisher, D.A. and Alt, B.T. (1985) A global oxygen isotope model-semi-empirical, zonally averaged. **Annals of Glaciology**, 7, 117-124.

Fisher, D.A. (1990) A zonally averaged isotope model coupled to a regional variable elevation stable isotope model. **Annals of Glaciology**, 14, 65-71. 1989 Seattle Symposium on **Ice and Climate**.

Fisher, D.A. in press. Remarks on the deuterium excess in precipitation in cold regions. **Tellus**.

Hammer, C.U., Clausen, H.B. and Tauber, H. (1986) Ice core dating of the Pleistocene/Holocene boundary applied to a calbration of the C-14 time scale. **Radiocarbon**, vol 28, 284-291.

Hooke, R.LeB. and Clausen, H.B. (1982) Wisconsin and Holocene S(^{18}O) variations, Barnes Ice Cap, Canada. **Geological Society of America Bulletin**, vol 93, 784-789.

IAEA/WMO, International Atomic Energy Agency/World Meteorological Organization. (1969-1986) **Environmental Isotope Data No. 1-8, World Survey of Isotope Concentration in Precipitation**. Technical reports series, Nos. 96,117,129,147,165,192,226 and 264. I.A.E.A. Vienna.

IAEA/WMO. (1981) **Statistical Treatment of Environmental Isotope Data in Precipitation**. Technical report series, No. 206. I.A.E.A, Vienna.

Jacka, J.H, Christou, L and Cook, B.J. (1985) Updating the sea ice and climate monitoring programs. **ANARE** research notes No. 28:59-62.

Jaeger, L. (1983) Monthly and areal patterns of mean global precipitation. **in** Street-Perrott A, Beran M, Ratcliffe R (eds) **Variations in The Global Water Budget**. Dordrecht etc, D. Reidel Publishing Co.

Jayaweera, K.O.L.F. and Ohtake, T. (1973) Concentration of ice crystals in Arctic stratus clouds. **J.Rech. Atmos.**, 7, 199-207.

Johnsen, S.J., Dansgaard, W. and White, J.W.C. (1989) The origin of Arctic precipitation under present and glacial conditions. **Tellus**. 41B, 4. 452-468.

Jouzel, J. and Merlivat, L.(1984) Dueterium and oxygen 18 in precipitation: modeling of the isotopic effects during snow formation. **J. Geophys. Res.**, 89(D7), 11749-11757.

Koerner, R.M. (1970) The mass balance of the Devon Island Ice Cap, Northwest Territories, Canada, 1961-66. **Journal of Glaciology**, vol 9 No 57, 325-336.

Koerner, R.M. and Russell, R.D. (1979) S(^{18}O) variations in snow on the Devon Island Ice Cap, Northwest Territories, Canada. **Canadian Journal of Earth Sciences**, vol 16, 1419-1427.

Koerner,R.M., Paterson,W.S.B. and Krouse,H.R. (1973) S(^{18}O) profile in ice formed between the equilibrium and firn lines. **NATURE**, vol 245 no 148, 137-140.

Koerner, R.M. and Fisher, D.A. (1990) A record of Holocene summer climate from a Canadian high-Arctic ice core. **NATURE**, vol 343, no 6259, 630-631.

Mason, B.J. (1975) **Clouds,Rain and Rain Making**, second edition. Cambridge University Press. Cambridge.

Merlivat, L. and Jouzel, J. (1979) Global climatic interpretation of deuterium-oxygen 18 relationship for precipitation. **Journal of Geophysical Research**. 84(C8), 5029-5033.

Mock, S.J. (1965) Glaciological studies in the vicinity of Camp Century Greenland. **CRREL Report**, 157.

Mock, S.J. (1967) Accumulation patterns on the Greenland Ice Sheet. **CRREL Report**, 233.

Oberhuber, J.M.(1988) **An Atlas Based on the 'COADS' Data Set: the Budgets of Heat, Buoyancy and Turbulent Kinetic Energy at the Surface of the Global Ocean**. Max-Planck-Inititute for Meteorology report No.15. Max-Planck-Institute, Hamburg.

Oort, A.H. (1983) **Global Atmospheric Circulation Statistics 1958-1973**. NOAA professional paper 14, pp 180 plus 36 micro-fiches.

Paterson, W.S.B., Koerner, R.M., Fisher, D., Johnsen, S.J., Clausen, H.B., Dansgaard, W., Bucher, P. and Oeschger, H. (1977) An oxygen-isotope climatic record from the Devon Island Ice Cap, Arctic Canada. **Nature**, 266(5602), 508-511.

Peixôto, J.P. and Oort, A.H. (1983) The atmospheric branch of the hydrological cycle and climate. **In** Street-Perrott A, Beran M, Ratcliffe R (eds) **Variations in the Global Water Budget**. Dordrecht etc, D. Reidel Publishing Co. 5-65.

Petit, J.R., White, J.W.C., Young, N.W., Jouzel, J. and Korotkevich, Y.S. (1991) Deuterium excess in recent Antarctic snow. **Journal of Geophysical Research**, 96 (3D), 5113-5122.

Redfield, A. and Friedman,I. (1965) Factors affecting the distribution of deuterium in oceans. **In: Symposium on Marine Geochemistry**. University of Rhode Island. Occas. Publication 3: 149-168.

Reeh, N., Thomsen, H.H., Frich, P. and Clausen, H.B. (1990) Stable isotope studies on ice margins in the Thule area, Northwest Greenland. **in** Svend Funder (ed) **Late Quaternary Stratigraphy in the Thule Area Northwest Greenland. Meddelelser om Gronland**. Geoscience 22.

Rozanski, K. and Pazdur, M.F. (1991), This workshop.

Sellers, W.D. (1967) **Physical Climatology. Second Edition**. Chicago and London, University of Chicago Press.

CLIMATIC CHANGES IN NORTHWEST AFRICA DURING THE LAST DEGLACIATION (16-7 ka BP)

F. Gasse and J. Ch. Fontes
Laboratoire d'Hydrologie et de Géochimie Isotopique
Bâtiment 504
Université de Paris-Sud
91405-Orsay Cedex, France.

ABSTRACT. This paper, based on the study of several lacustrine profiles, compares the climatic evolution of different climatic zones of northwest Africa during the period of the last deglaciation, in order to distinguish between large-scale climatic events, and regional patterns.

Methodological constraints for interpretation of palaeohydrological records in terms of palaeoclimate are first discussed. Attention is focussed on the selection of the palaeolakes, the establishement of a reliable chronology, and the interpretation of palaeohydrological indicators, namely mineralogy, stable isotope contents of carbonates, and biological remains.

Results are presented for 1) the central Sahel, 2) the central Sahara, 3) the northern Sahara, and 4) the Middle Atlas (Morocco). The major Arid-Humid Transitions (AHTs) appear to have been synchronous in all regions irrespective of whether they were submitted to tropical summer rainfall or to mid-latitude precipitation. The AHTs and reversal events are in good chronological agreement with the cold/warm fluctuations deduced from marine cores and high latitude continental records. Arid conditions prevailed during the Younger Dryas chronozone. Some regional differences are suggested, however, in the timing of optimal climatic conditions (maximum Precipitation-Evapotranspiration excesses), and the occurrence of a brief humid phase in central Sahel at ≈10.6-10.3 ka BP, the latter has not been found elsewhere.

Possible mechanisms for explaining palaeoclimatic covariability/regional differences are then discussed on the basis of studies on modern rainfall anomalies over Africa. Different types of factors are considered : 1) orbital forcing ; 2) atmospheric and oceanic circulation and Sea Surface Temperature anomalies; 3) land surface conditions. Emphasis is placed on the potential role of the soil moisture as an amplifying mechanism for squall line pluviogenesis.

NATO ASI Series, Vol. I 2
The Last Deglaciation: Absolute and Radiocarbon Chronologies
Edited by E. Bard and W. S. Broecker

INTRODUCTION

Within the past few years, attention was focussed on abrupt, sub-Milankovitch climatic changes (10 to 1000 years) which cannot be directly accounted for by orbital forcing. The two-step deglaciation model (Duplessy et al. 1981; Ruddiman and Duplessy, 1985) is now well documented from marine (Bard et al., 1987, 1989; Fairbanks, 1989; Kudrass et al., 1991) and from high and mid latitude continental records (see Berger and Labeyrie, 1987). The Late Glacial period is, however, poorly known at tropical and subtropical latitudes, where climate changes are mainly reflected by fluctuations in the water budget. The understanding of sudden events and related mechanisms require the study of high resolution sequences in sensitive sectors. Such favourable sequences can be offered by lacustrine deposits, despite the difficulty of obtaining long continuous records in arid and semi-arid areas.

This paper deals with climate-induced variations of water, salt and isotope balances of lakes and aquifers of northwest Africa during the last deglaciation. Lacustrine cores provide spatio-temporal information on timing and rate of sudden changes in the balance Precipitation-Evapotranspiration (P-E) which occurred in several different climatic zones (Fig. 1) of northwestern Africa : 1) the central Sahel, dry in winter and

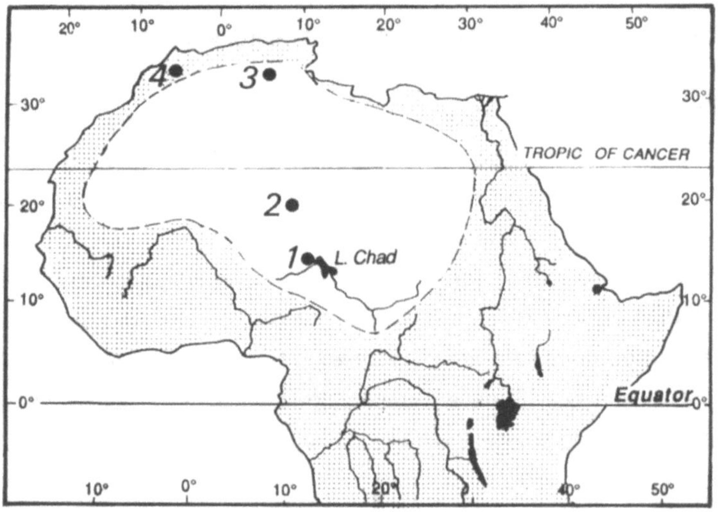

Fig. 1. Location of the investigated sites. The dashed line indicates the approximate area of present-day internal drainage. 1: Central Sahel (Bougdouma and Ari Koukouri, southern Niger ; 2 : central Sahara, eastern margin of the Aïr massif (Adrar Bous, northern Niger); 3: northern Sahara (Sebkha Mellala, Algeria); 4 : Middle Atlas (Tigalmamine, Morocco).

and with summer squalls and monsoonal rainfall; 2) the margin of a mountain massif of the central Sahara submitted to summer squall lines, and possibly to winter showers; 3) the northern Sahara, dominated by the Saharan anticyclone throughout the year, and receiving rare rainfall of northern origin; and 4) the Middle Atlas in the Maghreb, characterized by dry summers and winter precipitation from the mid-latitudes westerlies.

We compare below the evolution of regions with predominant tropical warm-season rainfall regimes (1 and 2), with poleward zones where winter rains are chiefly of extra-tropical origin (3 and 4).

It has been known for a long time that arid conditions prevailed over the northern African tropics during the Last Glacial Maximum (LGM), whereas the lower and/or mid-Holocene were much wetter than today (Gasse, 1977; Street and Grove, 1979; Talbot, 1980; Sarnthein et al., 1981; Servant, 1983). This general trend, which can be largely accounted for by seasonal fluctuations in solar radiation since 18 ka BP (Kutzbach & Street-Perrott, 1985), is compatible with the patterns of global energy balance as considered in the General Circulation Models (GCMs) (e.g. COHMAP Members, 1988; Street-Perrott et al., 1990). However, the Arid-Humid Transition (AHT) over the tropics has been discontinuous, as unexpected from the orbital control of insolation.

In the northern African tropics south of 15°N, lake records have clearly shown that the Arid-Humid Transition has been interrupted by at least two abrupt returns to dry conditions (Servant and Servant-Vildary, 1980; Servant, 1983; Talbot et al., 1984). The first one, dated at ≈11-9.5 ka BP, clearly coincides with the Younger Dryas chronozone (Gasse et al., 1990 Street-Perrott and Perrott, 1990) . The second one, ≈ 8-7 ka BP, is still difficult to correlate with marine records. These events have been tentatively linked (Street-Perrott and Perrott, 1990) to changes in the North Atlantic Deep Water (NADW) formation associated with the meltwater history of the Laurentide Ice Sheet, in agreement with an hypothesis proposed by Broecker et al. (1985, 1988, 1989). The effects of cold melt water on oceanic dynamics was invoked (Street-Perrott and Perrott, 1990) to explain the reduction in monsoon rainfall over the northern tropics, although the diminution of NADW formation during the Younger Dryas is still debated (Jansen & Veum, 1990).

In the central Sahara, north of 15°N, no evidence for a return to humid conditions prior to 9-8 ka BP had been reported from lake sediments until the recent results of the PALHYDAF project (Fontes and Gasse, 1991 a).

Few detailed records were available until the last few years from the northern Sahara and the Maghreb, and their chronology was sometimes poorly established (Fontes & Gasse, 1989 ; Street-Perrott et al., 1989). These areas, which mainly depend on Mediterranean and North Atlantic rainfall, are sometimes regarded as having been more humid than today during the LGM, and drier during the lower Holocene (Street-Perrott and Roberts, 1983 ; Rognon, 1987). It has been recently shown, however, that Saharan deposits attributed to the LGM on the basis of ^{14}C dates (Conrad, 1969) may be much older (Gibert, 1989 : Causse et al., 1988).

This paper shows clear similarities in the timing of the major hydroclimatic changes over the different climatic zones, although regional differences appear in the details of the evolution and in the periods of maximal (or minimal) Precipitation - Evapotranspiration (P-E) balance.

METHODS

Selection of the palaeolakes

It is obvious at a first glance that closed lakes supplied by surface-runoff could represent a favourable case for preservation of palaeoclimatic archives. However, the extent and the hydrogeological behaviour of the catchment area are major quality criteria for spatial reconstruction of rainfall fluctuations. For example, groundwater-fed palaeolakes of the Sahara and the Sahel are much more sensitive to local climatic changes that Lake Chad which integrates rainfall fluctuations over ≈10°N of latitude through the Chari river and its tributaries. Basic constraints are : (i) the recharge area of the aquifer must be restricted enough to allow climate reconstruction at a local scale, and (ii) the time of reaction to rainfall events must be short. Palaeolakes supplied by the largest confined aquifers of the Sahara, e.g. the Continental Intercalaire or the Nubian Sandstone aquifers, can provide good records of climatic events at a regional scale, provided flow patterns and distance to the recharge zone are known. The knowledge of chemical and isotope contents of groundwater is also necessary for accurate palaeohydrological interpretations. Lakes fed by groundwater which may have dissolved evaporites, e.g. NE of Lake Chad (Maglione, 1976) or Northern Mali (Fabre and Petit-Maire, 1988) might not be favourable because permanent high salinity may mask variations in hydrogeochemical balances and climatic conditions. Selection of the basins thus implies a thorough discussion of local and regional control of water level and water chemistry (Gasse et al. 1990; Fontes and Gasse, 1991 a).

Chronology of continental deposits

Causes of age distortion in continental sediments are more numerous than in marine deposits where bias seem to be limited to bioturbation, ageing of oceanic water masses, selective dissolution at depth (Broecker et al., 1991), and generation of old CO_2 by oxidation of organic matter from bottom sediments (until now, no evidence of diagenetic changes in the carbon isotope composition has been provided for oceanic sediments). For lake records, bioturbation is probably not a major problem because of the high sedimentation rates and limited development of benthic fauna. Ageing of waters and of aqueous inorganic carbon may be significant in lakes supplied by confined aquifers (Fontes et al., 1983). Oxidation of organic matter may contribute old CO_2 to the Total Dissolved Inorganic Carbon (TDIC), for example when decreases in environmental redox conditions allow sulphate reduction (Durand et al., 1984 ; Talbot and Kelts, 1986). A specific case of chemical ageing is the so called "hard water effect" where carbonate dissolution occurs in the catchment area, e.g. at Tigalmamine (Fig. 1). However, such dissolution does not necessarily means age distortion, as ^{14}C-dilution may be readily counteracted in lakes where isotope exchange reactions take place between the TDIC and the atmospheric CO_2 (Fontes, 1990). Such exchange is enhanced by high pH and low temperature conditions which increase the rate and the amount of CO_2 dissolution at the water-atmosphere interface. In that case, the ^{14}C content of the TDIC from surface waters reflects that of the atmosphere (Broecker and Peng, 1982). High mixing conditions, depending on thermic or saline gradient and/or wind activity, are then required for surface waters to reach the deepest parts of the system. Specially favourable conditions for $^{14}CO_2$ transit from the atmosphere into the bottom sediments occur in swampy lakes colonized by subaquatic vegetation (Fontes and Gasse, 1991a), e.g. Bougdouma (Fig. 1).

Diagenetic effects due to recrystallization through fluctuations of the groundwater table in the sediments after lake regression may give rise to further distortions towards older or younger ages depending on the origin of the TDIC from the aquifer.

The validation of continental ^{14}C chronology in lacustrine deposits implies several requirements. Although sometimes difficult, the detection of reworked material, e.g. terrestrial plant remains, detrital carbonates from the catchment, and even of sediments from previous highstands of the lake itself, is crucial. The authigeny of the dated material, and especially the idiomorphic shape of calcite crystals, has to be checked through optical and SEM observations. The ^{13}C content of the carbonate provides an additional criterion for discussion of ^{14}C dating but cannot be used as a simple validation index of carbonate ages. Dissolution of carbonates ("hard water effect") may be indicated by high ^{13}C contents if dissolved calcite is of marine origin. However, such dissolution may not change the ^{13}C content if the old calcite is of secondary origin (Fontes, 1990). Isotope equilibrium with atmospheric CO_2 (a case of age validity) also leads to high ^{13}C content of the carbonates (Durand et al, 1984 ; Fontes and Gasse, 1991 a). Low $\delta^{13}C$ values of carbonates indicate the relative importance of the contribution of biological CO_2 to the TDIC but do not imply in *per se* any age significance. The possible ageing of this biogenic CO_2 must then be discussed .

The "small" (generally <40 μm) fraction of calcite crystals which precipitate at the water surface, and therefore reflect close to equilibrium conditions with atmospheric CO_2 provide a valuable material for ^{14}C dating.

Attention must also be paid to the frequent unreliability of ^{14}C dates on organic matter, as compared to authigenic carbonates. In addition to the effects of water ageing, which equally affects carbonates and organic matter of aquatic origin, the organic matter of sediments percolated by circulating solutions can exchange with dissolved and younger organic coumpounds (Taylor, 1990). Even macrophyte debris may be contaminated by such percolations through adsorbtion of soil-derived organo-compounds (Téhet et al., 1990).

In conclusion, a quality assurance discussion of the ^{14}C dates is site specific and involves many criteria often absent in the literature. This discussion is easier to conduct for authigenic calcimicrite crystals than for organic rests. One of the major methodological results of the Palhydaf project (Fontes and Gasse, 1991 a, b) is a reappraisal of the validity of carbonate dating in lacustrine deposits.

Palaeohydrological indicators

Some of the variables sensitive to environmental changes in the studied profiles are considered below.

Mineralogy and stable isotope contents of carbonates

Oxygen-18 and carbon-13 contents are measured on authigenic carbonate separated by sieving. Under temperate and cold climates, oxygen-18 contents of lacustrine carbonates can often be intepreted in terms of isotopic composition of precipitation (Buchardt and

Fritz, 1980 : Siegenthaler et al., 1984). This requires that the residence time of waters in the lake is short and that evaporation plays a minor role in the lake balance. Thus, cold periods are reflected by low ^{18}O contents whereas climatic warmings give rise to higher $\delta^{18}O$ values in carbonates. Under tropical climates, an inversion of this pattern is expected as cold periods at high latitudes correspond to dry periods at low latitudes of the northern hemisphere and to a related increase in ^{18}O contents due to the influence of evaporation (Gasse et al., 1990). Conversely, a wet period may be accompanied by a lowering in the ^{18}O content of the carbonate due to a lesse influence of evaporation and lacustrine transgressions are marked by sudden decreases in ^{18}O contents from carbonates.

Two main differences with the marine environment should be noted : (i) the temperature is a minor factor of variation of the ^{18}O content of lacustrine carbonate, and (ii) humid/warm periods are reflected by significant increases in the calcitic fraction of the sediments, whereas dry (and probably cool) periods are favourable to the deposition of inorganic Mg-calcite, aragonite, and dolomite from solutions concentrated by evaporation. Therefore, the mineralogical composition of the sediment can be used as an additional criterion of palaeohydrological and palaeoclimatical variations.

The ^{13}C content of lacustrine carbonates provides an index of the origin of the TDIC. An interpretation of ^{13}C (and ^{14}C) contents in lakes has been proposed by Durand et al. (1984). As in marine environments, $\delta\,^{13}C$ values close to zero or slightly positive (up to + 2‰ or +3‰, if normalized with respect to the calcite) indicate a tendency towards equilibrium with the atmospheric reservoir of CO_2. This process is driven by high mixing rates, low alkalinity, and/or high residence time in the basin. In the latter case, evaporation leads to an increase in $\delta^{18}O$ values which is limited by the exchange with atmospheric moisture (Gonfiantini, 1986). Conversely, low ^{13}C contents, generally linked to low ^{18}O contents, are indicative of groundwater floods and of wet periods of high recharge rate. In these conditions, a ^{13}C and ^{18}O covariation is expected from low values during groundwater flood episodes to high values during stagnation episode of equilibrium with the atmosphere (Gasse et al., 1987). However, when the lake is supplied by river water, a flood with low ^{18}O contents (unvaporated waters) may be associated with high ^{13}C contents due to equilibrium with the atmosphere of a solution with a low buffer capacity (low alkalinity; Stumm and Morgan, 1984). Such may be the case on northwestern Sahara during the Holocene (Gasse et al., 1987). In the case of stagnant and warm waters like in Bougdouma, high organic activity may generate favourable conditions for methanogenesis (Durand et al.,

1984). The ^{13}C content of the carbonate reaches very high values due to the high enrichment which occurs between cogenetic CO_2 and CH_4 (Barker et al., 1979).

Biological remains

Amongst aquatic organisms preserved in the sediments, the most relevant in terms of palaeoconditions are diatoms, ostracods and charophytes. Emphasis is given here to diatoms which, in favourable cases, provide quantitative estimates of environmental changes.

There are generally marked differences between biological communities which occupy the littoral zones and those living offshore. The relative proportion of periphytic and planktonic diatoms thus informs on the water depth and distance from the shore. However, this pattern may be modified when currents move organisms from their original habitat, and when low transparency makes phytoplanktonic forms predominate in shallow open waters. The comparison between dated diatom assemblages from cores collected at the margin and at the centre of a lake may allow semi-quantitative estimate of the water depth fluctuations (e.g. Tigalmamine site ; El Hamouti et al., 1991).

Contemporary data sets on African diatoms allowed calibration functions to be developed that relate water salinity (or conductivity) to the composition of the communities. Diatom-inferred salinity can thus be calculated along stratigraphical profiles (e.g. southern Niger sites). In several cases, however, the abundance of species today unknown from Africa (e.g. *Cyclotella antiqua* at Tigalmamine, or *Cyclotella caspia* in central Sahara) precludes such statistical treatments.

RECORDED HYDROLOGICAL EVENTS

Central Sahel (southern Niger)

Previous work on the Chad basin (Durand and Mathieu, 1979-1980 ; Servant and Servant-Vildary, 1980 ; Maley, 1981; Servant, 1983) has already demonstrated that humid conditions re-established at ≈13-12 ka BP. During the last deglaciation, marked drops in the Precipitation-Evaporation balance occurred at 11-9.5 and 8-7 ka BP. South of the Sahel, lake Bosumtwi (Ghana) has experienced low lake-level before 13 ka BP, and during the intervals 11.2-10.1 and 8.2-7.1 ka BP (Talbot et al., 1984). Although Street-Perrott and Perrott (1990) regard the Younger Dryas chronozone (11-10 ka BP) as generally dry over Subsaharan Africa, our results suggest that the climatic evolution has been rather complex in the central Sahel during this interval.

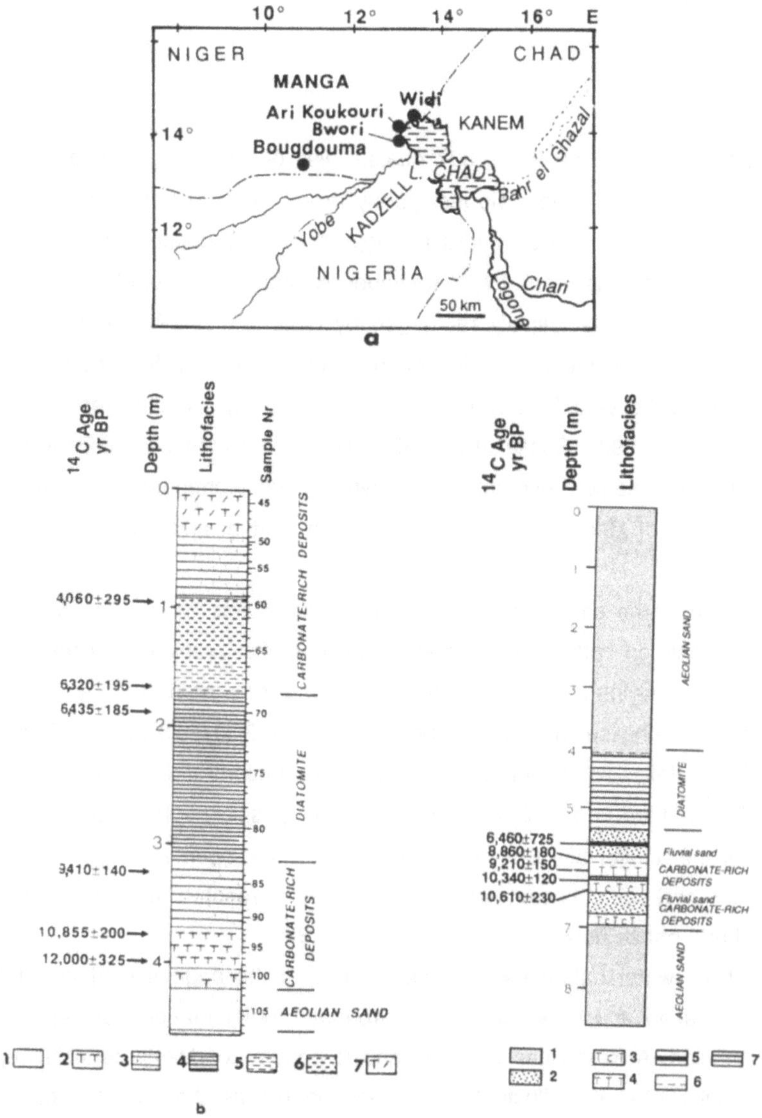

Fig. 2. Central Sahel (southern Niger). a) *Location of Bougdouma and Ari Koukouri.* b) *Bougdouma* - Simplified core stratigraphy; 1 : aeolian sand; 2: sandy carbonate; 3: diatomaceous carbonates; 4: carbonate-free diatomite; 5: calcareous clay; 6: calcareous sand; 7: carbonates and evaporites. c) *Ari Koukouri*- Simplified core stratigraphy; 1: aeolian sand; 2: fluvial sand; 3: shelly carbonate; 4: clayish carbonate; 6: clay; 7 palaeosoil.

The two sites selected here, Bougdouma and Ari Koukouri (Fig. 2), lie between ≈10 and 15°N, and 9 and 15°E., where summer squalls represents today 75% of the precipitation (Leroux, 1983). Both sites are sensitive to rainfall fluctuations which have occurred on a restricted catchment area.

Bougdouma

Bougdouma (13°19' N, 11°40' E, 337 m.a.s.l.) lies on the Manga Plateau (Fig. 2a), the surface of an old erg which slopes gently from ≈400 (northwest) to ≈300 m (southeast) over 120 km. The deepest interdunal depressions are occupied by permanent or seasonal waterbodies ranging from fresh to hyperalkaline, and supplied by the local unconfined aquifer which is exclusively recharged by local rainfall (Pirard, 1962).

A 4.58 m core from the depression of Bougdouma, almost dry today, provided a continuous profile from the Late Glacial to the Present (Fig. 2b). Detailed results are given elsewhere (Téhet et al., 1990 ; Gasse et al, 1990). During the Late Glacial-Early Holocene period, a step-wise hydrological evolution is shown by major changes in sedimentotogy, stable isotope contents of authigenic carbonates, diatom content and diatom assemblages (Figs. 2b, 3).

- At the base of the core, a calcareous diatomite reflects an alkaline-saline pond . Groundwater diagenesis on both carbonates and organic matter preclude any reliable dating (Téhet et al., 1990). Overlying aeolian sand indicates conditions more arid than today.

- Carbonate-bearing sediments are observed from >12 to ≈9.3 ka BP. Two sudden groundwater influxes, followed by phases of evaporative concentration, occurred at >12 ka BP and ≈12 ka BP, and led to a shallow saline waterbody, possibly ephemeral, until ≈ 11 ka BP. A new rise of the aquifer at ≈10.7 ka BP brought dilute, open water conditions in the depression. Isotope and salt concentration then increased to reach a maximum, indicating a short but marked arid phase between ≈10.3 and 10.0 ka BP.

- A positive water balance was re-established at the beginning of the Holocene, ≈9.7 ka BP, and peaked at ≈9.2 ka BP when carbonate sedimentation was replaced by a diatomite accumulation. A freshwater lake with anoxic bottom conditions persisted up to ≈8 ka BP, when the salinity began to fluctuate in response to increased evaporation rate.

- Calcite reappears at ≈ 6.4 ka BP (Fig. 2b) The upper part of the profile reflects unstable but generally saline, highly evaporative conditions interrupted, however, by a humid episode marked by a sudden rise of the aquifer at ≈4 ka BP.

The brief but clear humid phase dated from ≈10.7-10.3 ka BP is also found in isolated depressions in the Kadzell, along the northwestern margin of Lake Chad.

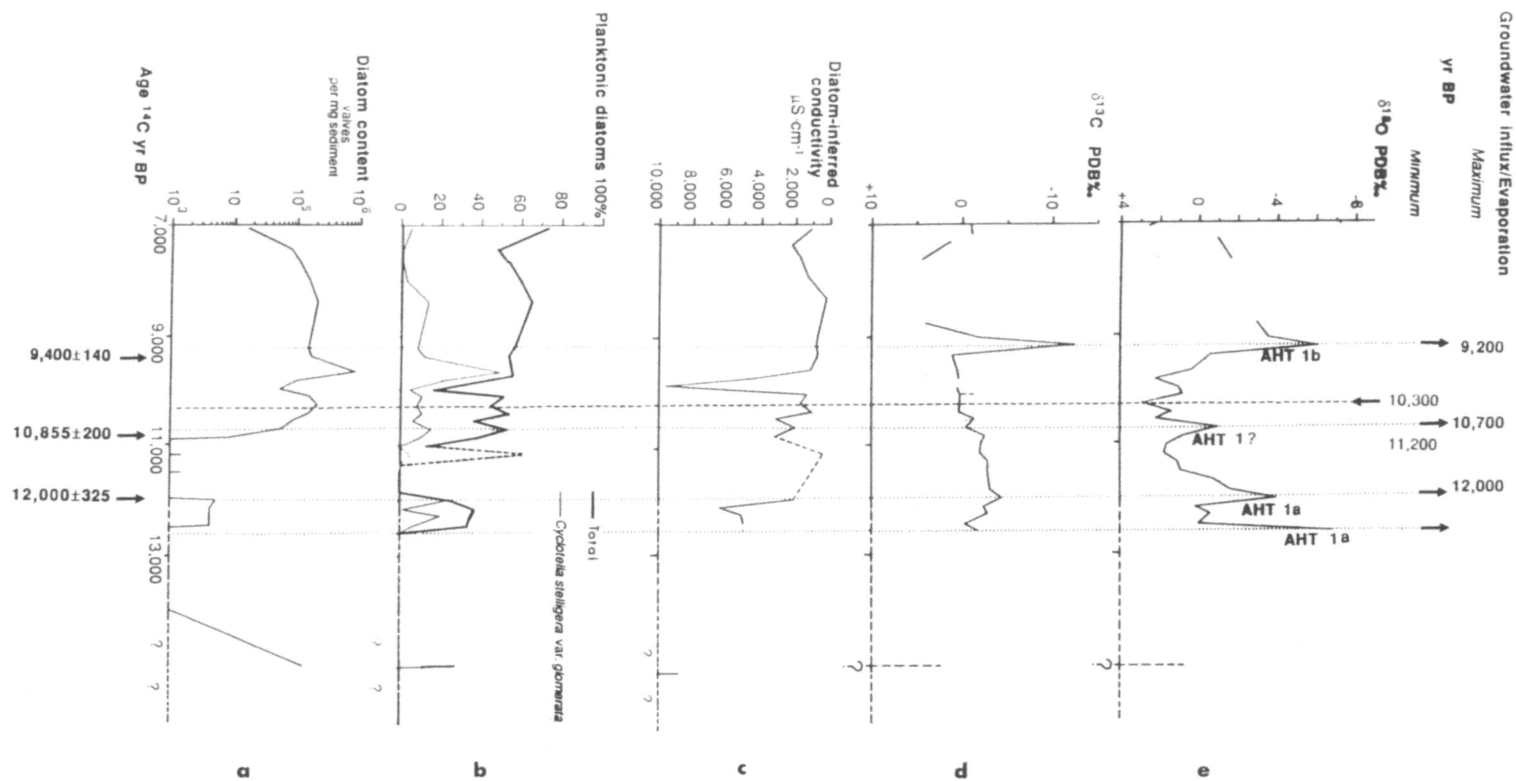

Fig. 3. *Bougdouma.* Palaeohydrological indicators plotted against ^{14}C ages (interval ≈15-7 ka BP). a) sediment diatom content; b) relative abundance of planktonic diatoms; c) diatom-inferred water conductivity; d, e) stable isotope contents of inorganic carbonates (fraction <8O µm). The results in b and c are uncertain between 12 and 11 ka BP due to partial selective dissolution. Arrows to the top mark the direction of the major hydrological events (after Gasse et al., 1990).

The Kadzell depressions

The Kadzell Plateau (Fig. 2a) represents the Holocene delta of the Yobe river, a small seasonal tributary of lake Chad which flows down from the Nigerian highlands and which represents ≈1% of the total water input to the lake (Roche, 1980). Natural exposures (e.g. Bwori, Widi; Durand et al., 1984, Fig. 2a), dug profiles from wells and hand-auger corings (Ari Koukouri; Durand et al., 1984; Gasse, 1987; this work) show that the small interdunal depressions have been occupied by Late Glacial-Early Holocene shallow waterbodies. Water was supplied either by flooding of the Yobe river and/or by rise of the Kadzell aquifer. The regions was then submerged by Lake Chad transgression (Durand et al., 1984), probably during the Mid-Holocene period (Gasse, 1987).

Despite the discontinuity of the profiles, correlations based on ^{14}C dates and sedimentary facies (Durand et al., 1984, and unpublished) allow three successive humid pulses to be distinguished during the last deglaciation, at ≈11.4, ≥10.6-10.3, and 9.4-9.2 ka BP. These humid episodes are separated by recession events registered by aeolian sands or a palaeosol (dated at 9,930 ± 240 ka BP at Bwori ; Durand, pers. comm.).

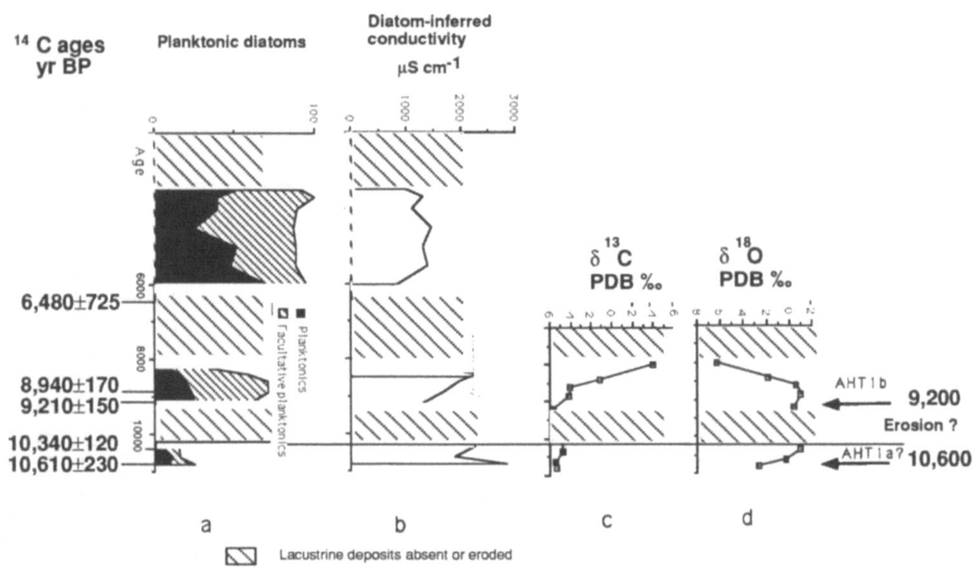

Fig. 4. *Ari Koukouri.* a-d) Palaeohydrological indicators plotted against ^{14}C ages. a) relative abundance of planktonic diatoms; c) diatom-inferred water conductivity; c,d) stable isotope contents of inorganic carbonates (fraction <80 µm). Arrows on the right mark the major freshwater inputs.

Attention is focussed here on the brief humid phase at ≥10.6-10.3. ka BP, with the example of Ari Koukouri (Figs. 2c, 4).

From >10.6 to ≈10.3 ka BP, sediments consist of Mg-rich calcite associated with abundant ostracods, mollusc shells and diatoms, reflecting a slightly saline, fluctuating shallow waterbody. The low content in detrital material suggests that the depression was supplied by the aquifer. A subsequeny regression is reflected by a sudden input of detrital particles (fluvial sand). The abrupt changes in facies and in diatom flora (Gasse, 1987), and the disappeance of calcareous shells between ≈10.3 and ≈9.2 ka BP suggest a sedimentary gap. At ≈9.2-8.9 ka BP, a new lacustrine phase is recorded by sandy marls, rich in freshwater diatoms of open water, deeper and more dilute than before 10.3 ka BP. Water then evaporated and the lake dried up. A palaeosol, dated at ≈6.4 ka BP, preceeds clayish, freshwater diatomites, the flora of which suggest a connection with the main lake Chad (Gasse, 1987). The top of the profile (4.2 m of aeolian sand) represents a palaeodune.

In the central Sahel, the Late Glacial is thus bracketed by two major and sudden increases in the P-E balance, at >12-11.4 and at ≈9.7-9.2 ka BP. The interval ≈11-10 ka BP, which is the Younger Dryas chronozone, corresponds to generally high evaporation conditions, and aridity maximum is recorded at ≈10.3-10 ka BP. However, this interval 11-10 ka BP is interrupted by a brief but clear humid phase found in several individual depressions, at ≈10.6-10.3 ka BP. This event, so far not observed from other regions, may characterize the central Sahel.

Central Sahara (Adrar Bous)

Although river-terrace studies suggested humid conditions on the Tibesti mountains during the Late Glacial (Jäkel, 1979), no lacustrine records were available for this period over the Central Sahara. Several works provide detailed information on outcropping lake sediments, of the Early and Mid-Holocene age (Hillaire-Marcel et al., 1983; Petit-Maire and Riser, 1981, 1983, 1987; Fabre et Petit-Maire, 1988; Dubar, 1988). The geographical origin of the lake water supplies is sometimes difficult to establish in this region because often depending on major confined or unconfined aquifers. However, the depression of Adrar Bous (Fig. 5), on the NE margin of the crystalline and volcanic Aïr massif, is believed to be mainly supplied by surface runoff and infiltration of water precipitated on the massif. Present rainfalls on the Aïr reliefs are relatively high (175 mm a^{-1}) in comparison with its eastern slopes (<50mm a^{-1}; Morel, 1985).

Detailed analyses of the mineralogy, stable isotope and trace elements of carbonates, supported by 48 ^{14}C dates (Dubar, 1988), and diatom study have been conducted on natural exposures and several hand-cores.

In the northern part of the the Adrar Bous basin, outcropping Early Holocene sediments overlain aeolian and/or fluvial sands crossed over 5-8 m in depth. These sands are interrupted by a layer of calcareous diatomite (≈50 cm thick, 2-2.5 m below the surface) rich in the freshwater planktonic species *Aulacosira ambigua* and *A. granulata* which reflects a first humid pulse. This level has been dated, on two cores, at 12900 ± 215, and 12,880±1440 ka BP (Dubar, 1988). In the southern part of depression, dating of "old" but recrystallized carbonates suggests that the water table was higher than today at 14-15 ka BP (Dubar, 1988; Fontes et Gasse, 1991 a).

The filling of the Holocene lake occurred rapidly prior to 9.7 ka BP and led to a shallow lake where carbonates accumulated until 8.4 ka BP. At the base, low stable istope contents reflect a rise of the water table. Waters then concentrated through evaporation, as indicated by the increase in ^{18}O contents and a diatom flora of the mesosaline type (El Hamouti, 1989), and the lake dried. The subsequent lacustrine optimum (≈8-≈7 ka BP), is recorded by a pure freshwater diatomite. The lake dried up after 6.7-6.5 ka BP. A return to palustral conditions is recorded at ≈4.9 ka BP (Dubar, 1988 ; Fontes and Gasse, 1991 b).

Fig. 5. Central Sahara. Location of the depression of *Adrar Bous* on the northeastern margin of the Aïr massif.

Although further investigations are needed for more precise dating of the Late Glacial events, the results indicate a step-wise establishment of wet conditions on the Central Sahara during the last deglaciation, with sudden increases in P-E balance at ≈13, 9.7 and ≈8 ka BP. Groundwater dating at the same latitude in central Mali also suggest that the recharge of the aquifers started during the Late Glacial, ≈13 ka BP (Fontes et al., 1991).

Northern Sahara (Sebkha Mellala, South Algeria)

The Sebkha Mellala (32°N, 5°12' E; ≈ 128 m.a.s.l. ; Fig. 6), dry today, lies in a hyperarid zone (mean rainfall 42 mm a^{-1}; mean evaporation 2000 mm a^{-1}; Dubief, 1959, 1963). It is a basin of internal drainage (playa) located in a zone of preferential discharge of the major aquifer of the Continental Intercalaire of the northern Sahara. Due to its

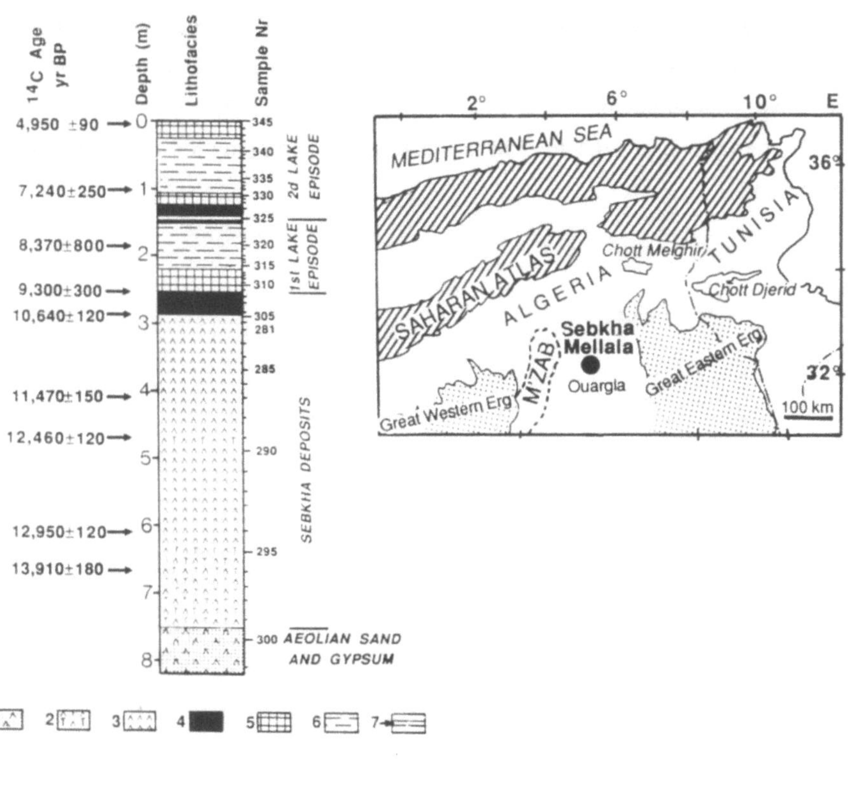

Fig. 6. Northern Sahara (south Algeria). Location of the *Sebkha Mellala*. a) Simplified stratigraphy of the core. 1: aeolian sand and gypsum; 2: massive carbonate-rich gypsum deposit; 3 : massive carbonate-poor gypsum deposit; 4 : laminated gypsum and carbonate; 5: laminated carbonate free of mollusc shells; 6 : alternation of shelly beds and of carbonate and gypsum layers; 7: palaeosol (after Gibert et al., 1990 and Gasse et al., 1990).

confinement, the aquifer responds without phase to any significant increase in rainfall on the closest zone of recharge situated some 60 km westwards on the M'zab heights. Late Quaternary deposits (Boyé et al. 1979; Gibert et al., 1990) accumulated in the basin and were partially deflated by wind erosion. Outcrops of these sediments provide good cross-sections and were complemented by hand auger coring at the foot of a hummock.

Evaporation of the groundwater supply allowed chemical sedimentation to take place with dry episodes marked by gypsum and dolomite precipitation associated with low amounts of calcite or Mg-calcite, while high water levels are characterized by the predominance of the calcite. Radiocarbon dating (AMS and conventional) and disequilibrium ($^{230}Th/^{234}U$) measurements performed on the sediment fraction smaller than 80 or 40 µm in size showed that the entire chronological interval from 15 ka to 5 ka was represented in the sedimentological column of about 8 m. The top of the sequence is wind deflated. A discussion of the validity of the chronology, and detailed results for the mineralogy, stable isotopes in inorganic and biogenic carbonates, element geochemistry on ostracod valves and biological remains are given elsewhere (Gibert et al., 1990; Gasse et al., 1990; Fontes et al., 1991). The major hydrological stages are summarized below (Figs. 6, 7).

- Dry conditions are recorded at the base of the profile by the dominance of gypsum and aeolian sand.

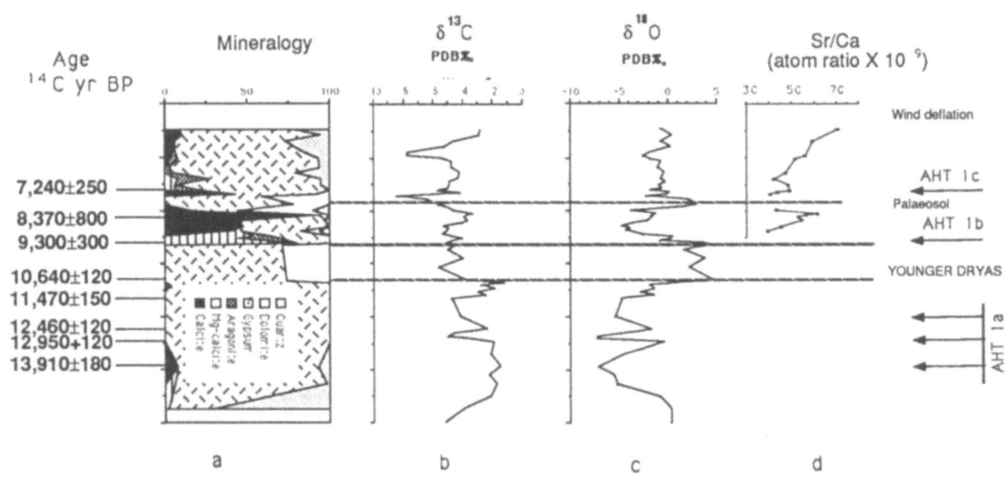

Fig. 7. Sebkha Mellala. Palaeohydrological indicators plotted against ^{14}C ages (inorganic carbonate fraction ≤80µm. a) Mineralogy (XRD peak areas) of bulk sediment (biological remains excluded). b, c) stable isotope contents of inorganic carbonates (fraction <80 µm) normalized to calcite. d) Sr/Ca ratio of ostracod valves, an indicator of water salinity. Arrows on the right mark the major freshwater inputs (AHT = Arid Humid Transition).

- The establishment of a wetter climate is marked, from ≈14.5 to 11 ka BP, by the precipitation of Mg-calcite and calcite depleted in ^{18}O. Substantial groundwater influxes are dated at ≈13.9, 12.7 and 12 ka BP.

- A rapid set back to arid conditions occurred from ≈11 to 10.6 ka BP, leading to the precipitation of post-gypsum dolomite from 10.6 to 9.3 ka BP. Even after correction for a specific isotope fractionation (Mc Kenzie, 1981), the high ^{18}O contents of this interval record the period of maximum local aridity in the profile.

- Arid conditions ended suddenly after 9.3 ka BP with the massive precipitation of carbonates and the appearance of biological remains (ostracods, molluscs, foraminifers and charophytes from brackish to hypersaline waters) from a lake which existed until ≈7.5 ka BP. The mineralogy (predominance of Mg-calcite associated to aragonite and gypsum, and even a narrow peak of dolomite ≈8.7 ka BP) suggests still concentrated waters until ≈8.7 ka BP. The minor discrepancies between high salinity and minimum ^{18}O values may be due to the leaching of salts deposited during the previous dry stage, as commonly occurs in playas (Gasse et al., 1987). Calcite with low ^{18}O content then indicates the dilution maximum in the basin. These wet, dilute conditions peaked at ≈8.3 ka BP.

- The lake dried up at ≈7.7 ka BP, as recorded by a palaeosol bracketed by dolomite-rich deposits devoid of shells.

- A new lacustrine phase is dated from ≈7.2 and ≤4.9 ka BP, with sudden influxes of dilute water ≈ 7.2 and 6 ka BP. Contents in ^{18}O and the amount of gypsum remain relatively high, suggesting a higher salinity and a lower P/E ratio than at 9-8 ka BP.

These results show that the last deglaciation in the northern Sahara is punctated by sudden pulses in the P-E ratio, dated at ≈14-12, ≈ 9. 6 and 7.2 ka BP. The period of maximal aridity coincides with the Younger Dryas chronozone (11-10 ka BP), and persisted to the beginning of the Early Holocene period.

Tigalmamine (Middle Atlas, Morocco)

In the Middle Atlas, numerous closed lakes formed by limestone dissolution provide good records of the influence of climatic changes on karstic circulation. Recent (20th century) variations in annual rainfall have significantly affected the magnitude of the water level fluctuations of L. Azigza, a natural hard-water upland lake (Flower and Foster, 1991). Tigalmamine lies in a very similar situation.

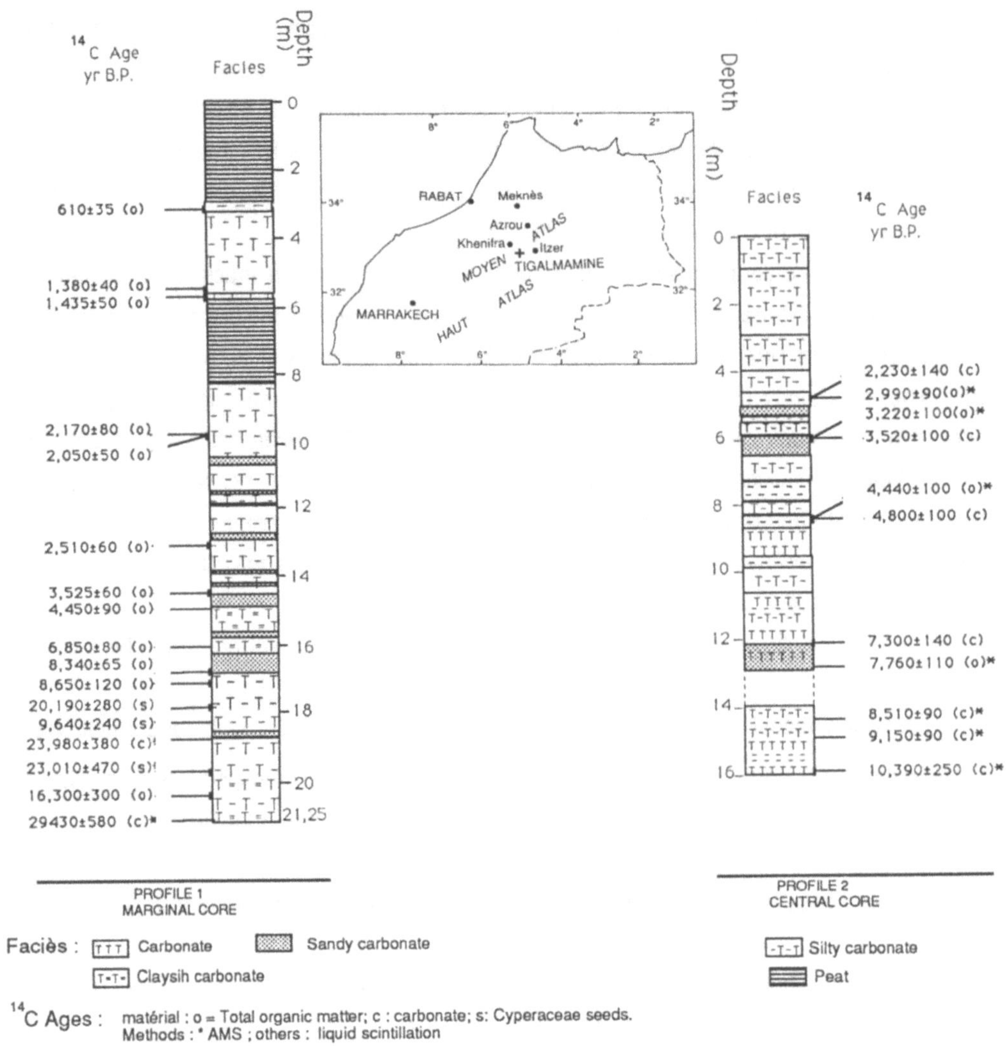

Fig. 8. Middle Atlas. *Tigalmamine* . Facies and chronology of the marginal (a) and the central (b) cores (after El Hamouti et al., 1991).

Tigalmamine (32°54'N, 5°21'E, 1626 m.a.s.l.) (Fig. 8) is a group of three small solution lakes in Lias dolomite. The climate (mean rainfall: 930 mm a^{-1}; mean annual air temperature: 10°C; Martin, 1981) shows large seasonal contrasts with most of the precipitation, of northern Atlantic origin, falling as snow in winter whilst summers are hot and dry.

The largest lake ($6X10^4$ m^2, 16 m deep; specific conductivity: 600-840 $\mu S.cm^{-1}$; pH ≈8,8), closed in summer, is supplied by local surface run-off and karstic circulations. Two piston-cores have been collected by H. Lamb from the lake margin (profile 1: 21.25 m deep; Lamb et al., 1989) and from the lake centre (under 16 m of water; profile 2: 16 m deep) respectively. Sediments consist of marls and calcareous silts rich in microorganisms, alternating with calcareous sands poor in biological remains, and with sedge peat levels in the littoral profile (Fig. 8). A previous investigation of the littoral core, dated at ≈16 ka BP at its base, suggested that the major climatic event after the LGM was a "sudden increase in atmospheric moisture" at ≈8.5 ka BP (Lamb et al., 1989). However, the pre-Holocene chronology of the littoral profile has not been ascertained (Fig. 8a), and several sedimentary gaps preclude age interpolation prior to≈ 9.6 ka BP (El Hamouti et al., 1991). We thus report here only on the Holocene period.

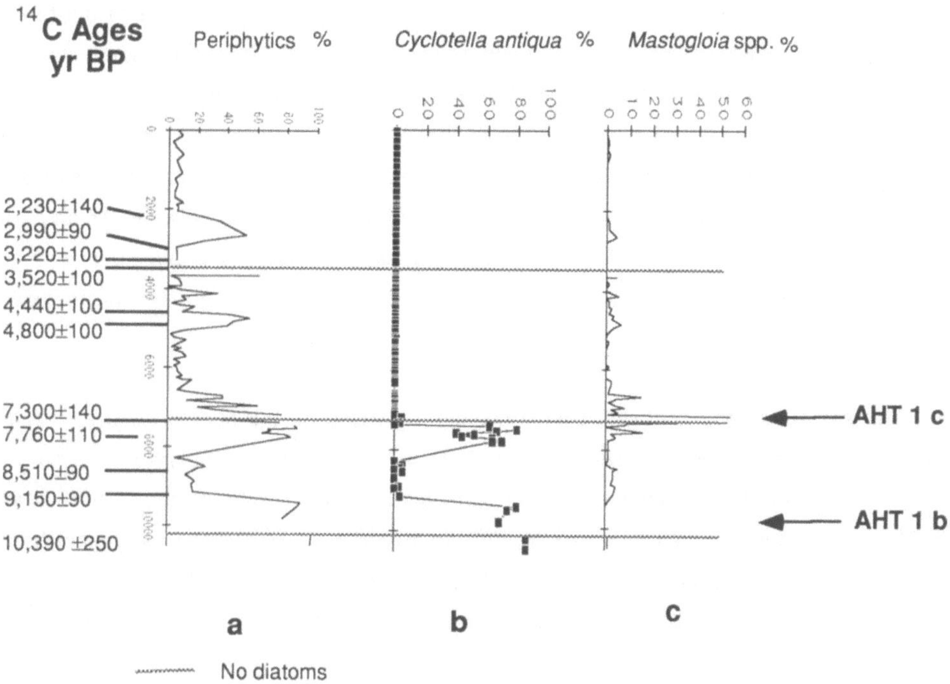

Fig. 9. *Tigalmamine.* Diatoms indicators of the central core. Frequencies of: a) periphytic diatoms, indicators of shallow water; b) *Cyclotella antiqua* , a littoral species of cool water; c) *Mastogloia elliptica* and M. *smithii,* littoral forms from saline waters.

The authigeny of the Holocene inorganic calcite was checked through scanning electron microscopy. The age consistency between carbonate and total organic matter supports the Holocene chronology. A percentage diatom study of the two cores provides the succession of the main hydroclimatic events (El Hamouti, 1989; El Hamouti et al., 1991) summarized below, and is illustrated with the example of the central core (Fig. 9). Three diatom indicators are considered (Fig. 9): (i) the percentage ratios: Planktonics/Periphytics which inform on the water-level fluctuations; (ii) the frequencies of *Cyclotella antiqua*, a littoral, aerophilous species from arctic-alpine regions and unknown so far in Africa. Its abundance (percentage ≥50%) is thus considered as an indicator of shallow environments cooler than today; (iii) the frequencies of the oligo- to mesohaline water species *Mastogloia smithii* and *M. elliptica* which developed during regressive stages as a result of evaporative concentration of dissolved salts.

At 10.4-10 ka B.P, the base of profile 2 registers swampy cold conditions which coincides with a gap (dry conditions) in profile 1. The diatom disappearance at ≈ 10.2 ka BP suggests a short episode of drought.

From ≈ 10 ka BP, the water-level rose rapidly until ≈8.5 ka BP, leading to a shallow, oligotrophic lake which submerged site 1 beneath a few metres of water.

A marked regression is recorded between ≈8 and 7 ka BP. The centre of the basin was occupied by a residual, slightly saline waterbody, as indicated by the predominance of the periphytics and peaks of *Mastogloia* . Site 1 was emerged. The absence of diatoms and the presence of coarse detrital material suggest that the lake dried up at ≈7.5 and ≈7.2 ka BP. This event coincides with the disappearance of the oligothermic form *Cyclotella antiqua* .

Highstands, recorded by planktonic flora in both cores, occurred between ≈7 and 4.8 ka BP. After a sudden drop at 4.8-4.5 ka BP, shallow environments prevailed at site 1. Planktonic flora similar to that of today were maintained at site 2, although two regressions are observed at ≈3.5-3.2 and ≈2.5 ka BP.

These results show that dry, cold conditions prevailed on the Middle Atlas at ≈10.4-10 ka BP. In contrast with previous hypotheses (Rognon, 1987), the return to humid conditions in the Maghreb, after the arid period of glacial times occurred much earlier than 5 ka BP, with sudden rainfall increase at ≈9.5 ka BP and ≈7 ka BP, as in the northern Sahara (see the previous case of the Sebkha Mellala record). In the Middle Atlas, however, optimal humidity (and probably temperature) conditions were reached after 7.5 ka BP.

DISCUSSION

The sites considered in this paper have been selected because of their hydrological setting which makes them representative of rainfall fluctuations on restricted catchment areas. Our results thus allow to distinguish between climatic events at the scale of Northwest Africa, and regional/local patterns at a given time. Although stratigraphical and chronological uncertainties still exist, some major statements can be made from our studies (Fig. 10).

Palaeoclimatic covariations and regional differences over tropical and extratropical Northwest Africa.

The last deglaciation in Northwest Africa was characterized by a series of rapid dry/wet oscillations which appear to have been synchronous in both regions which are submitted to tropical summer rainfall and to mid-latitudes precipitation, respectively. They may be summarized as :

-1) after the LGM, conditions wetter than today re-established during the Late Glacial (AHT 1a) over the Soudanian Africa (Talbot et al., 1984), central Sahel (Servant, 1983; Gasse et al., 1990; this work), central Sahara (Dubar, 1988 ; this work), and northern Sahara (Gibert et al., 1990; Gasse et al., 1990; this work). Detailed profiles show that this re-establishment occurred by steps, with at least two pulses at Bougdouma (at >12 and ≈12 ka BP respectively) and three pulses at the Sebkha Mellala (at ≈13.9, 12.7 and 12 ka BP);

-2) generally dry conditions prevailed during the interval ≈11-10 ka BP, which can be regarded as coeval with the Younger Dryas cooling event at high latitudes.;

-3) the dry episode synchronous with the Younger Dryas ended abruptly, probably over a few decades, by a sudden increase in the P-E balance at 9.7-9.2 ka BP (AHT 1b);

-4) after a regressive event at ≈8-7 ka BP, a new rise of the water level (AHT 1c) is recorded. The chronology of this oscillation has still to be defined more accurately.

Despite this strong tendency for climate covariability over a large area, regional differences are observed:

- 1) during the Younger Dryas chronozone, the central Sahel has experienced a short humid phase from ≈10.7 to 10.3 ka BP. This wet pulse has not been observed so far in the other regions, possibly because of the lack of precise chronology;

-2) during the Holocene, hydroclimatic optimum conditions (maximum P-E balances) seem to have been reached at 9-8 ka BP at Bougdouma and in the Sebkha Mellala, at 8-7 ka BP in the Aïr, and after 7 ka BP in the Middle Atlas, without latitudinal gradient.

The concept of a progressive return of wet conditions between the Equator and ≈30°N from 13-10 ka BP to 6 ka BP (Street-Perrott and Roberts, 1983), which was adopted by several authors (Rognon, 1987; Fabre and Petit-Maire 1988; Lézine and Casanova, 1989), is thus not supported by our results, possibly because Late Glacial sediments are accessible by coring only, whereas the former studies were mainly based on outcropping deposits of Holocene age.

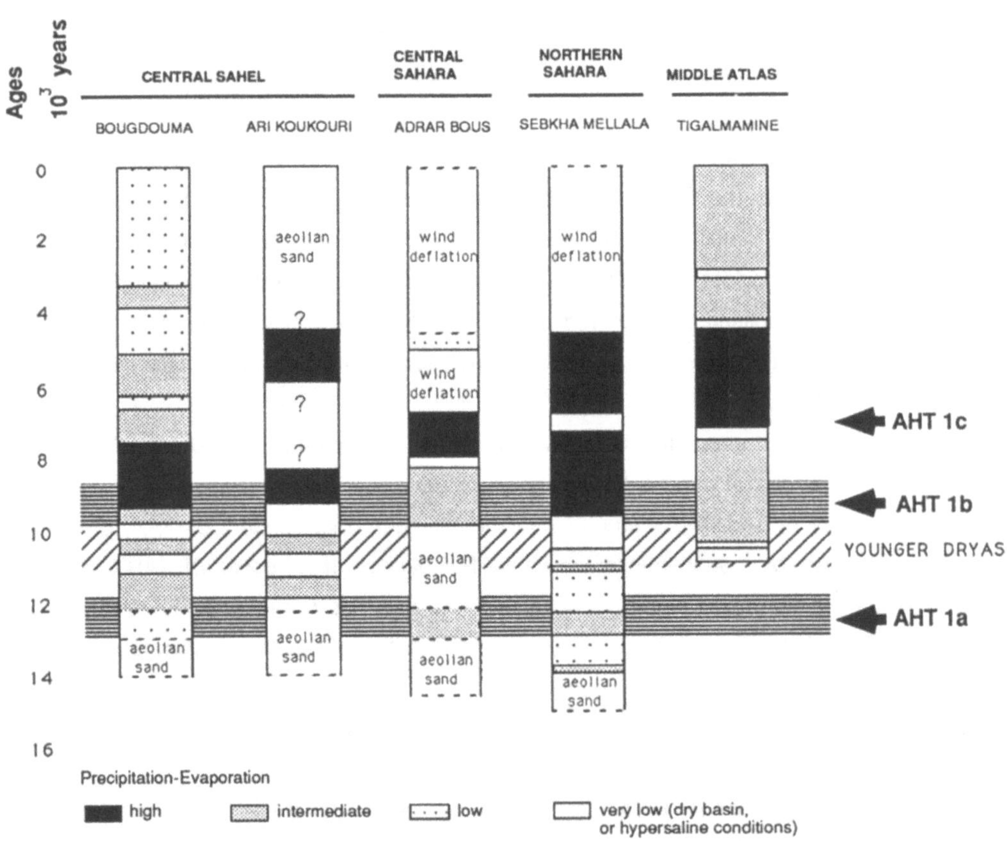

Fig. 10. Major hydrological changes recorded from the profiles selected for this paper.

Search for mechanisms

Our results show firstly a spatial coherence over Northwest Africa in the major Arid-Humid Transitions (AHTs) and reversal events, which are in good chronological agreement with the cold/warm fluctuations deduced from deep-sea sediments, ice cores and European climatolstratigraphy (Gasse et al., 1990).

Orbital forcing cannot account either for this step-wise climatic evolution, or for the timing of maximum P-E balances which post-dated the maximum in northern summer insolation by several thousands of years. During the Younger Dryas, the hypothesis of decreased monsoon rainfall resulting from a reduction of the NADW production (Street-Perrott and Perrott, 1990) does not explain the covariability between the Sahel and northern Sahara. Furthermore, the model of decreased NADW production as the cause of the Younger Dryas event (Broecker et al., 1985, 1988, 1989) is now contested (Fairbanks, 1989; Jansen and Veum, 1990; Broecker et al., 1990; Kudrass et al., 1991). Other possible global causes have been put forward to explain the abrupt climatic changes during the last deglaciation, e.g. 1) positive feedbacks in the response of the climate system to Milankovitch forcing from changes in albedo (Berger, 1990) ; 2) climatic induced fluctuations in atmospheric CO_2 (Berger, 1990; Kudrass et al., 1991) and/or methane (Lorius et al., 1990) ; 3) increased atmospheric dust introduced by volcanic activity (Berger, 1990); and 4) natural oscillations in the salinity budget of the Atlantic Ocean with approximate 2,000 years periods (Broecker *et al.*, 1990), a mechanism which could account for the abrupt regressions observed in Africa at 11-10, ≈8-7 (or ≈ 5) ka BP. No final conclusion is reached, possibly because of complex interferences between several mechanisms and perharps also because elegant hypotheses are still more numerous than ascertained facts.

The main difficulties in the interpretation of past climatic variations is that phenomenological covariations or codependences do not generally provide causal explanation. However, the configurations of the modern rainfall variability on the African Continent may help for interpreting past climatic patterns, although it is not established whether the series of dry years which occurred over the last century (at ≈1910, 1940, and from 1963) reflect a general tendency towards aridification or natural variability within a given type of climate.

Palaeoclimatic covariations and synchronisms in the sudden AHTs during the last deglaciation reflect rainfall "anomalies" of the same sign throughout nearly all the Northwest Africa, a pattern which is common today at the annual or pluriannual scale (Nicholson, 1986). Statistically-significant relationships have been established between modern rainfall on Soudano-Sahelian Africa and global anomalies in the atmospheric and oceanic circulation. Below-normal rainfall are associated with cold anomalies in boreal Sea Surface Temperature (SST), and warm anomalies in austral SST (Folland et al., 1986; Fontaine, 1990; Ward et al., 1990; Druyan, 1991). Another covariation may exist between low summer rainfall on the Sahel and an early northwards migration of the Intertropical Convergence Zone (ITCZ) in spring (Guillot et al., 1986). Pluriannual droughts on Soudano-Sahelian Africa are generally linked to a weaker Tropical Easterly Jet (TEJ), to a stronger mid-altitude African Easterly Jet (AEJ), to a stronger Hadley type overturning and to a weakening of the antithetical Walker circulation (Nicholson, 1986; Fontaine, 1989; Janicot, 1990). The synchronous occurrence of dry or wet conditions throughout Northwest Africa, including regions submitted to winter precipitation may be due to fluctuations in the intensity of the Hadley circulation over Africa (Nicholson, 1981, 1986; Nicholson and Chervin, 1983). However, these modern patterns can be regarded as energetical codependences rather than mechanistic hypotheses for explaining generally dry/wet conditions during several centuries. *Progresses in accounting for past regional differences have nevertheless to be sought from modern climatic studies which described several types of northern Africa rainfall anomalies on a regional scale.* In particular, a strong anticorrelation in rainfall departures (deviations from the mean) commonly occurs between equatorial-subequatorial (<10°N) and subtropical (>10°N) latitudes (Nicholson, 1986). This is related to regional SST anomalies (Guillot et al., 1986; Janicot, 1990): wet years south of 10°N corresponding to dry years in the Sahel are associated with positive SST anomalies in the Gulf of Guinea, and *vice versa* The brief humid phase recorded in the central Sahel but not in Soudanian Africa at 10.6-10.3 ka BP may be related to similar phenomena.

Rainfall fluctuations on Northwest Africa are also clearly linked to the land surface conditions. Sahel droughts are today associated with enhanced meridian gradients of surface temperature, albedo and humidity (Janicot, 1990), and significant positive feedbacks are as expected observed between increases in albedo and diminished rainfall (Charney et al., 1977; Mintz, 1984; Laval and Picon, 1986). Similarly, GCM results for 9 ka BP indicate that surface-albedo decrease in the northern tropics may have provide a significant positive feedback enhancing the direct climatic effects of Milankovitch forcing (Street-Perrott et al., 1990).

Emphasis must be given to the potential role of the water reservoir of vegetated soils in the amplification of non-monsoonal pluviogenesis, and in the possible feedback on the establishment of two distinct rainy seasons during wet periods. Local sources of water vapour represented by the soil moisture are very important in all regions of West Africa today (Druyan and Koster, 1989). The significant fraction of the total rainfall recycled by evapotranspiration from the soil horizons increases the precipitable vapour, and refills the squall lines along an east-west direction in the Sahel (Rowell and Blondin, 1990). This mechanism presently active at the daily scale can also be considered at the seasonal scale, and tentively extrapolated to the past. During wet periods of enhanced monsoonal influences, summer rains occurred over the northern Sahel and the central Sahara. Evapotranspiration during autumn and winter may have contributed vapour to the depressions driven by northern influences, and generated a second rainy season in these regions. Conversely, the winter refilling of the soil reservoir could have helped the successive monsoon to progress northwards. Studies of seasonality in central Sahara must be put on a high priority, with special reference to the expected occurrence of two rainy seasons of distinct origin. A winter recharge might thus have been the cause of the very low heavy isotope contents often observed in fossil groundwaters from the Sahara (Dray et al., 1983). *A full understanding of past climates thus requires the study of the aquifer recharge through space and time,* which complements surface palaeohydrological records (Fontes et al., 1991). In the most arid zones, the water table lies today at tens of metres, up to more than 100 m, below the land surface. Slight increases in the P-E balance may have been insufficient to provide the water which previously outcropped in the depressions. For instance, a P-E excess of about 20 mm a^{-1} in the net storage as compared to present conditions, would have generated an uprise of 6 m in 120 years for the water table of an unconfined aquifer with an average porosity of 40% (case of homogeneous sand). Observed fluctuations of groundwater-fed palaeolakes are usually greater than this estimate. Climatic variations recorded in groundwater-fed lakes were thus marked by P-E changes which were also greater. Such major variations in net infiltration rate would imply very large differences in rainfall. Detailed coupled investigations of aquifers and related palaeolakes are one of the ways to estimates P-E differences and, by comparison with modern references, to obtain evaluations of P.

Special attention must be paid to the potential role of methane as an amplifier factor for ATHs. In contrast to CO_2, CH_4 has no known sink except high atmosphere oxidation (see e.g. Quay, 1992). After its production (swamps, ruminants, termites...) CH_4 thus accumulates in the atmosphere with an average residence time of about one decade ($\approx$ 7 years for CO_2). Moreover, due to its high absorption power for long wave radiations (at least 20 times that of CO_2; Quay, 1992) the relative effect of CH_4 build up on atmospheric warming is probably higher than that of CO_2. The watering of natural wetlands may have been a triggering mechanism for atmospheric warming (Raynaud et al., 1988). This would have been especially especially the case for the flat tropical areas of Sahelian and Saharan Africa. In contrast to what occurs for CO_2 for which buffering mechanisms are not obvious, CH_4 production is automatically constrained by lake evolution towards either an increase in water level or a colonization by vegetation, inducing in both cases a decrease in CH_4 production. Methane production and induced warming would have labelled the beginning of humid phases. Attention has to be paid to the significance of burnt diatomites frequent in the Sahel and the Sahara, e.g. in the northern Mali (small depressions in the vicinity of Gargando, and Lake Faguibine cores) and in the northern Niger (Adrar Bous, Fig. 5).

Perspectives

Lacustrine deposits from tropical and extratropical Africa offer well-suited records to explore the sensitivity of the climate system. All the investigated sites provide a high resolution time, and more refined investigations could in most cases allow a better definition of the onset, course and termination of the hydroclimatic states. However, the results obtained so far on lacustrine cases are comparable to, or even more defined than those of deep-sea sediments. Our work identifies a number of specific types of palaeohydrologic, palaeoclimatic studies which will contribute to the understanding of the switching between wet and dry conditions over all northwest Africa or at a regional scale. Among short-term future surveys and expected results, one can list : 1) *a more precise determination of the timing of the rapid environmental changes*; 2) *the parallel studies of surface and groundwater palaeohydrology* which allow the influence of soil moisture to be investigated through space and time; 3) *the investigation of seasonal patterns of rainfall changes in various localities*; 4) *the comparison of the regional climatic patterns of the past, at a time scale of $\approx 10^2$ years, with the modern types of annual and pluriannual rainfall anomalies* ; 5) *the quantitative estimates of changes in the water budget of individual basins* selected as good recorders of regional rainfalls ; 6) *the evaluation of the flux of methane* which was produced in low lands by comparison with present swampy zones and its influence on global climate.

Such approaches are needed to estimate the relative role of orbital forcing, global/regional anomalies in oceanic and atmospheric circulation, land-atmosphere exchanges, and for discriminating among the different hypothetical causes of abrupt cliamtic changes.

ACKNOWLEDGEMENTS

This is a contribution to the PALHYDAF (*Paléohydrologie en Afrique*) programme, supported by the *Institut des Sciences de l'Univers, Centre National de la Recherche Scientifique* . We are grateful to Dr. R. Bonnefille and to the anonymous referee who made helpful comments in reviewing the manuscript. Thanks are due to Dr. J.N. Andrews for language revision.

REFERENCES

Bard E, Arnold M, Maurice P, Duprat J, Moyes J, Duplessy JC (1987) Retreat velocity of the North Altantic polar front during the last deglaciation determined by ^{14}C accelerator mass spectrometry. Nature 328: 791-794

Bard E, Labeyrie L, Arnold M, Labracherie M, Pichon JJ, Duprat J, Duplessy JC (1989) AMS ^{14}C in deep sea cores from the Southern Ocean : implications for sedimentation rates during isotope stage 2. Quat. Res. 31: 309-317

Barker JF, Fritz P, Brown RM (1979) Carbon-14 measurements in aquifers with methane. Isotope Hydrology 1978, IAEA Vienna vol. 2: 661-678

Berger WH (1990) The Younger Dryas cold spell - a quest for causes. Palaeogeogr. Palaeoclimatol. Palaeoecol. (Global and Planetary Change Section) 89: 219-237

Berger WH, Labeyrie LD (eds) (1987) Abrupt climatic change. Evidence and implications. Nato Asi Series, Ser. C, Mathematical and Physical Science 216: 425p

Boyé M, Marmier F, Nesson C, Trécolle G (1979) Les dépôts de la sebkha Mellala. Rev. Geomorph. Dyn. 28 (2-3): 49-62

Broecker WS, Peng TH (1982) Tracers in the sea. Eldigio Press Palisades

Broecker WS, Peteet DM, Rind P (1985) Does the ocean- atmosphere system have more than one stable mode of operation ? Nature 315: 21-26

Broecker WS, Andree M, Wölfli W, Oeschger H, Bonani G, Kennett J, Peteet D (1988) The chronology of the last deglaciation : implications to the cause of the Younger Dryas event. Paleoceanography 3 (1): 1-19

Broecker WS, Kennett JP, Flower BP, Teller JT, Trumbore S, Bonani G, Wolfli W (1989) Rounting of meltwater from the Laurentide ice sheet during the Younger Dryas cold episode. Nature 341: 318-321

Broecker WS, Bond G, Klas M (1990) A salt oscillator in the Glacial Atlantic ? 1. The concept. Paleoeceanography 5 (4): 469-477

Buchardt B, Fritz P (1980) Environmental isotope as environmental and climatological indicators. In Handbook of Environmental Isotope Geochemistry vol. I, The Terrestrial Environment, P Fritz, JCh Fontes (eds) Elsevier Amsterdam: 475-504

Causse C, Conrad G, Fontes JCh, Gasse F, Gibert E, Kassir A (1988) Le dernier "Humide" pléistocène du Sahara nord-occidental daterait de 80-100 000 ans. C.R. Acad. Sci. Paris 306 (II): 1459-1464

Charney J, Quirck WJ, Chow S, Kornfield J (1977) A comparative study of the effects of albedo change in drought in semi-arid regions. J. Atmos. Sci. 34 : 1366-1385

COHMAP Members (1988) Climatic changes of the last 18,000 years : observations and model simulations. Science 241: 1043-1052

Conrad G (1969). L'évolution continentale post-hercynienne du Sahara algérien. Centre de recherche sur les Zones Arides, Série Géol. 10 CNRS: 527 p

Delibrias G, Petit-Maire N, Fabre J (1984) Age des dépôts lacustres récents de la région de Taoudenni-Trhaza (Sahara malien). C. R. Acad. Sci. Paris 299: 1343-1346

Dray M, Gonfiantini R, Zuppi GM (1983) Isotopic composition of groundwater in the southern Sahara. In Palaeoclimates and palaeowaters : a collection of environmental isotope studies. I.A.E.A. Vienna 1980: 187-199

Druyan LM (1991) The sensitivity of Sub-Saharan precipitation to Atlantic SST. Climatic Change 18: 17-36

Druyan LM, Koster RD (1989) Sources of Sahel precipitation for simulated drought and rainy seasons. J. Climate 2 : 1438-1446

Dubar C (1988) Eléments de paléohydrologie de l'Afrique Saharienne : les dépôts quaternaires d'origine aquatique du Nord-Est de l'Aïr (Niger, Palhydaf site 3) Thesis Université Paris XI: 175 p

Dubief J (1959) Le climat du Sahara. Mém. Univ. Alger vol. I: 312 p

Dubief J (1963) Le climat du Sahara. Mém. Univ. Alger vol. II: 275 p

Duplessy JC, Delibrias G, Turon JL, Pujol C, Duprat J (1981) Deglacial warming of the northeastern Atlantic Ocean : correlation with the paleoclimatic evolution of the European continent. Palaeogeogr. Palaeoclimatol. Palaeoecol. 35: 121-144

Durand A, Mathieu P (1979-80) Le Quaternaire supérieur sur la rive sud du lac Tchad. Cah. ORSTOM série Géol XI (2): 189-209

Durand A, Fontes JCh, Gasse F, Icole M, Lang J (1984) Le nord-ouest du lac Tchad au Quaternaire : étude de paléoenvironnements alluviaux, éoliens, palustres et lacustres. Palaeoecology of Africa 16: 215-243

El Hamouti N (1989) Contribution à la reconstitution de la paléohydrologie et de la paléoclimatologie du Maghreb et du Sahara au Quaternaire supérieur à partir des diatomées. Thesis Université Parix XI: 299 p

El Hamouti N, Lamb H, Fontes JCh, Gasse F (1991) Changements hydroclimatiques abrupts dans le Moyen Atlas marocain depuis le dernier maximum glaciaire. C. R. Acad. Sci. Paris 313: 259-265

Fabre J, Petit-Maire N (1988) Holocene climate evolution at 22-23°N from two palaeolakes in the Taoudenni area (Northern Mali). Palaeogeogr. Palaeoclimatol. Palaeoecol. 65: 133-148

Fairbanks RG (1989) A 17,000 year glacio-eustatic sea level record : influence of glacial melting rates on the Younger Dryas event and deep-ocean circulation. Nature 342: 637-642

Flower R, Forster J (1991) Climatic implications of recent lake level changes at Lake Azigza, Morocco. Bull. Soc. géol. France in press

Folland CK, Parker TN, Parker DE (1986) Sahel rainfall and worldwide sea temperatures 1901-1985. Nature 320: 602-607

Fontaine B (1990) Etude comparée des moussons indienne et ouest africaine : caractéristiques, variabilité et téléconnexions. Thesis Université de Bourgogne Dijon vol. I : 233 p, vol. II : 200 figs

Fontes JCh (1990) Chemical and isotopic constraints on ^{14}C dating of groundwater. In Four decades of radiocarbon studies, Proceed. Lake Arrowhead conf. June 1991, in press

Fontes JCh, Gasse F (1989) On the ages of humid Holocene and Late Pleistocene phases in North Africa. Remarks on "Late Quaternary climatic reconstructions for the Maghreb (North Africa)" by P. Rognon. Palaeogeogr. Palaeoclimatol. Palaeoecol. 70: 393-398

Fontes JCh, Gasse F (1991a) PALHYDAF (Palaeohydrology in Africa). Objectives, methods, major results. Palaeogeogr. Palaeoclimatol. Palaeoecol. 84: 191-215

Fontes JCh, Gasse F (1991b) Chronology of the major Late Quaternary palaeohydrological events in NW Africa : PALHYDAF results. Hydrobiologia 214: 367-372

Fontes JCh, Coque R, Dever L, Filly A, Mamou A (1983) Paléohydrologie isotopique de l'Oued el Akarit (Sud Tunisie) au Pléistocène supérieur et à l'Holocène. Palaeogeogr. Palaeoclimatol. Palaeoecol. 43: 41-62

Fontes JCh, Andrews JN, Guerre A, Edmunds WM, Travi Y (1991a) Paleorecharge by the Niger River (Mali) deduced from groundwater geochemistry. Water Resources Research 27 (2): 199-214

Fontes JCh, Andrews JN, Causse C, Gibert E (1991b) A comparison of ^{14}C and Th/U ages on continental carbonates. 14th Int. Radiocarbon Conf. Tucson USA, May 1991, Proceedings in press

Gasse F (1977) Evolution of Lake Abhé (Ethiopia and T.F.A.I.) from 70,000 B.P.. Nature 2: 42-45

Gasse F (1987) Diatoms for reconstructing palaeoenvironments and palaeohydrology in tropical semi-arid zones. Example of some lakes from Niger since 12 000 BP. Hydrobiologia 154: 127-163

Gasse F, Fontes JCh, Plaziat JC, Carbonel P, Kaczmarska I, De Deckker P, Soulié-Marsche I, Callot Y, Dupeuble PA (1987) Biological remains, geochemistry and stable isotopes for the reconstruction of environmental and hydrological changes in the Holocene lakes from North Sahara. Palaeogeogr. Palaeoclimat. Palaeoecol. 60: 1-46

Gasse F, Téhet R, Durand A, Gibert E, Fontes JCh (1990) The Arid-Humid transition over the Sahara and the Sahel during the last deglaciation. Nature 346: 141-146

Gibert E (1989) Géochimie et paléohydrologie des bassins lacustres du nord-ouest saharien. Programme Palhydaf, site 2. Thesis Université Paris XI: 244 p

Gibert E, Arnold M, Conrad G, De Deckker P, Fontes JCh, Gasse F, Kassir A (1990) Retour des conditions humides au Tardiglaciaire au Sahara septentrional (sebkha Mellala, Algérie). Bull. Soc géol. France 8 (VI-3): 497-504

Gonfiantini, R., 1986. Environmental isotopes in lake studies. In P Fritz, JCh Fontes (Eds)Handbook of Environmental Isotope Geochemistry, Elsevier Amsterdam vol. 2: 113-168

Guillot B, Lahuec JP, Citeau J, Bellec B, Noyalet A (1986) Analyse de l'évolution climatique en Afrique de l'Ouest en 1983-1985 à l'aide de l'imagerie satellitaire et de données conventionnelles. INQUA/ 1986. Dakar Symposium "Changements globaux en Afrique" ORSTOM Bondy: 173-178

Hillaire-Marcel C, Riser J, Rognon P, Petit-Maire N, Rosso JC, Soulié-Märsche I (1983) Quaternary chronology of Holocene hydrologic changes in Northeastern Mali. Quat. Res. 20: 145-164

Jäkel D (1979) Run-off and fluvial processes in the Tibesti mountains as indicators of climatic history in the central Sahara during the last Pleistocene and Holocene. Palaeoecology of Africa 11: 13-44

Janicot S (1990) Variabilité des précipitations en Afrique de l'ouest et circulations quasi stationnaires durant une phase de transition climatique. Thesis Université Paris VI: 174 p

Jansen E, Veun T (1990) Evidence for two-step deglaciation and its impact on north Atlantic deep-water circulation. Nature 343: 612-616

Kudrass HR, Erlenkeuser H, Vollbrecht R, Weiss W (1991) Global nature of the Younger Dryas cooling event inferred from oyxgen isotope date from Sulu Sea cores. Nature 349: 406-408

Kutzbach JE, Street-Perrott F.A (1985) Milankovitch forcing of fluctuations in the level of

tropical lakes from 18 to 0 kyr B.P. Nature 317: 130-134
Lamb HF, Eicher U, Switsur V.R (1989) A 18,000-year record of vegetation, lake-level and climatic change from Tigalgamine, Middle Atlas, Morocco. Journ. of Biogeography 16: 65-74
Laval K, Picon L (1986) Effect of a change of the surface albedo of the Sahel on climate. J. Atmos. Sci. 43: 2418-2429
Leroux M (1983) Le climat de l'Afrique tropicale. Champion Paris, 2 vols.: 633 p
Lézine AM, Casanova J (1989) Pollen and hydrological evidence for the interpretation of past climates in tropical West Africa during the Holocene. Quat. Sci. Rev. 8: 45-55
Mc Kenzie JA (1981) Holocene dolomitization of calcium carbonate sediments from the coastal sabkhas of Aby Dhabi : a stable isotope study. J. Geol. 89: 185-198
Maglione G (1976) Géochimie des évaporites et silicates néoformés en milieu continental confiné. Les dépressions interdunaires du Tchad. Trav. Doc. ORSTOM 50: 335 pp
Maley J (1981) Etudes palynologiques dans le bassin du Tchad et paléoclimatologie de l'Afrique nord-tropicale de 30 000 ans à l'époque actuelle. Trav. Doc. ORSTOM Paris 129: 589 p
Martin J (1981) Le Moyen Atlas Central : étude géomorphologique. Notes et Mémoires du Service Géologique, 258, Service Géologique du Maroc Rabat
Mintz Y (1984) The sensitivity of numerically simulated climates to land-surface boundary conditions. In The Global Climate, JT Houghton (Ed) Cambridge University Press Cambridge: 79-105
Morel A (1985) Les hauts massifs de l'Aïr (Niger) et leurs piedmonts. Thesis Université de Grenoble: 404 p
Nicholson SE (1981) Rainfall and atmospheric circulation during drought and wetter periods in West Africa. Mon. Wea. Rev. 109: 2191-2208
Nicholson SE (1986) The spatial coherence of African rainfall anomalies : interhemispheric teleconnections. J. of Climate and Applied Meteorology 25: 1365-1381
Nicholson SE, Chervin RM (1983) Recent rainfall fluctuations in Africa - interhemispheric teleconnections. In : Variations in the global water budget, Street-Perrott et al. (Eds), Riedel Dordrecht: 221-238
Petit-Maire N, Riser J (eds) (1981) Holocene lake deposits and paleoenvironments in central Sahara northeastern Mali. Palaeogeogr. Palaeoclimatol. Palaeoecol. 35: 45-61
Petit-Maire N, Riser J (1983.) Sahara ou Sahel? Lamy Marseille: 473 p
Petit-Maire N,Riser J (1987) Holocene palaeohydrology of the Niger. Palaeoecology of Africa 18: 135-141
Petit-Maire N, Fabre J, Carbonel P, Schultz E, Aucour AM (1987) La dépression de Taoudenni (Sahara malien) à l'Holocène. Géodynamique 2 (2) :154-160
Pirard F (1962) Hydrologie du Niger Oriental. Rep. Bureau de Recherche Géologique et Minière DAK 62 A4, Dakar: 32 p
Quay PD (1992) Radiocarbon in trace gases in the atmosphere. In Radiocarbon after four decades : an interdisciplinary perspective, Springler-Verlag New-York, in press
Raynaud D, Chapellaz J, Barnola JM, Korotkevich YS, Lorius C (1988) Climatic and CH_4 cycle implications of glacial-interglacial CH_4 change in the Vostok ice core. Nature 333: 655-657
Riser J, Petit-Maire N (1986) Paléogéographie du bassin d'Araouane à l'Holocène. Rev. Géol. Dyn. Géogr. Phys. 27 (3-4) : 205-212
Roche M.A (1980.)Traçage naturel salin et isotopique des eaux du système hydrologique du lac Tchad. Trav. Doc. ORSTOM 117: 389 p
Rognon P (1987) Late Quaternary climatic reconstruction for the Maghreb (North Africa). Palaeogeogr.Palaeoclimatol. Palaeoecol. 58: 11-34

Rowell DP, Blondin C (1990) The influence of soil wetness distribution on short-range rainfall forecasting in the West African Sahel. Quart. J.R. Met. Soc. 116 : 1471-1485

Ruddiman WF, Duplessy JC (1985) Conference on the Last Deglaciation : timing and mechanism. Quat. Res. 23: 1-17

Sarthein M, Tetzlaff G, Koopman B, Wolter K, Pflaumann U (1981) Glacial and interglacial wind regimes over Atlantic and north-west Africa. Nature 293: 193-196

Servant M (1983) Séquences continentales et variations climatiques : évolution du bassin tchadien au Cénozoïque supérieur. Trav. Doc. ORSTOM PARIS 159: 573 p

Servant M, Servant-Vildary S (1980) L'environnement quaternaire du bassin du Tchad. In MAJ Williams, H Faure (eds). The Sahara and the Nile, Balkema Rotterdam: 133-162

Siegenthaler U, Eicher U, Oeschger H, Dansgaard W (1984) Lake sediments as continental $\delta^{18}O$ records from the Glacial/Post Glacial transition. Annals of Glaciology 5: 149-152

Street FA, Grove AT (1979) Global maps of lake-level fluctuations since 30,000 B.P. Quat. Res. 12: 83-118

Street-Perrott FA, Perrott FA (1990) Abrupt climatic fluctuations in the tropics. An Atlantic ocean feed back mechanism. Nature 343: 1-26

Street-Perrott FA, Robert N (1983) Fluctuations in closed-basin lakes as an indicator of past atmospheric circulation patterns. In : Variations in the global water budget, Street-Perrott et al. (Eds) Riedel Dordrecht: 331-345

Street-Perrott FA, Marchand DS, Roberts N, Harrison SP (1989) Global lake level variations from 18,000 to 0 years ago : a palaeoclimatic analysis. In CO_2, Rep. DOE/ER/60304 H1, National Technical Information Service Springfield: 213 pp

Street-Perrott FA, Mitchell JFB, Marchand DS, Brunner JS (1990) Milankovitch and albedo forcing ot the tropical monsoons ; a comparison of geological evidence and numerical simulations for 9000 yrsBP. Trans. Roy. Soc. Edinburgh : Earth Sciences 81: 407-427

Stumm W, Morgan JJ (1984) Aquatic chemistry. Wiley-Interscience Inc. New-York 2nd ed.

Talbot M (1980) Environmental responses to climatic change in the West African Sahel over the past 20,000 years. In : The Sahara and the Nile MAJ Williams, H Faure (Eds) Balkema Rotterdam: 37-62

Talbot MR, Kelts K (1986) Primary and diagenetic carbonates in the anoxic sediments of Lake Bosumtwi Ghana Geology 14 (11): 912-916

Talbot MR, Livingstone DA, Palmer DG, Maley J, Melack JM, Delibrias G, Gulliksen S (1984) Preliminary results from sediment cores from lake Bosumtwi Ghana. Palaeoecology of Africa 16: 173-192

Taylor RE (1990) Personnal communication

Téhet R, Gasse F, Durand A, Schroeter P, Fontes JCh (1990) Fluctuations climatiques du Tardiglaciaire à l'Actuel au Sahel (Bougdouma, Niger méridional). C.R. Acad. Sci. Paris 311, II: 253-258

Ward MN, Folland CK, Maskell K, Rowell D, Colman, A (1990) Understanding and predicting seasonal rainfall in subsaharian Africa. WMO Tropical Meteorology Research Programme Report Series 36: 157-161

THE PALYNOLOGICAL EXPRESSION AND TIMING OF THE YOUNGER DRYAS EVENT - EUROPE VERSUS EASTERN NORTH AMERICA

Dorothy M. Peteet
NASA/Goddard Institute
2880 Broadway, NY NY 10025
&
Lamont-Doherty Geological Observatory
Palisades, NY 10964

ABSTRACT

A comparison of the distribution, magnitude and timing of the Younger Dryas event in Europe and eastern North America shows several striking similarities and differences. The European event is documented throughout the region from Scandinavia to the Mediterranean while the North American oscillation is most evident in the maritime Canadian provinces and southern New England. The pollen signal in Europe and Canada typically shows a drop in tree pollen percentage and increase in herbs (ie. Artemisia) while the southern New England stratigraphy shows the replacement of a mixed boreal conifer-hardwood forest with boreal conifer forest alone. Bulk radiocarbon stratigraphy places the Younger Dryas on both sides of the Atlantic approximately 11-10,000 ^{14}C yr BP, but available AMS dates from both regions suggest the timing may be closer to 10,600-10,000 ^{14}C yr BP. Uncertainties in ^{14}C production rate and/or partitioning in the carbon cycle suggest that establishing the true age and duration of the Younger Dryas may be quite difficult to ascertain.

NATO ASI Series, Vol. I 2
The Last Deglaciation: Absolute and Radiocarbon Chronologies
Edited by E. Bard and W. S. Broecker

INTRODUCTION

In attempting to understand the nature and timing of climate change from terrestrial records during the Younger Dryas interval, it is useful to compare the information given by pollen stratigraphy on both sides of the North Atlantic (Fig. 1). The term Younger Dryas is used in this paper as a biostratigraphic zone, although it has also been defined by Mangerud et al. (1974) as a chronostratigraphic zone approximately 11-10 ka. While the European Younger Dryas biozone has been documented for over half a century (Jessen, 1938) the coastal North American evidence for this climatic change has only recently been re-evaluated and identified regionally in the maritime provinces of Canada (Mott et al., 1986) and the coastal regions of the northeastern United States (Peteet et al., 1990). The questions of distribution, magnitude, and rapidity of the change are all essential for comparison of the pollen stratigraphy with independent data from glacial moraines, ice cores, and marine isotopic and faunal records.

Three questions are important in the assessment of whether or not a palynological signal reflects a climatic cooling which is correlative with the European Younger Dryas. These are: 1) Does the palynologic signal depict a regional expression? 2) Does the palynological change reflect a real vegetational change? 3) Does the change occur between approximately 11-10,000 ^{14}C yr BP?

DISTRIBUTION

Europe

The stratigraphic change identified as the Younger Dryas event has been documented from hundreds of sites in Europe (see Watts, 1980; Rind et al., 1986; Pons et al., 1987) ranging from Scandanavia south to Spain and eastward into Hungary and Poland (Rozanski and Pazdur, this volume) (Fig. 1). The available bulk radiocarbon dating indicates that the change occurred roughly between 11 and 10,000 ^{14}C yr BP, but the range is considerable. For example, Price (1983) notes that inception of the Younger Dryas ranges in Scotland alone between 11,300 and 10,600

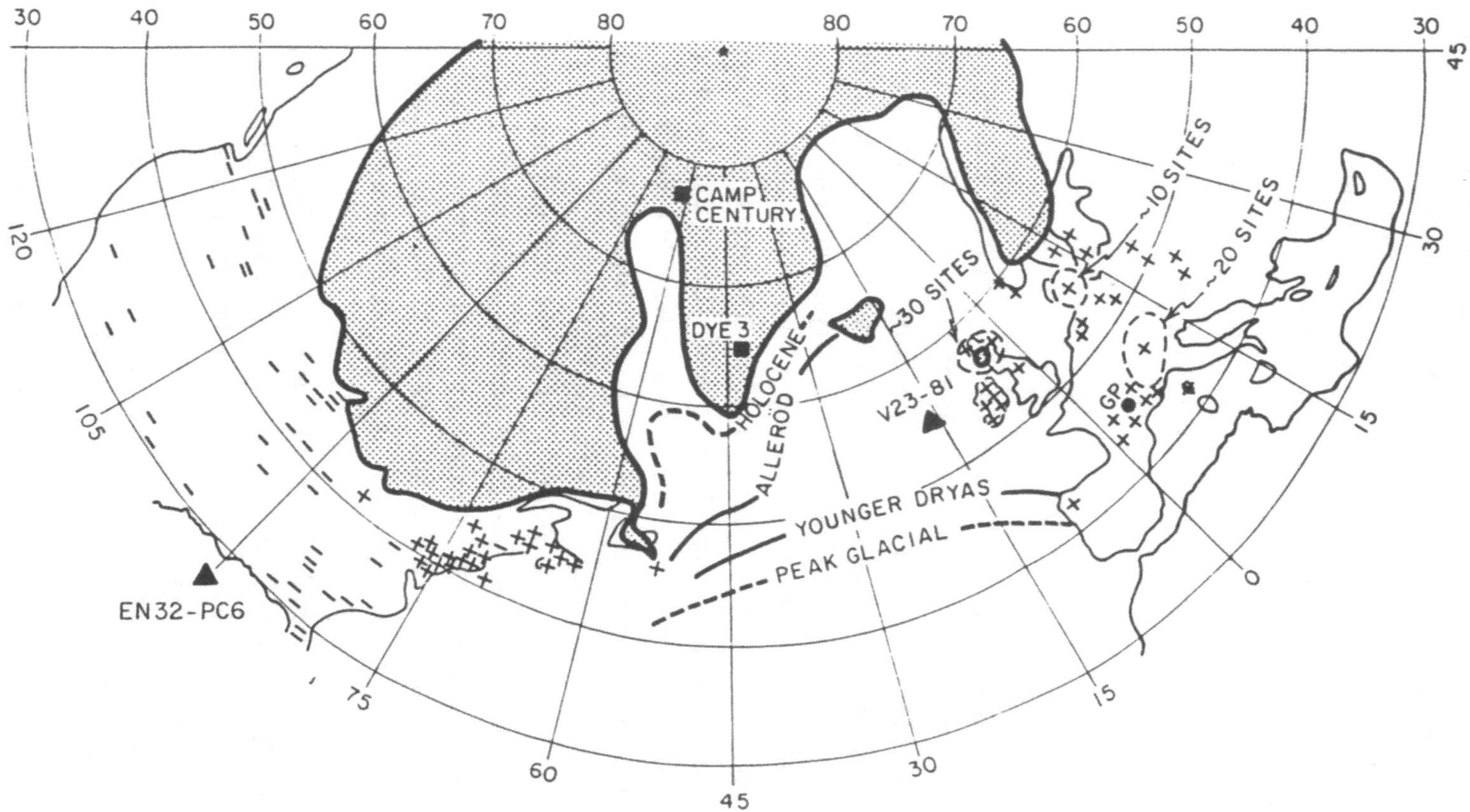

Figure 1. Location maps showing sites where the Younger Dryas has been identified on land (+) and is absent (-). Triangles represent marine sites, and squares indicate ice core sites.

^{14}C yr BP. However, the correlation between biostratigraphic zones is assumed because of the well-known abundance of sites and is strengthened because most of these sites have a lithostratigraphic change which parallels the biostratigraphic event. In addition, glacial moraines, stable isotopes, and insect fossils all attest to the presence of this cold event in Europe.

As recent attention has increasingly focused upon the Younger Dryas event, southern European sites have been added to expand the range to include the Mediterranean region (Pons et al., 1987), including Corsica.

Eastern North America

A climatic shift correlative with the Younger Dryas of Europe, has been documented from over twenty sites in eastern maritime Canada (Mott et al., 1986) (Fig. 1). These North American maritime sites are similar to many of the European pollen sites in that a lithological shift to inorganic sediments characterizes the Younger Dryas interval. In northern Maine, Gould Pond (Fig. 5C) shows a spruce and alder peak which may be indicative of this cooling as well. The apparent lack of this reversal in other northern New England sites (Davis and Jacobson, 1986) is puzzling, and may involve the difficulty with low deposition rate late-glacial cores, interpretation of the pollen data, or a different climatic regime due to the influence of the Gulf of Maine.

Further to the south, over fifteen sites in the northeastern United States also record a late-glacial oscillation (Peteet, 1987) which has recently been correlated with the Younger Dryas in three locations using AMS dates on macrofossils (Peteet et al., 1990). In these sites, however, the sedimentary change is negligible, as it often occurs in gyttja with very high deposition rates.

To the west, in the mid-continental region, two authors point to a late-glacial climatic reversal correletive with the Younger Dryas chronozone. From Rattle Lake in northwestern Ontario Bjorck (1985) shows the replacement of a mixed spruce-ash-elm forest prior to 11,200 yr BP with willow, sedge, grass, and composites, a climatic deterioration which lasted until 10,200 yr BP, when pine markedly increased. A similar pattern is reflected in the Ohio area where Shane (1987) identified the rapid rise of boreal conifers after the presence of a mixed hardwood-conifer assemblage. Although it is tempting to correlate these reversals with the coastal sites and the European event (Wright, 1989) the correlation is controversial because of possible regional temperature depression near proglacial lakes (Lewis and Anderson, 1989).

MAGNITUDE

Europe

The strength of the palynologic and lithologic signal in Europe appears to decrease as one moves inland from the N. Atlantic (Watts, 1980). Throughout Europe, increases in Artemisia, Poaceae and other non-arboreal pollen mark the climatic deterioration. The British Isles records exhibit a pronounced decline in tree birch and juniper and increases in Artemisia and Empetrum. At sites which were drier and colder during the Younger Dryas, Artemisia plays a larger role, as illustrated in areas adjacent to the Scottish highlands (Macpherson, 1980) (Fig. 2). In Scandinavia, pine, birch and Empetrum pollen of the Allerod decrease as an increase in Artemisia, chenopods, grass, and dwarf birch mark the Younger Dryas. This pattern characterizes the pollen stratigraphy of Germany, the Netherlands, and Belgium as well (Bohncke et al., 1988). However, these authors conclude that differences in precipitation (snow cover) to a great extent determine the inferred differences in vegetational development. In several lakes on the Swiss plateau, a 2% increase in Artemisia is accompanied by increases in pine pollen concurrent with decreases in birch, increased clastic mineral input, and an oxygen isotope drop of 2% (Lotter and Zbinden, 1989; Amman and Lotter, 1989). The role that Artemisia plays during the Younger Dryas at several sites in France is shown in Fig. 3 (Pons et al. (1987)

Attempts to relate the strength of the palynological signal to temperature and/or precipitation is quite difficult, as has been noted by many authors (Watts, 1980; Pons et al., 1987; Rind et al., 1986). Inherent problems involve the difference in seasonality compared with conditions today, as well as an actual assessment of vegetational change from the pollen percentage changes. Nevertheless, estimates range from summer cooling of 2 to 7 degrees C. These estimates are less

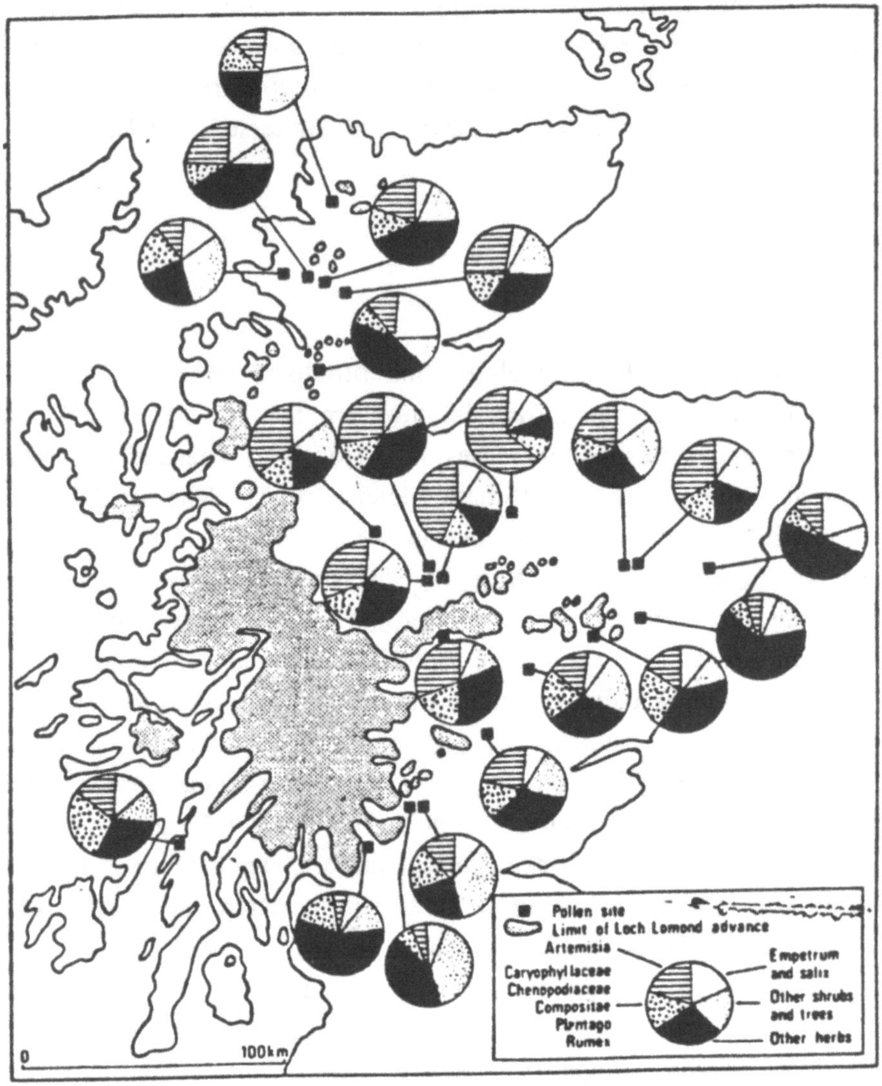

Figure 2. Pollen spectra at <u>Artemisia</u> maximum from sites in and adjacent to the Scottish highlands (Macpherson 1980).

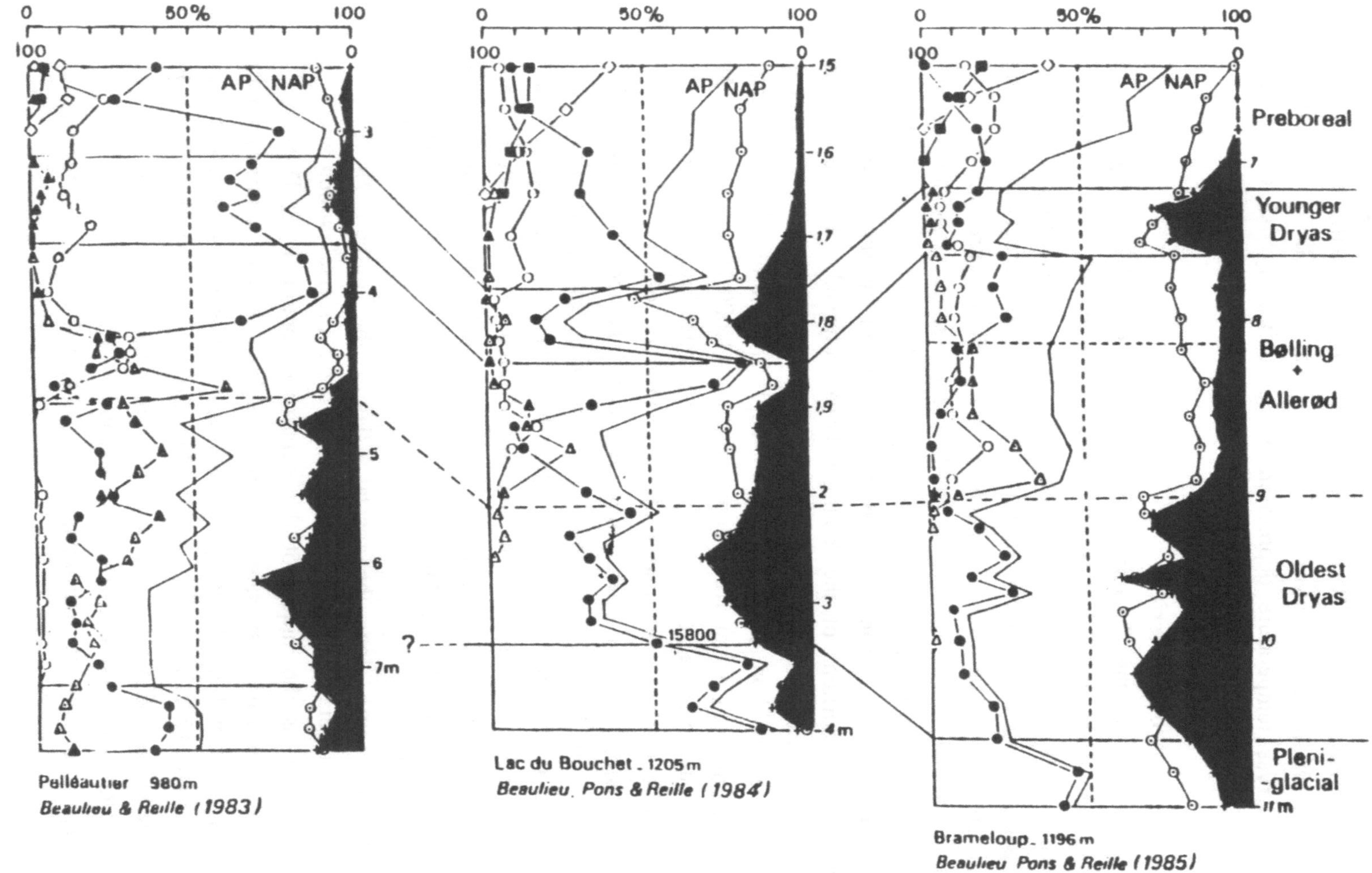

Figure 3. Variations in arboreal pollen percentages (AP) and the role of Artemisia (black) during the Younger Dryas from three sites in France (Pons et al., 1987).

than glacial or fossil beetle estimates of temperature change, which are summer decreases of 8 to 10 degrees (see summary in Rind et al., 1986; Mangerud, 1987). In southwestern France, Pons et al (1987) estimate a Younger Dryas temperature depression of approximately 4-5 C and a decrease in precipitation of 400 to 500 mm.

Eastern North America

The sites on the Canadian Atlantic were begun by Livingstone and Livingstone (1958) at Gillis Lake on Cape Breton Island. Since that time, numerous sites have been investigated, primarily concerning two types of deposits (Mott et al., 1986) (Fig. 4). The first are lateglacial organic layers sandwiched between inorganic layers. They represent the remains of plants which grew in shallow depressions after the warming following the deglaciation, and generally date between 12 and 11,000 ^{14}C yr BP (Fig. 4). These soil layers were buried by mineral deposits when colder conditions of the Younger Dryas returned. The second type of deposit investigated are lake sediments, which show a palynological change generally from trees (spruce or poplar) to shrubs (willow) and herbs (Artemisia, Carex) from approximately 11,000-10,000 ^{14}C yr BP. Often the palynological change is relatively subtle, but the lithologic change quite distinct. Estimates of the magnitude of the inferred climatic change are not quantitative, but certainly the landscape and vegetation were strongly affected. Mass wasting and solifluction were widespread as indicated from lake sites (Mott et al., 1986) and renewed glacial activity in some areas is well-documented (Stea and Mott, 1989). Spruce trees were decimated at several sites, and shrub birch reverted to willow dominance, suggesting temperature decreases of at least several degrees.

Further south along the coastal Atlantic, the southern New England late-glacial palynology exhibits a clear reversal which was noted as early as the middle of this century (Leopold, 1956; Deevey, 1958). In over fifteen sites, exhibiting a regional signal, a mixed conifer-hardwood assemblage (pine-spruce-oak-ash-hornbeam) is evident at approximately 12-11,000 ^{14}C yr BP. (Fig. 5A, Alpine Swamp, NJ and Fig. 5B, Rogers Lake, Conn). This mixed assemblage declines as the boreal forest components (spruce-fir-larch) expand. The hardwoods subsequently expand at approximately 10,000 yr BP. Whether or not

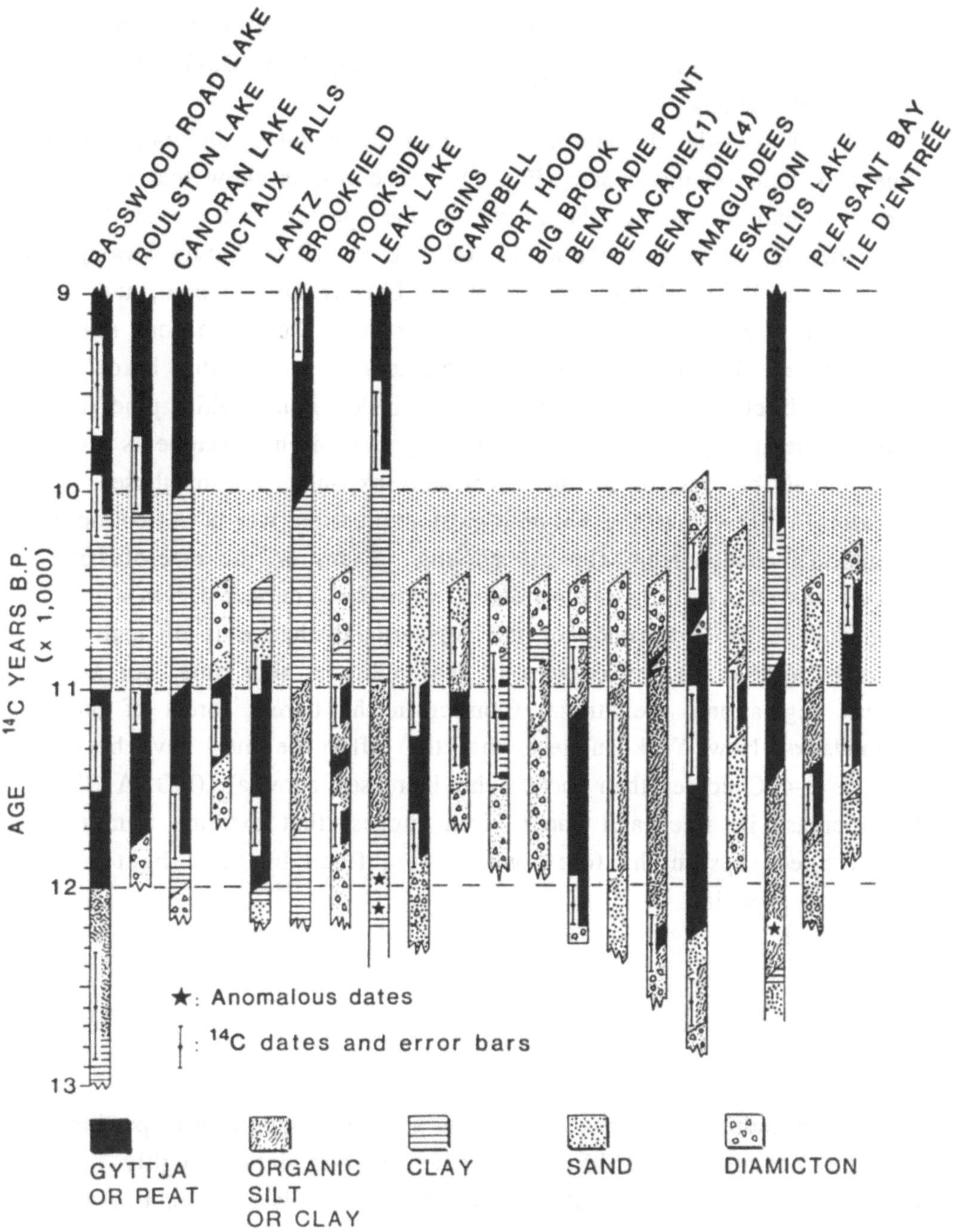

Figure 4. Schematic diagram showing the stratigraphy and radiocarbon dates for 14 buried organic and 6 lake sediment sites. Stippled pattern denotes period of climatic cooling (Mott et al., 1986).

this reversal was indeed a cooling-induced vegetational change or an expansion of northward migration of these taxa as climate warmed has been controversial for over fifty years (Leopold, 1956; Davis, 1969; Peteet et al., 1990).

Accelerator-dated macrofossils from three lakes sediments shed some new light on the debate (Fig. 6). The results show that the boreal increase cannot be due to the arrival of the species to the area, because spruce needles occur as early as 12,300 ^{14}C yr BP at Alpine Swamp, NJ (Peteet et al., 1990), 12,600 ^{14}C yr BP at Linsley Pond, Conn., and 12,300 ^{14}C yr BP at Allamuchy Pond, NJ. Paper birch seeds and cone scales accompanying the birch pollen rise indicate that it was this boreal species of birch that also expanded. At Linsley Pond, white pine (a more temperate species than spruce or paper birch) disappears during this interval, but returns as the hardwoods expand. All of these indicators suggest that the palynological expression was a reflection of a vegetational change from climatic cooling. Dates on the macrofossils indicate that the cooling is correlative with the Younger Dryas chronozone. The magnitude of the vegetational change is difficult to assess because of the differences in seasonality at this time, but a modern vegetational elevational transect in the boreal forest of the Adirondacks, New York suggests that the difference may have been as great as 3-4 C colder than today with increased snowfall (NOAA, 1974). The increase in alder and paper birch suggest that extreme winters may have created gaps in the forest which were filled by these light-demanding species.

TIMING

Europe

Bulk radiocarbon stratigraphy in sites throughout Europe indicates that the Younger Dryas biostratigraphic reversal dates roughly between 11,000 and 10,000 ^{14}C yr BP (Watts, 1980). This is consistent with correlation by Mangerud et al. (1984) of the Vedde Ash bed in peat bogs of western Norway at 10,600 ^{14}C yr BP (mid-Younger Dryas chronozone) with ash zones found in the North Atlantic and the Norwegian Sea, thus providing a common Younger Dryas marker for

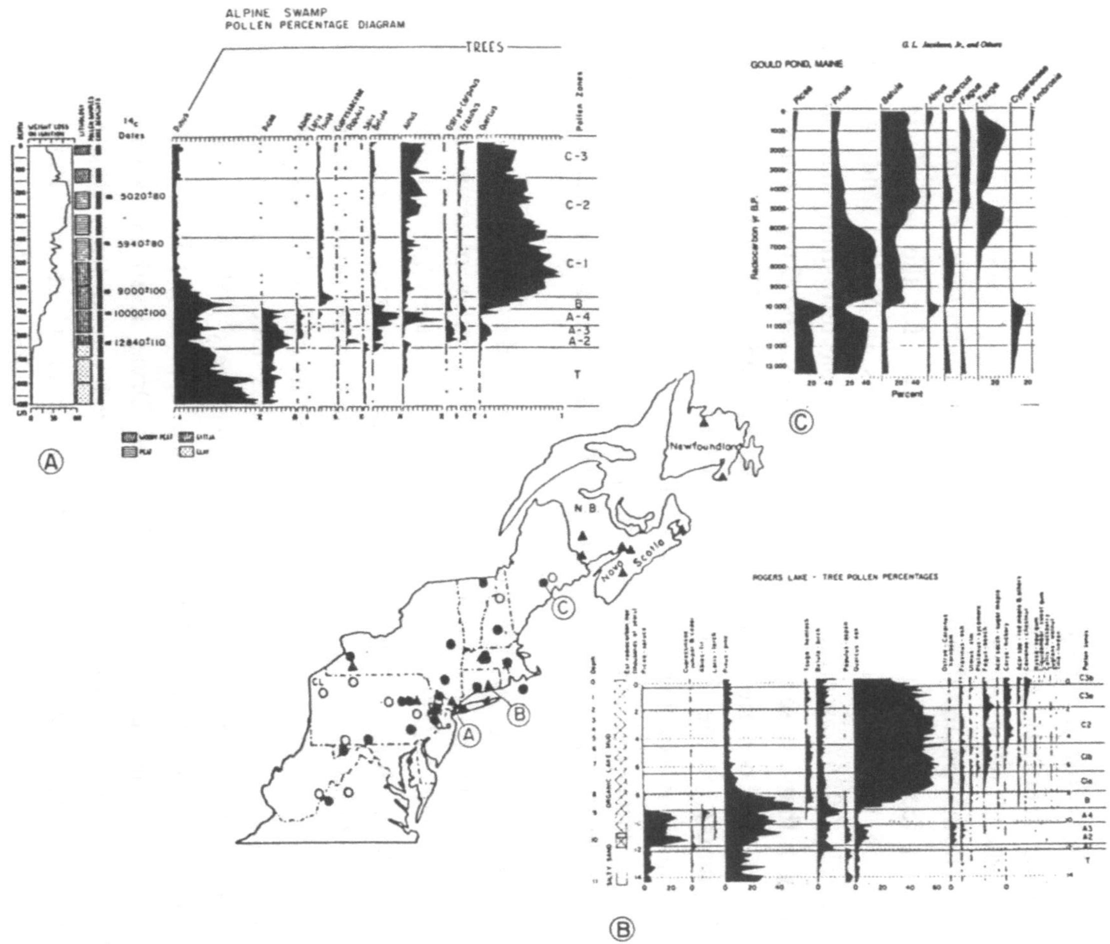

Figure 5. Map showing northeastern sites where a YD oscillation is present(triangle and full circle) or absent (open circle). Pollen stratigraphy from Alpine Swamp,NJ, Rogers Lake, Conn., and Gould Pond, Maine showing the boreal expansion in the A-4 stage, 11-10,000 yr BP

both land and sea. Problems with recycled older carbon, stratigraphic resolution, and the effects of contamination by mineral carbon are responsible for much of the scatter visible in ages beyond these ranges (Lowe et al., 1988). In some areas, the Younger Dryas pollen zone can further be subdivided into several subzones, as Bjorck and Moller (1987) found in southeastern Sweden. Summarizing bulk radiocarbon dates from eleven sites from southwestern Europe, Pons et al. (1987) show that recent dates for the palynological Younger Dryas biozone indicate its occurrence between 10,700 and 10,300 ^{14}C yr BP.

The advent of accelerator mass spectrometry (AMS) ^{14}C dating has enhanced the potential for dating this event carefully on land, ocean, and in ice. The first terrestrial AMS dates for the late-glacial were those on macrofossils from lake sediments in Switzerland, where Andree et al. (1986) ascertained 1) that dates on terrestrial macrofossils were younger than those on carbonate and gyttja, and 2) a plateau of constant ^{14}C age occurred about 12,415 yr BP and 9922 + 97 yr BP, and 3) the beginning of the Younger Dryas occurred around 11,000 ^{14}C yr BP. Explanations for these plateaus include higher sedimentation rates during these intervals, decreasing ^{14}C production rates, or dilution of atmospheric ^{14}C with carbon of lower ^{14}C concentration (Andree et al., 1986).

In a study using 90 AMS dates from two lakes in lowland Switzerland, Ammann and Lotter (1989) found similar plateaus at 12,700 yr BP and 10,000 ^{14}C yr BP. If these plateaus are not due to some lake depositional process, it means that correlation between sites is impossible within the age plateaus, and that more refined dating within these intervals is not possible. Lotter (this workshop) presents evidence from varve chronology from the Swiss lakes that the length of the 10,000 yr BP plateau is actually approximately 350 years, that the Allerod/Younger Dryas boundary took place approximately 10,700 ^{14}C yr BP, and that the duration of the Younger Dryas itself in varve years is 700. In contrast to this record is the varved record from Gosciaz Lake, Poland (Pazdur and Rozanski, this volume) which indicates a Younger Dryas duration of 1250 years. One available AMS date for the Younger Dryas/Preboreal boundary is 10,080 + 120 ^{14}C yr BP; a mid-Younger Dryas date is 10,450 + 140 ^{14}C yr BP. A third varved record of the Younger Dryas from a lake in Germany (Zolitchka, this volume) showed a Younger Dryas duration of 370 varved years. Thus additional AMS research is necessary to resolve the Allerod/Younger Dryas chronology.

Few other AMS macrofossil studies of lake sediments are available, primarily due to funding limitations. Lowe et al. (1988) obtained AMS ages for the beginning of the Younger Dryas as 11,750 ^{14}C yr BP and its close at 10,400 ^{14}C yr BP from one site in North Wales. These dates appear anomalously old for the cooling episode, and the authors discuss several possible reasons for the discrepancy between AMS and bulk radiocarbon samples. Cwynar and Watts (1989) obtained dates of 10,600 ^{14}C yr BP for the Younger Dryas inception, and 10,070 ^{14}C yr BP for its close in Ireland, more in keeping with the Swiss chronology.

The German oak/pine tree ring chronology also supports a plateau scenario for changes in ^{14}C production, with 400 tree ring years that date to approximately 10,000 yr BP (Becker and Kromer, this volume). The disparity between ^{14}C and U/Th ages on Barbados corals (Bard et al., 1990) suggests that ^{14}C production rates or ocean related ^{14}C distribution between reservoirs affected the late-glacial ^{14}C of the atmosphere. Differences in the three varve chronologies, the 14C and U/Th scale, and the tree ring chronologies indicates that the true age of the Younger Dryas is as yet an elusive entity.

Eastern North America

Bulk radiocarbon dates from the maritime provinces of Canada (Fig. 4) suggest that the age range for the Younger Dryas inception and conclusion is similar to the situation from bulk radiocarbon chronology in Europe (Mott et al., 1986). To date, AMS ages are available from two sites only, Basswood Road Lake in New Brunswick and Lac a Majie in Nova Scotia. Both of these sites indicate that the Younger Dryas inception was approximately 10,600 ^{14}C yr BP and its close 10,000 ^{14}C yr BP (Mayle and Cwynar, 1990 and personal communication).

Three lake sediments in eastern United States coastal region have recently been investigated for the timing of the Younger Dryas. AMS ages from these lakes are shown in Figure 6. These sites are Alpine Swamp, NJ (Peteet et al., 1990), Allamuchy Swamp, NJ, and Linsley Pond, Conn (Peteet et al., in prep.). The three sites indicate that

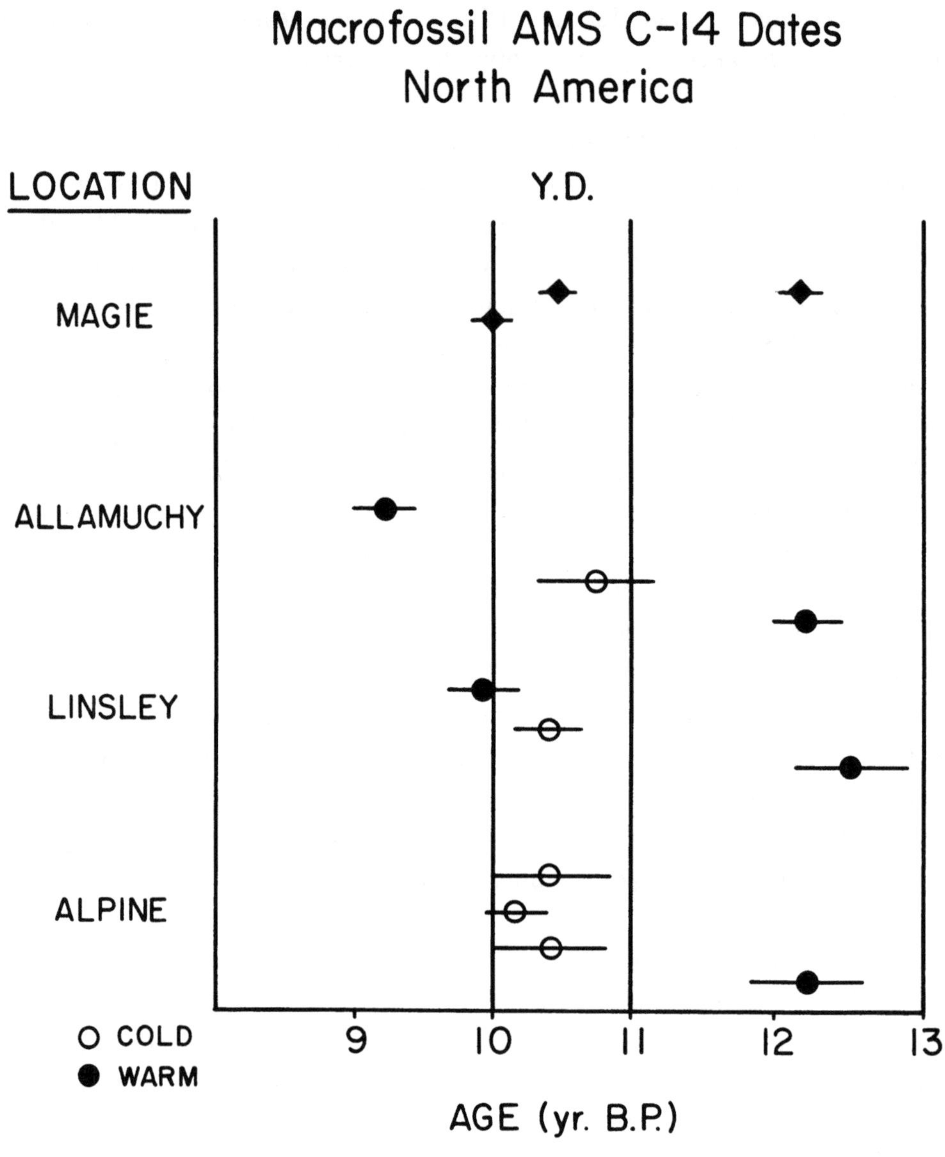

Figure 6. Available AMS ^{14}C dates on the Younger Dryas interval from four lake sites in North America suggests that the cooling took place between 11 and 10,000 yr BP

a Younger Dryas interval occurred between approximately 10,600 and 10,000 ^{14}C yr BP. However, many additional dates in high resolution studies, both with varves and macrofossils are needed in North America to parallel the European studies, and to investigate whether or not a plateau in ^{14}C age is apparent.

Concluding Remarks

In summary, attention over the last decade to late-glacial palynological stratigraphy has shown the regional European extent of the Younger Dryas biozone to be approximately the same as described by Watts (1980). In contrast to this scenario in Europe, recent research in eastern North America has demonstrated a Younger Dryas extent from northeastern Canada southward as far as New Jersey. Estimates of the magnitude of the climatic change range from the vegetational change are 2 to 7 C in Europe, and 3 to 4 in the United States. The timing of the Younger Dryas roughly is between 11 and 10,000 ^{14}C yr BP, but plateaus in the best AMS chronologies suggest that the timing of the inception and conclusion of the Younger Dryas may be very difficult to ascertain.

Acknowledgments

The research was supported by NSF Grant ATM-8604397 to Columbia University.

References

Ammann B, and Lotter, AF (1989) Late-glacial radiocarbon and palynostratigraphy on the Swiss Plateau. Boreas 18: 109-126.

Andree M, Oeschger H, Siegenthaler U, Riesen T, Moell M, Ammann B, and Tobolski K (1986). C-14 dating of plant macrofossils in lake sediment. Radiocarbon 28 (2A): 411-416.

Bard E, Hamelin, B, Fairbanks R, Zindler, A (1990) Calibration of the 14C timescale over the past 30,000 years using mass spectrometric U-Th ages from Barbados corals. Nature 345: 405-410.

Bohncke S, Wijmstra L, van der Woode J, and Sohl H (1988) The late-glacial infill of three lake successions in The Netherlands: Regional developments. Boreas 17: 385-402.

Bjorck S. (1985). Deglaciation chronology and revegetation in northwestern Ontario. Can. J. Earth Sci. 22: 850-871.

Bjorck, S and Moller, P (1987) Late Weichselian environmental history in southeastern Sweden during the deglaciation of the Scandinavian ice sheet. Quaternary Research 28: 1-37.

Cwynar LC and Watts WA (1989) Accelerator-mass spectrometer ages for the late-glacial events at Ballybetagh, Ireland. Quaternary Research 31: 377-380.

Davis MB (1969) Climatic changes in southern Connecticut recorded by pollen deposition at Rogers Lake. Ecology 50: 409-422.

Davis, RB and Jacobson GL (1986) Late-glacial and early Holocene landscapes in northern New England and adjacent areas of Canada. Quaternary Research 23: 341-368.

Deevey, ES (1958) Radiocarbon-dated pollen sequences in eastern North America. Veroffentilichungen des Geobotanischen Institutes Rubel in Zurich 34: 30-37.

Jacobson GL, Webb T III, and Grimm EC (1987) Patterns and rates of vegetational change during the deglaciation of eastern North America. In: Ruddiman WF and Wright Jr HE (eds) North America and Adjacent Oceans During the Last Deglaciation, pp 277-288, Geology of North America Vol K-3, Boulder, Colo.

Jessen K (1938) Some west Baltic pollen diagrams. Quartar 1: 124-139.

Leopold, EB (1956) Two late-glacial deposits in southern Connecticut. Proceedings of the National Academy of Sciences 42: 863-867.

Lewis, CFM and Anderson TW (1989) Oscillations of levels and cool phases of the Laurentian Great Lakes caused by inflows from glacial Lakes Agassiz and Barlow-Ojibway. Journal of Paleolimnology 2: 99-146.

Livingstone DA and Livingstone BGR (1958) Late-glacial and postglacial vegetation from Gillis Lake in Richmond County, Cape Breton Island, Nova Scotia. American Journal of Science 256: 341-359.

Lotter AF and Zbinden H (1989) Late-glacial pollen analysis, oxygen-isotope record, and radiocarbon stratigraphy from Rotsee (Lucerne), Central Swiss Plateau. Eclogae geol. Helv. 82/1: 191-202.

Lowe JJ, Lowe S, Fowler AJ, Hedges REM, and Austin TJF (1988) Comparison of accelerator and radiometric radiocarbon measurements obtained from Late Devensian Lateglacial lake sediments from Llyn Gwernan, North Wales, UK. Boreas 17: 355-369.

Macpherson JB (1980) Environmental change during the Loch Lomond stadial: evidence from a site in the Upper Spey Valley, Scotland. In: Lowe, JJ, Gray, JM and Robinson JE (eds) Studies in the Late-Glacial of Northwest Europe, 89-102. Pergamon Press, Oxford.

Mangerud, J (1987) The Allerod/Younger Dryas boundary. In: Berger WH and Labeyrie LD (eds) Abrupt Climatic Change, Reidel Publishing Co., Dordrecht, pp. 163-172.

Mangerud J, Lie SE, Furnes H, Kristiansen IL, and Lomo L (1984) A Younger Dryas ash bed in western Norway, and its Possible Correlations with tephra in cores from the Norwegian Sea and the North Atlantic. Quaternary Research 21: 85-104.

Mayle F, and Cwynar LC 1990. Accelerator-mass spectometer C-14 dates for the Younger Dryas at Lac a Magie, Nova Scotia. AMQUA Abstract, Waterloo, Ontario , p. 25.

Mott RJ, Grant DR, Stea R, and Occhietti S (1986) Late-glacial climatic oscillation in Atlantic Canada equivalent to the Allerod-Younger Dryas event. Nature 323: 247-250.

NOAA (National Oceanic and Atmospheric Administration) (1974). "Climates of the States;", Water Information Center, Inc., Port Washington, New York.

Peteet, DM (1987) Younger Dryas in North America - modeling, data analysis, and re-evaluation. In: Berger WH and Labeyrie LD (eds)

Abrupt Climatic Change, Reidel Publishing Co., Dordrecht, pp. 185-193.

Peteet, DM, Vogel, JS, Nelson, DE, Southon, JR, Nickmann RJ, and Heusser LE (1990) Younger Dryas climatic reversal in northeastern USA? AMS ages for an old problem. Quaternary Research 33: 219-230.

Pons A, de Beaulieu JL, Guiot J and Reille M (1987) The Younger Dryas in southwestern Europe: an abrubt climatic change as evidenced from pollen records. In: Berger WH and Labeyrie LD (eds) Abrupt Climatic Change, Reidel Publishing Co., Dordrecht, pp. 195-208.

Price, RJ (1983) Scotland's environment during the last 30,000 years. Scott. Academic Press, Edinburgh.

Rind, D, Peteet, D, Broecker, W, McIntyre, A, and Ruddiman, W (1986). The impact of cold North Atlantic sea surface temperatures on climate: implications for the Younger Dryas cooling (11-10 k). Climate Dynamics 1: 3-33.

Shane LCK (1987) Late-glacial vegetational and climatic history of the Allegheny Plateau and the Till Plains of Ohio and Indiana. Boreas 16: 1-20.

Stea, RR and Mott RJ (1989) Deglaciation environments and evidence for glaciers of Younger Dryas age in Nova Scotia, Canada. Boreas 18: 169-187.

Watts, WA (1980) Regional variation in the response of vegetation of Lateglacial climatic events in Europe. In: Lowe, JJ, Gray, JM, and Robinson, JE Studies in the LateGlacial of Northwest Europe, eds. Pergamon Press, Oxford, pp. 1-22.

Wright, H.E. Jr. (1989) The amphi-Atlantic distribution of the Younger Dryas paleoclimatic oscillation. Quaternary Science Reviews 8: 295-306.

NATO ASI Series I

Vol. 1: **Global Environmental Change.**
Edited by R. W. Corell and P. A. Anderson. 278 pages. 1991.

Vol. 2: **The Last Deglaciation: Absolute and Radiocarbon Chronologies.**
Edited by E. Bard and W. S. Broecker. 358 pages. 1992.

MIX
Papier aus verantwortungsvollen Quellen
Paper from responsible sources
FSC® C105338

If you have any concerns about our products,
you can contact us on
ProductSafety@springernature.com

In case Publisher is established outside the EU,
the EU authorized representative is:
Springer Nature Customer Service Center GmbH
Europaplatz 3, 69115 Heidelberg, Germany

Printed by Libri Plureos GmbH
in Hamburg, Germany